ADVANCES IN CODING THEORY AND CRYPTOGRAPHY

Series on Coding Theory and Cryptology

Editors: Harald Niederreiter *(National University of Singapore, Singapore)* and
San Ling *(Nanyang Technological University, Singapore)*

Published

Vol. 1 Basics of Contemporary Cryptography for IT Practitioners
B. Ryabko and A. Fionov

Vol. 2 Codes for Error Detection
by T. Kløve

Vol. 3 Advances in Coding Theory and Cryptography
eds. T. Shaska et al.

Series on Coding Theory and Cryptology – Vol. 3

ADVANCES IN CODING THEORY AND CRYPTOGRAPHY

Editors

T. Shaska
Oakland University, USA

W. C. Huffman
Loyola University, USA

D. Joyner
US Naval Academy, USA

V. Ustimenko
The University of Maria Curie Sklodowska, Poland

NEW JERSEY • LONDON • SINGAPORE • BEIJING • SHANGHAI • HONG KONG • TAIPEI • CHENNAI

Published by

World Scientific Publishing Co. Pte. Ltd.
5 Toh Tuck Link, Singapore 596224
USA office: 27 Warren Street, Suite 401-402, Hackensack, NJ 07601
UK office: 57 Shelton Street, Covent Garden, London WC2H 9HE

Library of Congress Cataloging-in-Publication Data
Advances in coding theory and cryptography / editors T. Shaska ... [et al.].
p. cm. -- (Series on coding theory and cryptology ; vol. 3)
Includes bibliographical references.
ISBN-13: 978-981-270-701-7
ISBN-10: 981-270-701-8
1. Coding theory--Congresses. 2. Cryptography--Congresses. I. Shaska, Tanush. II. Vlora Conference in Coding Theory and Cryptography (2007 : Vlorë, Albania) III. Applications of Computer Algebra Conference (2007 : Oakland University)
QA268.A38 2007
003'.54--dc22

2007018079

British Library Cataloguing-in-Publication Data
A catalogue record for this book is available from the British Library.

Printed in Singapore.

PREFACE

Due to the increasing importance of digital communications, the area of research in coding theory and cryptography is broad and fast developing. In this book there are presented some of the latest research developments in the area. The book grew as a combination of two research conferences organized in the area: the *Vlora Conference in Coding Theory and Cryptography* held in Vlora, Albania during May 26-27, 2007, and the special session on coding theory as part of the *Applications of Computer Algebra* conference, held during July 19-22, Oakland University, Rochester, MI, USA.

The Vlora Conference in Coding Theory and Cryptography is part of Vlora Conference Series which is a series of conferences organized yearly in the city of Vlora sometime in the period April 25 - May 30. The conference is 3-4 days long and focuses on some special topic each year. The topic of the 2007 conference was coding theory and cryptography. The Vlora conference series will host a Nato Advanced Study Institute during the year 2008 with the theme *New Challenges in Digital Communications*. More information of the conferences organized by the Vlora group can be found at `http://www.albmath.org/vlconf`.

Applications of Computer Algebra (ACA) is a series of conferences devoted to promoting the applications and development of computer algebra and symbolic computation. Topics include computer algebra and symbolic computation in engineering, the sciences, medicine, pure and applied mathematics, education, communication and computer science. Occasionally the ACA conferences have special sessions on coding theory and cryptography.

I especially want to thank A. Elezi who shared with me the burdens of organizing the *Vlora Conference in Coding Theory and Cryptography*, the participants of the conference in Vlora, and the Department of Mathematics and Informatics at the Technological University of Vlora for helping host the conference.

Also, my thanks go to the Department of Mathematics and Statistics at Oakland University for hosting the *Applications of Computer Algebra* conference. Without their financial and administrative support such a conference would not be possible. My special thanks go to J. Nachman for sharing with me all the burdens of organizing such a big conference. I want to thank also the co-organizers of the coding theory session D. Joyner and C. Shor and all the participants of this session.

There are fourteen papers in this book which cover a wide range of topics and 26 authors from institutions across North America and Europe. I want to thank all the authors for their contributions to this volume. Finally, my special thanks go to my co-editors W. C. Huffman, D. Joyner, and V. Ustimenko for their continuous support and excellent editorial job. It was their efforts which made the publication of this book possible.

T. Shaska

LIST OF AUTHORS

T. L. Alderson	– University of New Brunswick,
M. Borges-Quintana	– Universidad de Oriente, Santiago de Cuba, Cuba
M. A. Borges-Trenard	– Universidad de Oriente, Santiago de Cuba, Cuba Saint John, NB., E2L 4L5, Canada
I. G. Bouykliev	– Institute of Mathematics and Informatics, Veliko Tarnovo, Bulgaria
J. Brevik	– California State University, Long Beach, CA, USA
D. Coles	– Bloomsburg University, Bloomsburg PA, USA
M. J. Jacobson, Jr.	– University of Calgary, Calgary, Canada
D. Joyner	– US Naval Academy, Annapolis, ML, USA
X. Hou	– University of South Florida, Tampa, FL, USA
J. D. Key	– Clemson University, Clemson SC, USA
J. L. Kim	– University of Louisville, Louisville, KY, USA
A. Ksir	– US Naval Academy, Annapolis, ML, USA
J. B. Little	– College of the Holy Cross, Worcester, MA, USA
E. Martinez-Moro	– Universidad de Valladolid, Valladolid, Spain
K. Mellinger	– University of Mary Washington, Fredericksburg, VA, USA
M. E. O'Sullivan	– San Diego State University, San Diego, CA, USA
E. Previato	– Institut Mittag-Leffler, Djursholm, Sweden Boston University, Boston, MA USA
R. Scheidler	– University of Calgary, Calgary, Canada
P. Seneviratne	– Clemson University, Clemson, SC, USA
T. Shaska	– Oakland University, Rochester, MI, USA
C. Shor	– Bates College, Lewiston, ME, USA
A. Stein	– University of Wyoming, Laramie, WY, USA
W. Traves	– US Naval Academy, Annapolis, ML, USA
V. Ustimenko	– The University of Maria Curie-Sklodowska, Lublin, POLAND
H. N. Ward	– University of Virginia, Charlottesville, VA, USA
R. Wolski	– University of California, Santa Barbara, CA, USA

CONTENTS

The key equation for codes from order domains

John B. Little

Department of Mathematics and Computer Science,
College of the Holy Cross,
Worcester, MA 01610, USA
E-mail: little@mathcs.holycross.edu

We study a sort of analog of the *key equation* for decoding Reed-Solomon and BCH codes and identify a key equation for all codes from order domains which have finitely-generated value semigroups (the field of fractions of the order domain may have arbitrary transcendence degree, however). We provide a natural interpretation of the construction using the theory of Macaulay's *inverse systems* and duality. O'Sullivan's generalized Berlekamp-Massey-Sakata (BMS) decoding algorithm applies to the duals of suitable evaluation codes from these order domains. When the BMS algorithm does apply, we will show how it can be understood as a process for constructing a collection of solutions of our key equation.

Keywords: order domain, key equation, Berlekamp-Massey-Sakata algorithm

1. Introduction

The theory of error control codes constructed using ideas from algebraic geometry (including the geometric Goppa and related codes) has undergone a remarkable extension and simplification with the introduction of codes constructed from *order domains*. This development has been largely motivated by the structures utilized in the Berlekamp-Massey-Sakata decoding algorithm with Feng-Rao-Duursma majority voting for unknown syndromes.

The order domains, see [1–4], form a class of rings having many of the same properties as the rings $R = \cup_{m=0}^{\infty} L(mQ)$ underlying the one-point geometric Goppa codes constructed from curves. The general theory gives a common framework for these codes, n-dimensional cyclic codes, as well as many other Goppa-type codes constructed from varieties of dimension > 1. Moreover, O'Sullivan has shown in [5] that the Berlekamp-Massey-Sakata decoding algorithm (abbreviated as the BMS algorithm in the following) and the Feng-Rao procedure extend in a natural way to a suitable class of codes in this much more general setting.

For the Reed-Solomon codes, the Berlekamp-Massey decoding algorithm can be phrased as a method for solving a *key equation*. For a Reed-Solomon code with minimum distance $d = 2t + 1$, the key equation has the form

$$fS \equiv g \bmod \langle X^{2t} \rangle. \tag{1}$$

Here S is a known univariate polynomial in X constructed from the error syndromes, and f, g are unknown polynomials in X. If the error vector e satisfies $wt(e) \leq t$, there is a unique solution (f, g) with $\deg(f) \leq t$, and $\deg(g) < \deg(f)$ (up to a constant multiple). The polynomial f is known as the *error locator* because its roots give the *inverses* of the error locations; the polynomial g is known as the *error evaluator* because the error values can be determined from values of g at the roots of f, via the Forney formula.

O'Sullivan has introduced a generalization of this key equation for one-point geometric Goppa codes from curves in [6] and shown that the BMS algorithm can be modified to compute the analogs of the error-evaluator polynomial together with error locators.

Our main goal in this article is to identify an analog of the key equation Eq. (1) for codes from general order domains, and to give a natural interpretation of these ideas in the context of Macaulay's *inverse systems* for ideals in a polynomial ring (see [7–10]) and the theory of duality. We will only consider order domains whose value semigroups are finitely generated. In these cases, the ring R can be presented as an affine algebra $R \cong \mathbb{F}[X_1, \ldots, X_s]/I$, where the ideal I has a Gröbner basis of a very particular form (see [3]). Although O'Sullivan has shown how more general order domains arise naturally from valuations on function fields, it is not clear to us how our approach applies to those examples. On the positive side, by basing all constructions on algebra in polynomial rings, all codes from these order domains can be treated in a uniform way, Second, we also propose to study the relation between the BMS algorithm and the process of solving this key equation in the cases where BMS is applicable.

Our key equation generalizes the key equation for n-dimensional cyclic codes studied by Chabanne and Norton in [12]. Results on the algebraic background for their construction appear in [13]. See also [14] for connections with the more general problem of finding shortest linear recurrences, and [15] for a generalization giving a key equation for codes over commutative rings.

The present article is organized as follows. In Section 2 we will briefly review the definition of an order domain, evaluation codes and dual evaluation codes. Section 3 contains a quick summary of the basics of Macaulay inverse systems and duality. In Section 4 we introduce the key equation and relate the BMS algorithm to the process of solving this equation.

2. Codes from Order Domains

In this section we will briefly recall the definition of order domains and explain how they can be used to construct error control codes. We will use the following formulation.

Definition 2.1. Let R be a $\mathbb{F}_q$-algebra and let $(\Gamma, +, \succ)$ be a well-ordered semigroup. We assume the ordering is compatible with the semigroup operation in the sense that if $a \succ b$ and c is arbitrary in Γ, then $a+c \succ b+c$. An *order function* on R is a surjective mapping $\rho : R \rightarrow \{-\infty\} \cup \Gamma$ satisfying:

(1) $\rho(f) = -\infty \Leftrightarrow f = 0$,
(2) $\rho(cf) = \rho(f)$ for all $f \in R$, all $c \neq 0$ in $\mathbb{F}_q$,
(3) $\rho(f+g) \preceq \max_{\succ}\{\rho(f), \rho(g)\}$,
(4) if $\rho(f) = \rho(g) \neq -\infty$, then there exists $c \neq 0$ in $\mathbb{F}_q$ such that $\rho(f) \prec \rho(f - cg)$,
(5) $\rho(fg) = \rho(f) + \rho(g)$.

We call Γ the *value semigroup* of ρ.

Axioms 1 and 5 in this definition imply that R must be an integral domain. In the cases where the transcendence degree of R over $\mathbb{F}_q$ is at least 2, a ring R with one order function will have many others too. For this reason an *order domain* is formally defined as a pair (R, ρ) where R is an $\mathbb{F}_q$-algebra and ρ is an order function on R. However, from now on, we will only use one particular order function on R at any one time. Hence we will often omit it in refering to the order domain, and we will refer to Γ as the value semigroup of R. Several constructions of order domains are discussed in [3] and [4].

The most direct way to construct codes from an order domain given by a particular presentation $R \cong \mathbb{F}_q[X_1, \ldots, X_s]/I$ is to generalize Goppa's construction in the case of curves.

Let X_R be the variety $V(I) \subset \mathbb{A}^s$ and let

$$X_R(\mathbb{F}_q) = \{P_1, \ldots, P_n\}$$

be the set of $\mathbb{F}_q$-rational points on X_R. Define an evaluation mapping

$$\begin{aligned} ev : R &\rightarrow \mathbb{F}_q^n \\ f &\mapsto (f(P_1), \ldots, f(P_n)) \end{aligned}$$

Let $V \subset R$ be any finite-dimensional vector subspace. Then the image $ev(V) \subseteq \mathbb{F}_q^n$ will be a linear code in $\mathbb{F}_q^n$. One can also consider the dual code $ev(V)^{\perp}$.

Of particular interest here are the codes constructed as follows (see [5]). Let R be an order domain whose value semigroup Γ can be put into order-preserving one-to-one correspondence with $\mathbb{Z}_{\geq 0}$. We refer to such Γ as *Archimedean* value semigroups because it follows that for all nonconstant $f \in R$ and all $g \in R$ there is some $n \geq 1$ such that $\rho(f^n) \succ \rho(g)$. This property is equivalent to saying that the corresponding valuation of $K = QF(R)$ has *rank 1*. O'Sullivan gives a necessary and sufficient condition for this property when $\succ$ is given by a monomial order on $\mathbb{Z}^r_{\geq 0}$ in [2], Example 1.3. Let Δ be the ordered basis of R with ordering by ρ-value. Let $\ell \in \mathbb{N}$ and let V_ℓ be the span of the first ℓ elements of Δ. In this way, we obtain evaluation codes $Ev_\ell = ev(V_\ell)$ and dual codes $C_\ell = Ev_\ell^{\perp}$ for all ℓ.

O'Sullivan's generalized BMS algorithm is specifically tailored for this last class of codes from order domains with Γ Archimedean. If the C_ℓ codes are used to encode messages, then the Ev_ℓ codes describe the parity checks and the syndromes used in the decoding algorithm.

3. Preliminaries on Inverse Systems

A natural setting for our formulation of a key equation for codes from order domains is the theory of inverse systems of polynomial ideals originally introduced by Macaulay. There are several different versions of this theory. For modern versions using the language of differentiation operators, see [9, 10]. Here, we will summarize a number of more or less well-known results, using an alternate formulation of the definitions that works in any characteristic. A reference for this approach is [8].

Let k be a field, let $S = k[X_1, \ldots, X_s]$ and let T be the formal power series ring $k[[X_1^{-1}, \ldots, X_s^{-1}]]$ in the inverse variables. T is an S-module under a mapping

$$\begin{aligned} c : S \times T &\to T \\ (f, g) &\mapsto f \cdot g, \end{aligned}$$

sometimes called *contraction*, defined as follows. First, given monomials X^α in S and $X^{-\beta}$ in T, $X^\alpha \cdot X^{-\beta}$ is defined to be $X^{\alpha-\beta}$ if this is in T, and 0 otherwise. We then extend by linearity to define $c : S \times T \to T$.

Let $Hom_k(S, k)$ be the usual linear dual vector space. It is a standard fact that the mapping

$$\phi : Hom_k(S,k) \to T$$
$$\Lambda \mapsto \sum_{\beta \in \mathbb{Z}^s_{\geq 0}} \Lambda(X^\beta)X^{-\beta}$$

is an isomorphism of S-modules, if we make $Hom_k(S,k)$ into an S-module in the usual way by defining $(q\Lambda)(p) = \Lambda(qp)$ for all polynomials p, q in S. In explicit terms, the k-linear form on S obtained from an element $g \in T$ is a mapping Λ_g defined as follows. For all $f \in S$,

$$\Lambda_g(f) = (f \cdot g)_0,$$

where $(t)_0$ denotes the constant term in $t \in T$. In the following we will identify elements of T with their corresponding linear forms on S.

The theory of inverse systems sets up a correspondence between ideals in S and submodules of T. All such ideals and submodules are finitely generated and we will use the standard notation $\langle f_1, \ldots, f_t \rangle$ for the ideal generated by a collection of polynomials $f_i \in S$.

For each ideal $I \subseteq S$, we can define the annihilator, or *inverse system*, of I in T as

$$I^\perp = \{\Lambda \in T : \Lambda(p) = 0, \ \forall \, p \in I\}.$$

It is easy to check that $I^\perp$ is an S-submodule of T under the module structure defined above. Similarly, given an S-submodule $H \subseteq T$, we can define

$$H^\perp = \{p \in S : \Lambda(p) = 0, \ \forall \, \Lambda \in H\},$$

and $H^\perp$ is an ideal in S. The key point in this theory is the following duality statement.

Theorem 3.1. *The ideals of S and the S-submodules of T are in inclusion-reversing bijective correspondence via the constructions above, and for all I, H we have:*

$$(I^\perp)^\perp = I, \quad (H^\perp)^\perp = H.$$

See [8] for a proof.

We will be interested in applying Theorem 3.1 when I is the ideal of some finite set of points in the n-dimensional affine space over k (e.g. when $k = \mathbb{F}_q$ and I is an error-locator ideal arising in decoding – see Section 4 below). In the following, we will use the notation m_P for the maximal ideal of S corresponding to the point $P \in k^s$.

Theorem 3.2. *Let $P_1, \ldots, P_t$ be points in k^s and let*

$$I = m_{P_1} \cap \cdots \cap m_{P_t}.$$

The submodule of T corresponding to I has the form

$$H = I^{\perp} = (m_{P_1})^{\perp} \oplus \cdots \oplus (m_{P_t})^{\perp}.$$

Proof. In Proposition 2.6 of [11], Geramita shows that $(I \cap J)^{\perp} = I^{\perp} + J^{\perp}$ for any pair of ideals. The idea is that $I^{\perp}$ and $J^{\perp}$ can be constructed degree by degree, so the corresponding statement from the linear algebra of finite-dimensional vector spaces applies. The equality $(I + J)^{\perp} = I^{\perp} \cap J^{\perp}$ also holds from linear algebra (and no finite-dimensionality is needed). The sum in the statement of the Lemma is a direct sum since $m_{P_i} + \cap_{j \neq i} m_{P_j} = S$, hence $(m_{P_i})^{\perp} \cap \Sigma_{j \neq i} (m_{P_j})^{\perp} = \{0\}$. □

We can also give a concrete description of the elements of $(m_P)^{\perp}$.

Theorem 3.3. *Let $P = (a_1, \ldots, a_s) \in \mathbb{A}^s$ over k, and let L_i be the coordinate hyperplane $X_i = a_i$ containing P.*

(1) $(m_P)^{\perp}$ is the cyclic S-submodule of T generated by

$$h_P = \sum_{u \in \mathbb{Z}^s_{\geq 0}} P^u X^{-u},$$

where if $u = (u_1, \ldots, u_s)$, P^u denotes the product $a_1^{u_1} \cdots a_s^{u_s}$ (X^u evaluated at P).

(2) $f \cdot h_P = f(P) h_P$ for all $f \in S$, and the submodule $(m_P)^{\perp}$ is a one-dimensional vector space over k.

(3) Let I_{L_i} be the ideal $\langle X_i - a_i \rangle$ in S (the ideal of L_i). Then $(I_{L_i})^{\perp}$ is the submodule of T generated by $h_{L_i} = \sum_{j=0}^{\infty} a_i^j X_i^{-j}$.

(4) In T, we have $h_P = \prod_{i=1}^{s} h_{L_i}$.

Proof. (1) First, if $f \in m_P$, and $g \in S$ is arbitrary then

$$\Lambda_{g \cdot h_P}(f) = (f \cdot (g \cdot h_P))_0 = ((fg) \cdot h_P)_0 = f(P)g(P) = 0.$$

Hence the S-submodule $\langle h_P \rangle$ is contained in $(m_P)^{\perp}$. Conversely, if $h \in (m_P)^{\perp}$, then for all $f \in m_P$,

$$0 = \Lambda_h(f) = (f \cdot h)_0.$$

An easy calculation using all f of the form $f = x^{\beta} - a^{\beta} \in m_P$ shows that $h = c h_P$ for some constant c. Hence $(m_P)^{\perp} = \langle h_P \rangle$.

(2) The second claim follows by a direct computation of the contraction product $f \cdot h_p$.

(3) Let $f \in I_{L_i}$ (so f vanishes at all points of the hyperplane L_i), and let $g \in S$ be arbitrary. Then

$$\begin{aligned}\Lambda_{g \cdot h_{L_i}}(f) &= (f \cdot (g \cdot h_{L_i}))_0 = ((fg) \cdot h_{L_i})_0 \\ &= f(0, \ldots, 0, a_i, 0, \ldots, 0) g(0, \ldots, 0, a_i, 0, \ldots, 0) = 0,\end{aligned}$$

since the only nonzero terms in the product $((fg) \cdot h_{L_i})$ come from monomials in fg containing only the variable X_i. Hence $\langle h_{L_i} \rangle \subset T$ is contained in $I_{L_i}^{\perp}$. Then we show the other inclusion as in the proof of (1).

(4) We have $m_P = I_{L_1} + \cdots + I_{L_s}$. Hence $(m_P)^{\perp} = (I_{L_1})^{\perp} \cap \cdots \cap (I_{L_s})^{\perp}$, and the claim follows. We note that a more explicit form of this equation can be derived by the formal geometric series summation formula:

$$h_P = \sum_{u \in \mathbb{Z}^s_{\geq 0}} P^u X^{-u} = \prod_{i=1}^{s} \frac{1}{1 - a_i/X_i} = \prod_{i=1}^{s} h_{L_i}. \qquad \square$$

Both the polynomial ring S and the formal power series ring T can be viewed as subrings of the field of formal Laurent series in the inverse variables,

$$K = k((X_1^{-1}, \ldots, X_s^{-1})),$$

which is the field of fractions of T. Hence the (full) product fg for $f \in S$ and $g \in T$ is an element of K. The contraction product $f \cdot g$ is a projection of fg into $T \subset K$. We can also consider the projection of fg into $S_+ = \langle X_1, \ldots, X_s \rangle \subset S \subset K$ under the linear projection with kernel spanned by all monomials not in S_+. We will denote this by $(fg)_+$.

4. The Key Equation and its Relation to the BMS Algorithm

Let C be one of the codes $C = ev(V)$ or $ev(V)^{\perp}$ constructed from an order domain $R \cong \mathbb{F}_q[X_1, \ldots, X_s]/I$. Consider an error vector $e \in \mathbb{F}_q^n$ (where entries are indexed by the elements of the set $X_R(\mathbb{F}_q)$). In the usual terminology, the *error-locator ideal* corresponding to e is the ideal $I_e \subset \mathbb{F}_q[X_1, \ldots, X_s]$ defining the set of error locations:

$$I_e = \{f \in \mathbb{F}_q[X_1, \ldots, X_s] : f(P) = 0, \ \forall\ P \ s.t.\ e_P \neq 0\}.$$

We will use a slightly different notation and terminology in the following because we want to make a systematic use of the observation that this ideal depends only on the support of e, not on the error values. Indeed, many different error vectors yield the same ideal defining the error locations. For this reason we will introduce $\mathcal{E} = \{P : e_P \neq 0\}$, and refer to the error-locator ideal for any e with $supp(e) = \mathcal{E}$ as $I_{\mathcal{E}}$.

For each monomial $X^u \in \mathbb{F}_q[X_1, \ldots, X_s]$, we let

$$E_u = \langle e, ev(X^u) \rangle = \sum_{P \in X_R(\mathbb{F}_q)} e_P P^u \tag{2}$$

be the corresponding syndrome of the error vector. (As in Theorem 3.3, P^u is shorthand notation for the evaluation of the monomial X^u at P.)

In the practical decoding situation, of course, for a code $C = ev(V)^{\perp}$ where V is a subspace of R spanned by some set of monomials, only the E_u for the X^u in a basis of V are initially known from the received word.

In addition, the elements of the ideal $I + \langle X_1^q - X_1, \ldots, X_s^q - X_s \rangle$ defining the set $X_R(\mathbb{F}_q)$ give relations between the E_u. Indeed, the E_u for u in the ordered basis Δ for R with all components $\leq q-1$ determine all the others, and these syndromes still satisfy additional relations. Thus the E_u are, in a sense, highly redundant.

To package the syndromes into a single algebraic object, following [12], we define the *syndrome series*

$$\mathcal{S}_e = \sum_{u \in \mathbb{Z}^s_{\geq 0}} E_u X^{-u}$$

in the formal power series ring $T = \mathbb{F}_q[[X_1^{-1}, \ldots, X_s^{-1}]]$. (This depends both on the set of error locations $\mathcal{E}$ and on the error values.) As in Section 3, we have a natural interpretation for $\mathcal{S}_e$ as an element of the dual space of the ring $S = \mathbb{F}_q[X_1, \ldots, X_s]$.

The following expression for the syndrome series $\mathcal{S}_e$ will be fundamental. We substitute from Eq. (2) for the syndrome E_u and change the order of summation to obtain:

$$\begin{aligned} \mathcal{S}_e &= \sum_{u \in \mathbb{Z}^n_{\geq 0}} E_u X^{-u} = \sum_{u \in \mathbb{Z}^n_{\geq 0}} \sum_{P \in X_R(\mathbb{F}_q)} e_P P^u X^{-u} \\ &= \sum_{P \in X_R(\mathbb{F}_q)} e_P \sum_{u \in \mathbb{Z}^n_{\geq 0}} P^u X^{-u} = \sum_{P \in X_R(\mathbb{F}_q)} e_P h_P, \end{aligned}$$

where h_P is the generator of $(m_P)^{\perp}$ from Theorem 3.3. The sum here taking the terms with $e_P \neq 0$, gives the decomposition of $\mathcal{S}_e$ in the direct sum expression for $I_{\mathcal{E}}^{\perp}$ as in Theorem 3.2.

The first statement in the following Theorem is well-known; it is a translation of the standard fact that error-locators give linear recurrences on the syndromes. But to our knowledge, this fact has not been considered from exactly our point of view in this generality (see [16] for a special case).

Theorem 4.1. *With all notation as above,*

(1) $f \in I_{\mathcal{E}}$ if and only if $f \cdot \mathcal{S}_e = 0$ for all error vectors e with supp(e) $= \mathcal{E}$.
(2) For each e with supp(e) $= \mathcal{E}$, $I_{\mathcal{E}} = \langle \mathcal{S}_e \rangle^{\perp}$ in the duality from Theorem 3.1.
(3) If e, e' are two error vectors with the same support, then $\langle \mathcal{S}_e \rangle = \langle \mathcal{S}_{e'} \rangle$ as submodules of T.

Proof. For (1), we start from the expression for $\mathcal{S}_e$ from Eq. (3). Then by Theorem 3.3, we have

$$f \cdot \mathcal{S}_e = \sum_{P \in \mathcal{E}} e_P (f \cdot h_P) = \sum_{P \in \mathcal{E}} e_P f(P) h_P.$$

If $f \in I_{\mathcal{E}}$, then clearly $f \cdot \mathcal{S}_e = 0$ for all choices of error values e_P. Conversely, if $f \cdot \mathcal{S}_e = 0$ for all e with *supp*(e) $= \mathcal{E}$, then $f(P) = 0$ for all $P \in \mathcal{E}$, so $f \in I_{\mathcal{E}}$.

Claim (2) follows from (1).

The perhaps surprising claim (3) is a consequence of (2). Another way to prove (3) is to note that there exist $g \in R$ such that $g(P)e_P = e'_P$ for all $P \in \mathcal{E}$. We have

$$g \cdot \mathcal{S}_e = \sum_{P \in \mathcal{E}} e_P (g \cdot h_P) = \sum_{P \in \mathcal{E}} e_P g(P) h_P = \sum_{P \in \mathcal{E}} e'_P h_P = \mathcal{S}_{e'}.$$

Hence $\langle \mathcal{S}_{e'} \rangle \subseteq \langle \mathcal{S}_e \rangle$. Reversing the roles of e and e', we get the other inclusion as well, and (3) follows. □

The following explicit expression for the terms in $f \cdot \mathcal{S}_e$ is also useful. Let $f = \sum_m f_m X^m \in S$. Then

$$f \cdot \mathcal{S}_e = (\sum_m f_m X^m) \cdot (\sum_{u \in \mathbb{Z}^s_{\geq 0}} E_u X^{-u}) = \sum_{r \in \mathbb{Z}^s_{\geq 0}} (\sum_m f_m E_{m+r}) X^{-r}.$$

Hence $f \cdot \mathcal{S}_e = 0 \Leftrightarrow \sum_m f_m E_{m+r} = 0$ for all $r \geq 0$.

The equation $f \cdot \mathcal{S} = 0$ from (1) in Theorem 4.1 is the prototype, so to speak, for our generalizations of the key equation to codes from order domains, and we will refer to it as the *key equation* in the following. It also naturally generalizes all the various key equations that have been developed in special cases, as we will demonstrate shortly. Before proceeding with that, however, we wish to make several comments about the form of this equation.

Comparing the equation $f \cdot \mathcal{S}_e = 0$ with the familiar form Eq. (1), several differences may be apparent. First, note that the syndrome series $\mathcal{S}_e$ will not be entirely known from the received word in the decoding situation. The same is true in the Reed-Solomon case, of course. The polynomial S in the congruence in Eq. (1) involves only the known syndromes, and Eq. (1) is derived by accounting for the other terms in the full syndrome series. With a truncation of $\mathcal{S}_e$ in our situation we would obtain a similar type of congruence (see the discussion following Eq. (8) below, for instance). It is apparently somewhat rare, however, that the portion of $\mathcal{S}_e$ known from the received word suffices for decoding up to half the minimum distance of the code.

Another difference is that there is no apparent analog of the error-evaluator polynomial g from Eq. (1) in the equation $f \cdot \mathcal{S}_e = 0$. The way to obtain error evaluators in this situation is to consider the "purely positive parts" $(f\mathcal{S}_e)_+$ for certain solutions of our key equation.

We now turn to several examples that show how our key equation relates to several special cases that have appeared in the literature.

Example 4.1. We begin by providing more detail on the precise relation between Theorem 4.1, part (1) in the case of a Reed-Solomon code and the usual key equation from Eq. (1). These codes are constructed from the order domain $R = \mathbb{F}_q[X]$ (where $\Gamma = \mathbb{Z}_{\geq 0}$ and ρ is the degree mapping). The key equation Eq. (1) applies to the code $Ev_\ell = ev(V_\ell)$, where $V_\ell = Span\{1, X, X^2, \ldots, X^{\ell-1}\}$, and the evaluation takes place at all $\mathbb{F}_q$-rational points on the affine line, omitting 0.

Our key equation in this case is closely related to, but not precisely the same, as Eq. (1). The reason for the difference is that Theorem 4.1 is applied to the dual code $C_\ell = Ev_\ell^\perp$ rather than Ev_ℓ. Starting from Eq. (3) and using the formal geometric series summation formula as in Theorem 3.3 part (4), we can write:

$$\mathcal{S}_e = \sum_{P \in \mathcal{E}} e_P \sum_{u \geq 0} P^u X^{-u} = X \frac{\sum_{P \in \mathcal{E}} e_P \prod_{Q \in \mathcal{E}, Q \neq P} (X - Q)}{\prod_{P \in \mathcal{E}} (X - P)}.$$

Hence, in this formulation, $\mathcal{S}_e = Xq/p$, where p is the generator of the actual error locator ideal (not the ideal of the inverses of the error locations). Moreover if we take $f = p$ in Theorem 4.1, then

$$(p\mathcal{S}_e)_+ = Xq \tag{3}$$

gives an analog of the error evaluator. There are no "mixed terms" in the products $f\mathcal{S}_e$ in this one-variable situation.

Example 4.2. The key equation for s-dimensional cyclic codes introduced by Chabanne and Norton in [12] has the form

$$\sigma \mathcal{S}_e = \left(\prod_{i=1}^{s} X_i \right) g, \tag{4}$$

where $\sigma = \prod_{i=1}^{s} \sigma_i(X_i)$, and σ_i is the univariate generator of the elimination ideal $I_{\mathcal{E}} \cap \mathbb{F}_q[X_i]$. Our version of the Reed-Solomon key equation from Eq. (3) is a special case of Eq. (4). Moreover, Eq. (4) is clearly the special case of Theorem 4.1, part (1) for these codes where $f = \sigma$ is the particular error locator polynomial $\prod_{i=1}^{s} \sigma_i(X_i) \in I_{\mathcal{E}}$. For this special choice of error locator, $\sigma \cdot \mathcal{S}_e = 0$, and $(\sigma \mathcal{S}_e)_+ = (\prod_{i=1}^{s} X_i)\, g$ for some polynomial g. We see that $\mathcal{S}_e$ can be written as

$$\mathcal{S}_e = \sum_P e_P h_P = \left(\prod_{i=1}^{s} X_i \right) \sum_P e_P \frac{1}{\prod_{i=1}^{s} (X_i - X_i(P))},$$

and the product $\sigma \mathcal{S}_e = (\sigma \mathcal{S}_e)_+$ reduces to a polynomial (again, there are no "mixed terms").

Example 4.3. We now turn to the key equation for one-point geometric Goppa codes introduced by O'Sullivan in [6]. Let $\mathcal{X}$ be a smooth curve over $\mathbb{F}_q$ of genus g, and consider one-point codes constructed from $R = \cup_{m=0}^{\infty} L(mQ)$ for some point $Q \in \mathcal{X}(\mathbb{F}_q)$, O'Sullivan's key equation has the form:

$$f \omega_e = \phi. \tag{5}$$

Here ω_e is the syndrome differential, which can be expressed as

$$\omega_e = \sum_{P \in \mathcal{X}(\mathbb{F}_q)} e_P \omega_{P,Q},$$

where $\omega_{P,Q}$ is the differential of the third kind on Y with simple poles at P and Q, no other poles, and residues $res_P(\omega_{P,Q}) = 1, res_Q(\omega_{P,Q}) = -1$. For any $f \in R$, we have

$$res_Q(f\omega_e) = \sum_P e_P f(P),$$

the syndrome of e corresponding to f. (We only defined syndromes for monomials above; taking a presentation $R = \mathbb{F}_q[X_1, \ldots, X_s]/I$, however, any $f \in R$ can be expressed as a linear combination of monomials and the syndrome of f is defined accordingly.) The right-hand side of Eq. (5) is also a differential. In this situation, Eq. (5) furnishes a key equation in the following sense: f is an error locator (i.e. f is in the ideal of R corresponding to $I_{\mathcal{E}}$) if and only if ϕ has poles only at Q. In the special case that $(2g-2)Q$ is a canonical divisor (the divisor of zeroes of some differential of the first kind ω_0 on $\mathcal{X}$), Eq. (5) can be replaced by the equivalent equation $fo_e = g$, where $o_e = \omega_e/\omega_0$ and $g = \phi/\omega_0$ are rational functions on $\mathcal{X}$. Since ω_0 is zero only at Q, the key equation is now that f is an error locator if and only if Eq. (5) is satisfied for some $g \in R$.

For instance, when $\mathcal{X}$ is a smooth plane curve $V(F)$ over $\mathbb{F}_q$ defined by $F \in \mathbb{F}_q[X,Y]$, with a single smooth point Q at infinity, then it is true that $(2g-2)Q$ is canonical. O'Sullivan shows in Example 4.2 of [6] (using a slightly different notation) that

$$o_e = \sum_{P \in \mathcal{X}(\mathbb{F}_q)} e_P H_P, \tag{6}$$

where if $P = (a,b)$, then $H_P = \frac{F(a,Y)}{(X-a)(Y-b)}$. This is a function with a pole of order 1 at P, a pole of order $2g-1$ at Q, and no other poles.

To relate this to our approach, note that we may assume from the start that $Q = (0:1:0)$ and that F is taken in the form

$$F(X,Y) = X^\beta - cY^\alpha + G(X,Y)$$

for some relatively prime $\alpha < \beta$ generating the value semigroup at Q. Every term in G has (α,β)-weight less than $\alpha\beta$. First we rearrange to obtain

$$H_P = \frac{F(a,Y)}{(X-a)(Y-b)} = \frac{(a^\beta - X^\beta) + F(X,Y) + (G(a,Y) - G(X,Y))}{(X-a)(Y-b)}$$

The $F(X,Y)$ term in the numerator does not depend on P. We can collect those terms in the sum Eq. (6) and factor out the $F(X,Y)$. We will see shortly that those terms can in fact be ignored. The $G(a,Y) - G(X,Y)$ in the numerator furnish terms that go into the error evaluator g here. The remaining portion is

$$\frac{-(X^\beta - a^\beta)}{(X-a)(Y-b)} = -\frac{X^{\beta-1}}{Y} \sum_{i=0}^{\beta-1} \sum_{j=0}^{\infty} \frac{a^i b^j}{X^i Y^j}.$$

The sum here looks very much like that defining our h_P from Theorem 3.3, except that it only extends over the monomials in complement of $\langle LT(F) \rangle$. Call this last sum h'_P. As noted before the full series h_P (and consequently $\mathcal{S}$) are redundant. For example, every ideal contained in m_P (for instance the ideal $I = \langle F \rangle$ defining the curve), produces relations between the coefficients. From the duality theorem, Theorem 3.1, we have that $I \subset m_P$ implies $(m_P)^\perp \subset I^\perp$, so $F \cdot h_P = 0$.

The relation $F \cdot h_P = 0$ says in particular that the terms in h'_P are sufficient to determine the whole series h_P. Indeed, we have

$$h_P = \sum_{i=0}^{\infty} \left(\frac{(cY^\alpha - G)}{X^\beta} \right)^i h'_P = \left(\frac{X^\beta}{F} \right) h'_P.$$

It follows that O'Sullivan's key equation and ours are equivalent.

We now turn to the precise relation between solutions of our key equation and the polynomials generated by the BMS decoding algorithm applied to the $C_\ell = Ev_\ell^\perp$ codes from order domains R. We will see that the BMS algorithm systematically produces successively better approximations to solutions of $f \cdot \mathcal{S}_e = 0$, so that in effect, *the BMS algorithm is a method for solving the key equation for these codes.*

For our purposes, it will suffice to consider the *"Basic Algorithm"* from §3 of [5], in which all needed syndromes are assumed known and no sharp stopping criteria are identified. The *syndrome mapping* corresponding to the error vector e is

$$\begin{aligned} Syn_e : R &\to \mathbb{F}_q \\ f &\mapsto \sum_{P \in \mathcal{E}} e_P f(P), \end{aligned}$$

where as above $\mathcal{E}$ is the set of error locations. The same reasoning used in the proof of our Theorem 4.1 shows

$$f \in I_{\mathcal{E}} \Leftrightarrow Syn_e(fg) = 0, \forall g \in R. \tag{7}$$

From Definition 2.1 and Geil and Pellikaan's presentation theorem, we have an ordered monomial basis of R:

$$\Delta = \{X^{\alpha(j)} : j \in \mathbb{N}\},$$

whose elements have distinct ρ-values. As in the construction of the Ev_ℓ codes, we write $V_\ell = Span\{1 = X^{\alpha(1)}, \ldots, X^{\alpha(\ell)}\}$. The V_ℓ exhaust R, so for $f \neq 0 \in R$, we may define

$$o(f) = \min\{\ell : f \in V_\ell\},$$

and (for instance) $o(0) = -1$. In particular the semigroup Γ in our presentation carries over to a (nonstandard) semigroup structure on $\mathbb{N}$ defined by the addition operation

$$i \oplus j = k \Leftrightarrow o(X^{\alpha(i)} X^{\alpha(j)}) = k.$$

Given $f \in R$, one defines

$$span(f) = \min\{\ell : \exists g \in V_\ell \text{ s.t. } Syn_e(fg) \neq 0\}$$
$$fail(f) = o(f) \oplus span(f).$$

When $f \in I_\mathcal{E}$, $span(f) = fail(f) = \infty$.

The BMS algorithm, then, is an iterative process which produces a Gröbner basis for $I_\mathcal{E}$ with respect to a certain monomial order $>$. The strategy is to maintain data structures for all $m \geq 1$ as follows. The Δ_m are an increasing sequence of sets of monomials, converging to the monomial basis for $I_\mathcal{E}$ as $m \to \infty$, and δ_m is the set of maximal elements of Δ_m with respect to $>$ (the "interior corners of the footprint"). Similarly, we consider the complement Σ_m of Δ_m, and σ_m, the set of minimal elements of Σ_m (the "exterior corners"). For sufficiently large m, the elements of σ_m will be the leading terms of the elements of the Gröbner basis of $I_\mathcal{E}$, and Σ_m will be the set of monomials in $LT_>(I_\mathcal{E})$.

For each m, the algorithm also produces collections of polynomials $F_m = \{f_m(s) : s \in \sigma_m\}$ and $G_m = \{g_m(c) : c \in \delta_m\}$ satisfying:

$$o(f_m(s)) = s, \quad fail(f_m(s)) > m$$
$$span(g_m(c)) = c, \quad fail(g_m(c)) \leq m.$$

In the limit as $m \to \infty$, by Eq. (7), the F_m yield the Gröbner basis for $I_\mathcal{E}$.

We record the following simple observation.

Theorem 4.2. *With all notation as above, suppose $f \in R$ satisfies $o(f) = s$, $fail(f) > m$. Then*

$$f \cdot \mathcal{S}_e \equiv 0 \bmod W_{s,m},$$

where $W_{s,m}$ is the $\mathbb{F}_q$-vector subspace of the formal power series ring T spanned by the $X^{-\alpha(j)}$ such that $s \oplus j > m$.

Proof. By the definition, $fail(f) > m$ means that $Syn_e(fX^{\alpha(k)}) = 0$ for all k with $o(f) \oplus k \leq m$. By the definitions of $\mathcal{S}_e$ and the contraction product, $Syn_e(fX^{\alpha(k)})$ is exactly the coefficient of $X^{-\alpha(k)}$ in $f \cdot \mathcal{S}_e$. $\square$

The subspace $W_{s,m}$ in Theorem 4.2 depends on $s = o(f)$. In our situation, though, note that if $s' = \max\{o(f) : f \in F_m\}$, then Theorem 4.2 implies

$$f \cdot \mathcal{S}_e \equiv 0 \bmod W_{s',m} \tag{8}$$

for all $f = f_m(s)$ in F_m. Moreover, only finitely many terms from $\mathcal{S}_e$ enter into any one of these congruences, so Eq. (8) is, in effect, a sort of general analog of Eq. (1).

The $f_m(s)$ from F_m can be understood as approximate solutions of key equation (where the goodness of the approximation is determined by the subspaces $W_{s',m}$, a decreasing chain, tending to $\{0\}$ in T, as $m \to \infty$). The BMS algorithm thus systematically constructs better and better approximations to solutions of the key equation. O'Sullivan's stopping criteria (see [5]) show when further steps of the algorithm make no changes. The Feng-Rao theorem shows that any additional syndromes needed for this can be determined by the majority-voting process when $wt(e) \leq \lfloor \frac{d_{FR}(C_\ell)-1}{2} \rfloor$.

We conclude by noting that O'Sullivan has also shown in [6] that, for codes from curves, the BMS algorithm can be slightly modified to compute error locators and error evaluators simultaneously in the situation studied in Example 4.3. The same is almost certainly true in our general setting, although we have not worked out all the details.

Acknowledgements

Thanks go to Mike O'Sullivan and Graham Norton for comments on an earlier version prepared while the author was a visitor at MSRI. Research at MSRI is supported in part by NSF grant DMS-9810361.

References

[1] T. Høholdt, R. Pellikaan, and J. van Lint, Algebraic Geometry Codes, in: *Handbook of Coding Theory*, W. Huffman and V. Pless, eds. (Elsevier, Amsterdam, 1998), 871-962.

[2] M. O'Sullivan, New Codes for the Berlekamp-Massey-Sakata Algorithm, *Finite Fields Appl.* **7** (2001), 293-317.

[3] O. Geil and R. Pellikaan, On the Structure of Order Domains, *Finite Fields Appl.* **8** (2002), 369-396.

[4] J. Little, The Ubiquity of Order Domains for the Construction of Error Control Codes, *Advances in Mathematics of Communications* **1** (2007), 151-171.

[5] M. O'Sullivan, A Generalization of the Berlekamp-Massey-Sakata Algorithm, preprint, 2001.

[6] M. O'Sullivan, The key equation for one-point codes and efficient error evaluation, *J. Pure Appl. Algebra* **169** (2002), 295-320.

[7] F.S. Macaulay, *Algebraic Theory of Modular Systems*, Cambridge Tracts in Mathematics and Mathematical Physics, v. 19, (Cambridge University Press, Cambridge, UK, 1916).

[8] D.G. Northcott, Injective envelopes and inverse polynomials, *J. London Math. Soc. (2)* **8** (1974), 290-296.

[9] J. Emsalem and A. Iarrobino, Inverse System of a Symbolic Power, I, *J. Algebra* **174** (1995), 1080-1090.

[10] B. Mourrain, Isolated points, duality, and residues *J. Pure Appl. Algebra* **117/118** (1997), 469-493.

[11] A. Geramita, Inverse systems of fat points, Waring's problem, secant varieties of Veronese varieties and parameter spaces for Gorenstein ideals, *The Curves Seminar at Queen's (Kingston, ON)* **X** (1995), 2–114.

[12] H. Chabanne and G. Norton, The n-dimensional key equation and a decoding application, *IEEE Trans. Inform Theory* **40** (1994), 200-203.

[13] G.H. Norton, On n-dimensional Sequences. I, II, *J. Symbolic Comput.* **20** (1995), 71-92, 769-770.

[14] G.H. Norton, On Shortest Linear Recurrences, *J. Symbolic Comput.* **27** (1999), 323-347.

[15] G.H. Norton and A. Salagean, On the key equation over a commutative ring, *Designs, Codes and Cryptography* **20** (2000), 125-141.

[16] J. Althaler and A. Dür, Finite linear recurring sequences and homogeneous ideals, *Appl. Algebra. Engrg. Comm. Comput.* **7** (1996), 377-390.

A Gröbner representation for linear codes

M. Borges-Quintana* and M. A. Borges-Trenard**

Departamento de Matemáticas, Universidad de Oriente,
Santiago de Cuba, Cuba
* *E-mail: mijail@csd.uo.edu.cu*
** *E-mail: mborges@csd.uo.edu.cu*

E. Martínez-Moro

Departamento de Matemática Aplicada,
Universidad de Valladolid
Valladolid, Spain
E-mail: edgar@maf.uva.es

This work explains the role of Möller algorithm and Gröbner technology in the description of linear codes. We survey several results of the authors about FGLM techniques applied to linear codes as well as some results concerning the structure of the code.

Keywords: Linear code; Möller algorithm; Gröbner representation.

1. Introduction

In this paper we survey several results of the authors about the nice role of Gröbner bases technology and Möller FGLM techniques (FGLM stands for Faugère, Gianni, Lazard and Mora, see [7]) applied to linear codes over finite fields. This work is intended as an attempt to clarify and summarize as well as unify several previous works of the authors [3–5]. We follow Teo Mora's approach for the presentation of Gröbner bases theory [15] and study how this theory can describe several combinatorial properties of linear codes. Section 2 contains a brief summary of Möller algorithm and related concepts. In the third section we set up the notation and terminology of the structures associated to a linear code. In section 4 we will look more closely at those structures and we will indicate the resemblances with the Gröbner bases technology, with special emphasis on the binary case. Although it is not the main goal of this survey, in section 5 we point out several directions of how these techniques can be used to derive solutions for several coding theory problems such that gradient decoding, combinatorial problems, minimal codeword bases, etc.

2. Möller's algorithm

No attempt has been made here to develop the whole theory of Möller algorithm. We will touch only a few aspects of the theory useful for our paper. For a thorough treatment of Gröbner bases we refer the reader to [15] and for a recent survey on Möller algorithm and FGLM techniques we refer to [16].

As usual we will denote by X the finite set of variables $\{x_1, \ldots, x_n\}$ and if $\mathbf{a} = (a_1, \ldots, a_n) \in \mathbb{N}^n$ we will denote $\mathbf{x}^{\mathbf{a}} = x_1^{a_1} \ldots x_n^{a_n}$. Let $\mathcal{P} = \mathbb{F}[X]$ the polynomial ring over the field $\mathbb{F}$ and $\mathcal{T} = \{\mathbf{x}^{\mathbf{a}} \mid \mathbf{a} \in \mathbb{N}^n\}$ the set of terms. Let $\prec$ be a Notherian semigroup ordering on the set $\mathcal{T}$ (this is called either term ordering or admissible ordering), for each $f = \sum_{\tau \in \mathcal{T}} c(f, \tau)\tau \in \mathcal{P}$ we write $\mathrm{T}(f) = \max_{\prec}\{\tau \in \mathcal{T} \mid c(f, \tau) \neq 0\}$ and $\mathrm{lc}(f) = c(f, \mathrm{T}(f))$ for the leading term and leading coefficient of f respectively.

If $F \subseteq \mathcal{P}$ then $\mathrm{T}(F) = \{\mathrm{T}(f) \mid f \in F\}$ and for each ideal $\mathcal{I} \subset \mathcal{P}$ we consider the semigroup ideal $\mathrm{T}(\mathcal{I})$ and the Gröbner éscalier $\mathrm{N}(\mathcal{I}) = \mathcal{I} \backslash \mathrm{T}(\mathcal{I})$. It is well known that $\mathcal{P} \cong \mathcal{I} \bigoplus \mathrm{span}_{\mathbb{F}}(\mathrm{N}(\mathcal{I}))$ as $\mathbb{F}$-vector spaces, which in turn gives a unique canonical form for each element $f \in \mathcal{P}$

$$\mathrm{Can}(f, \mathcal{I}, \prec) = \sum_{\tau \in \mathrm{N}(\mathcal{I})} c(f, \tau, \prec)\tau \in \mathrm{span}_{\mathbb{F}}(\mathrm{N}(\mathcal{I})) \tag{1}$$

such that $f - \mathrm{Can}(f, \mathcal{I}, \prec) \in \mathcal{I}$.

Let $\mathrm{G}_{\prec}(\mathcal{I})$ denote the unique minimal basis of $\mathrm{T}_{\prec}(\mathcal{I})$, a set $\mathcal{G} \subseteq \mathcal{I}$ is said to be a Gröbner basis of the ideal $\mathcal{I}$ with respect to (w.r.t. for short) the ordering $\prec$ if the set $\mathrm{T}_{\prec}(\mathcal{G})$ generates $\mathrm{T}_{\prec}(\mathcal{I})$ as a semigroup ideal. The reduced Gröbner basis of the ideal $\mathcal{I}$ w.r.t. $\prec$ is the set

$$\mathrm{Red}_{\prec}(\mathcal{I}) = \{\tau - \mathrm{Can}(\tau, \mathcal{I}, \prec) \mid \tau \in \mathrm{G}_{\prec}(\mathcal{I})\}, \tag{2}$$

and the border basis of $\mathcal{I}$ w.r.t. $\prec$ is

$$\mathrm{Bor}_{\prec}(\mathcal{I}) = \{\tau - \mathrm{Can}(\tau, \mathcal{I}, \prec) \mid \tau \in \mathrm{B}_{\prec}(\mathcal{I})\}, \tag{3}$$

thus the border basis of the ideal $\mathcal{I}$ is a Gröbner basis of $\mathcal{I}$ that contains the reduced Gröbner basis.

An ideal $\mathcal{I} \subset \mathcal{P}$ is zero-dimensional if $\dim_{\mathbb{F}}(\mathcal{P}/\mathcal{I}) < \infty$ where $\dim_{\mathbb{F}}$ denotes the dimension as $\mathbb{F}$ vector space. From now on we will make the assumption that our ideal $\mathcal{I}$ is zero-dimensional. The following representation will play a central role in the paper

Definition 2.1 (Gröbner representation). *Let $\mathcal{I} \subset \mathcal{P}$ be a zero-dimensional ideal and $s = \dim_{\mathbb{F}}(\mathcal{P}/\mathcal{I})$. A Gröbner representation of $\mathcal{I}$ is the assignment of*

(1) a set $N = \{\tau_1, \ldots, \tau_s\} \subseteq \mathrm{N}_{\prec}(\mathcal{I})$
(2) and a set of square matrices

$$\phi = \left\{ \phi(r) = \left(a_{ij}^r\right)_{i,j=1}^s \mid r = 1, \ldots, n \,, a_{ij}^r \in \mathbb{F} \right\}$$

such that

$$\mathcal{P}/\mathcal{I} = \mathrm{span}_{\mathbb{F}}(N), \qquad \tau_i x_r \equiv_{\mathcal{I}} \sum_{j=1}^{s} a_{ij}^r \tau_j \quad \forall 1 \le i \le s, 1 \le r \le n.$$

We call ϕ the matphi structure and $\phi(r)$ the matphi matrices. They first appear in [7] in a procedure to describe the multiplication structure in the quotient algebra $\mathcal{P}/\mathcal{I}$. Note that ϕ is independent of the particular set N of representatives of $\mathcal{P}/\mathcal{I}$ we have chosen. For each $f \in \mathcal{P}$ the Gröbner description of f in terms of the Gröbner representation (N, ϕ) is

$$\mathrm{Rep}(f, N) = (\gamma(f, \tau_1), \ldots, \gamma(f, \tau_s)) \in \mathbb{F}^s$$

such that $f - \sum_{i=1}^{s} \gamma(f, \tau_i)\tau_i \in \mathcal{I}$.

We write $\mathcal{P}^* = \mathrm{Hom}_{\mathbb{F}}(\mathcal{P}, \mathbb{F})$ to denote the vector space of all linear functionals $\ell : \mathcal{P} \to \mathbb{F}$. $\mathcal{P}^*$ is a $\mathcal{P}$-module defined by the product

$$(\ell \cdot f)(g) = \ell(fg) \quad \ell \in \mathcal{P}^*, f, g \in \mathcal{P}$$

where

$$\ell(f) = \sum_{\tau \in \mathcal{T}} c(f, \tau)\ell(\tau).$$

Two ordered sets $L = \{\ell_1, \ldots, \ell_r\} \subset \mathcal{P}^*$, $q = \{q_1, \ldots, q_s\} \subset \mathcal{P}$ are said to be triangular if $r = s$ and $\ell_i(q_j) = 0$ for all $i < j$. For each $\mathbb{F}$-vector space $L \subseteq \mathcal{P}^*$ we define the ideal $\mathfrak{P}(L) = \{g \in \mathcal{P} \mid \ell(g) = 0, \forall \ell \in L\}$.

Proposition 2.1 (Möller's theorem). *Let $\prec$ be any term ordering and $L = \{\ell_1, \ldots, \ell_s\} \subset \mathcal{P}^*$ be a set of functionals such that $\mathcal{I} = \mathfrak{P}(L)$ is a zero-dimensional ideal. Then there are a $r \in \mathbb{N}$, an order ideal $N = \{\tau_1, \ldots, \tau_r\} \subset \mathcal{T}$ and two ordered subsets*

$$\Lambda = \{\lambda_1, \ldots, \lambda_r\} \subset L, \quad \mathbf{q} = \{q_1, \ldots, q_r\} \subset \mathcal{P}$$

such that

(1) $r = \deg(\mathcal{I}) = \dim_{\mathbb{F}}(\mathrm{span}_{\mathbb{F}}(L))$.

(2) $\mathrm{N}_\prec(\mathcal{I}) = N$.
(3) $\mathrm{span}_\mathbb{F}(L) = \mathrm{span}_\mathbb{F}(\Lambda)$.
(4) $\mathrm{span}_\mathbb{F}(\tau_1, \ldots, \tau_\nu) = \mathrm{span}_\mathbb{F}(q_1, \ldots, q_\nu)$ *for all* $\nu \leq r$.
(5) *The sets* $\{\lambda_1, \ldots, \lambda_\nu\}$ *and* $\{q_1, \ldots, q_\nu\}$ *are triangular for all* $\nu \leq r$.

Möller's algorithm [7, 13, 14] is a procedure that returns the data stated in the proposition above given a set of linear functionals L such that $\mathfrak{P}(L)$. As a byproduct of Möller algorithm one can compute a Gröbner representation of the ideal. We will give in next section a modified version of such algorithm adapted to the setting of linear codes.

3. Gröbner representation of a linear code

We will touch only a few aspects of the theory of linear codes over finite fields, the reader is expected to be familiar with basic algebraic coding theory (see [12] for a basic account). Just to fix the notation we will give a few notions of linear codes in the following paragraphs.

Let $\mathbb{F}_q$ a finite field with q elements ($q = p^m$, p a prime and $m \in \mathbb{N}$). A linear code $\mathcal{C}$ of length n and dimension k ($k < n$) is the image of an injective linear mapping $c : \mathbb{F}_q^k \to \mathbb{F}_q^n$. All codes in this paper are linear and from now on will write code for linear code.

The set

$$\mathcal{C}^\perp = \{\ell \in (\mathbb{F}_q^n)^* \mid \ell(\mathbf{c}) = 0 \text{ for all } \mathbf{c} \in \mathcal{C}\} \tag{4}$$

is a $\mathbb{F}_q$-linear subspace of $(\mathbb{F}_q^n)^* = \mathrm{Hom}(\mathbb{F}_q^n, \mathbb{F}_q)$ of dimension $n - k$, thus $\mathcal{C}^\perp$ can be seen as a code of length n and dimension $n - k$ over the field $\mathbb{F}_q$ called dual code of $\mathcal{C}$ (just fixing coordinates in $(\mathbb{F}_q^n)^*$) . A generator matrix of the code $\mathcal{C}$ is a $k \times n$ matrix such that its rows span $\mathcal{C}$ as a $\mathbb{F}_q$-linear space. If we consider the dual standard basis in $(\mathbb{F}_q^n)^*$ the generator matrix H of $\mathcal{C}^\perp$ fulfills $H \cdot \mathbf{c} = \mathbf{0}$ for all $\mathbf{c} \in \mathcal{C}$ and is called parity check matrix.

The Hamming weight of a vector $\mathbf{v} \in \mathbb{F}_q^n$ is the number of non-zero entries in $\mathbf{v}$ and will be denoted by $\mathrm{w}_h(\mathbf{v})$. The Hamming distance between two vectors $\mathbf{u}$ and $\mathbf{v}$ is defined as $\mathrm{d}_h(\mathbf{u}, \mathbf{v}) = \mathrm{w}_h(\mathbf{u} - \mathbf{v})$ and the minimum distance d of the code $\mathcal{C}$ is the minimum Hamming weight among all its non-zero codewords. The error correcting capacity of a code is[a] $t = \lfloor \frac{d-1}{2} \rfloor$.

Let $\mathcal{C}$ be a code and H be its parity check matrix, the syndrome of a vector $\mathbf{u} \in \mathbb{F}_q^n$ is $H \cdot \mathbf{u}$. Two vectors belong to the same coset if and only if they have the same syndrome. The *weight of a coset* is the smallest weight

[a] $\lfloor \cdot \rfloor$ denotes the floor function.

of a vector in the coset and any vector of smallest weight in the coset is called a *coset leader*. Every coset of a weight at most t has a unique coset leader thus the equation

$$H \cdot \mathbf{u} = H \cdot \mathbf{e}$$

has a unique minimal weight solution $\mathbf{e}$ among the coset leaders of the code $\mathcal{C}$ for each $\mathbf{u} \in B(\mathcal{C}, t)$ called the *error vector* of $\mathbf{u}$ where

$$B(\mathcal{C}, t) = \left\{ \mathbf{u} \in \mathbb{F}_q^n \mid \exists \mathbf{c} \in \mathcal{C} \text{ such that } \mathrm{d}_h(\mathbf{u}, \mathbf{c}) \leq t \right\}.$$

If we fix α a root of an irreducible polynomial of degree m over $\mathbb{F}_p$ we can represent any element of $\mathbb{F}_q$ as $a_0 + a_1\alpha + \cdots + a_{m-1}\alpha^{m-1}$ with $a_i \in \mathbb{F}_p$ for all i. Let $\mathcal{T}$ be the set of terms, i.e., the free commutative monoid generated by the nm variables $X = \{x_{11}, \ldots, x_{1m}, \ldots, x_{n1}, \ldots, x_{nm}\}$, and consider the morphism of monoids from $\mathcal{T}$ onto $\mathbb{F}_q^n$:

$$\begin{aligned} \psi : \mathcal{T} &\to \mathbb{F}_q^n \\ x_{ij} &\mapsto (0, \ldots, 0, \underbrace{\alpha^{j-1}}_{i}, 0, \ldots, 0) \end{aligned}$$

and, by morphism extension,

$$\prod_{i=1}^{n} \prod_{j=1}^{m} x_{ij}^{\beta_{ij}} \mapsto \left(\left(\textstyle\sum_{j=1}^{m} \beta_{1j}\alpha^{j-1} \right), \ldots, \left(\textstyle\sum_{j=1}^{m} \beta_{nj}\alpha^{j-1} \right) \right) \tag{5}$$

We say that $\prod_{i=1}^{n} \prod_{j=1}^{m} x_{ij}^{\beta_{ij}} \in \mathcal{T}$ is in standard representation if $\beta_{ij} < p$ for all i, j.

A code $\mathcal{C}$ defines an equivalence relation $R_{\mathcal{C}}$ in $\mathbb{F}_q^n$ given by

$$(\mathbf{u}, \mathbf{v}) \in R_{\mathcal{C}} \Leftrightarrow \mathbf{u} - \mathbf{v} \in \mathcal{C}. \tag{6}$$

This relation can be translated to $\mathbf{x}^{\mathbf{a}}, \mathbf{x}^{\mathbf{b}} \in \mathcal{T}$ as follows

$$\mathbf{x}^{\mathbf{a}} \equiv_{\mathcal{C}} \mathbf{x}^{\mathbf{b}} \Leftrightarrow (\psi(\mathbf{x}^{\mathbf{a}}), \psi(\mathbf{x}^{\mathbf{b}})) \in R_{\mathcal{C}} \Leftrightarrow \xi_{\mathcal{C}}(\mathbf{x}^{\mathbf{a}}) = \xi_{\mathcal{C}}(\mathbf{x}^{\mathbf{b}}) \tag{7}$$

where $\xi_{\mathcal{C}}(\mathbf{x}^{\mathbf{a}}) = H \cdot \psi(\mathbf{x}^{\mathbf{a}})$ is the transition from the monoid $\mathcal{T}$ to the set of syndromes associated to the word u through ψ.

The support of $\mathbf{x}^{\mathbf{a}} \in \mathcal{T}$ will be the set of variables in X that divide $\mathbf{x}^{\mathbf{a}}$ and is denoted by $\mathrm{supp}(\mathbf{x}^{\mathbf{a}})$ whereas the index[b] of $\mathbf{x}^{\mathbf{a}}$ is defined as

$$\mathrm{ind}(\mathbf{x}^{\mathbf{a}}) = \{ i \mid \exists j \in \{1, \ldots, m\} \text{ such that } x_{ij} \in \mathrm{supp}(\mathbf{x}^{\mathbf{a}}) \}. \tag{8}$$

For the sake of simplicity in notation from now on we write the set of nm variables as x_k, where $k = (i-1)m + j$ instead of x_{ij}.

[b]Note that this definition for elements in $\mathcal{T}$ corresponds to the definition of support of the corresponding vector in $\mathbb{F}_q^n$.

Definition 3.1 (Error vector ordering). *We say that* $\mathbf{x}^{\mathbf{a}}$ *is less than* $\mathbf{x}^{\mathbf{b}}$ *w.r.t. the error-vector ordering, and denote it by* $\mathbf{x}^{\mathbf{a}} \prec_e \mathbf{x}^{\mathbf{b}}$*, if one of the following conditions holds:*

(1) $|\mathrm{ind}(\mathbf{x}^{\mathbf{a}})| < |\mathrm{ind}(\mathbf{x}^{\mathbf{b}})|$.
(2) $|\mathrm{ind}(\mathbf{x}^{\mathbf{a}})| = |\mathrm{ind}(\mathbf{x}^{\mathbf{b}})|$ *and* $\mathbf{x}^{\mathbf{a}} \prec_{ad} \mathbf{x}^{\mathbf{b}}$*, where* $\prec_{ad}$ *denotes an arbitrary but fixed admissible ordering on* $\mathcal{T}$.

Note that the error vector ordering is a total degree compatible ordering on $\mathcal{T}$ but in general it is not admissible (the multiplicative property of admissible orderings sometimes fails). For example let the vector space $\mathbb{F}_3^7$ and $\prec_{ad}$ be the degree reverse lexicographical ordering, we have

$$x_1x_5 \prec_e x_3x_7 \quad \text{but} \quad x_1x_5x_7 \succ_e x_3x_7^2.$$

Anyway the error vector ordering still shares two important properties of admissible orderings

(1) $1 \prec_e \mathbf{x}^{\mathbf{a}}$ for all $\mathbf{a} \neq \mathbf{0}$.
(2) $\mathbf{x}^{\mathbf{a}} \prec_e \mathbf{x}^{\mathbf{a}}x_i$ for all $i = 1, \ldots, n$.

The two properties above will allow us to construct a Gröbner representation of a code using a sort of Möller algorithm as an analogue of the Gröbner representation of a zero-dimensional ideal in Definition 2.1.

Definition 3.2 (Gröbner representation of a code). *Let* $\mathcal{C}$ *be a* $\mathbb{F}_q$*-linear code of dimension* k*. A Gröbner representation of* $\mathcal{C}$ *is the assignment of*

- *a set* $N = \{\tau_1, \ldots, \tau_{q^{n-k}}\} \subseteq \mathcal{T}$
- *and a function* $\phi : N \times X \to N$ *(the function Matphi)*

such that

(1) $1 \in N$.
(2) If $\tau_1, \tau_2 \in N$ *and* $\tau_1 \neq \tau_2$ *then* $\xi_{\mathcal{C}}(\tau_1) \neq \xi_{\mathcal{C}}(\tau_2)$.
(3) For all $\tau \in N \setminus \{1\}$ *there exist* $x \in X$ *such that* $\tau = \tau' x$ *and* $\tau' \in N$.
(4) $\xi_{\mathcal{C}}(\phi(\tau, x_i)) = \xi_{\mathcal{C}}(\tau x_i)$.

Note that N has as many elements as different syndromes has code $\mathcal{C}$ and condition (2) states that two different elements in N have different syndrome. The function ϕ gives us a multiplicative structure that is independent of the particular set N of representative elements of the cosets determined by the code (i.e. ϕ can be seen as a function on the cosets of the code).

The following algorithm is an instance of the general Möller algorithm. Note that in the case of codes we can specify a system of generators of $C^{\perp}$ just by giving the parity check matrix H of the code. For our purpose we are just interested in computing a Gröbner representation of the code.

Algorithm 3.1 (Möller's algorithm for codes).

Input: *The parity check matrix of a linear code* $\mathcal{C}$ *over* $\mathbb{F}_q$ *and* m *such that* $p^m = q$, p *a prime number.*

Output: N, ϕ *for* $\mathcal{C}$ *as in Definition 3.2.*

1: List $\leftarrow \{1\}$, $N \leftarrow \emptyset$, $r \leftarrow 0$
2: ***while*** List $\neq \emptyset$ ***do***
3: $\quad \tau \leftarrow$ NextTerm[List], $\mathbf{v} \leftarrow \xi_{\mathcal{C}}(\tau)$
4: $\quad j \leftarrow$ Member[$\mathbf{v}, \{\mathbf{v}_1, \ldots, \mathbf{v}_r\}$]
5: $\quad$ ***if*** $j \neq$ false ***then***
6: $\quad\quad$ ***for*** k *such that* $\tau = \tau' x_k$ *with* $\tau' \in N$ ***do***
7: $\quad\quad\quad \phi(\tau', x_k) = \tau_j$
8: $\quad\quad$ ***end for***
9: $\quad$ ***else***
10: $\quad\quad r \leftarrow r+1$, $\mathbf{v}_r \leftarrow \mathbf{v}$, $\tau_r \leftarrow \tau$, $N \leftarrow N \cup \{\tau_r\}$
11: $\quad\quad$ List $\leftarrow$ InsertNext[τ_r, List]
12: $\quad\quad$ ***for*** k *such that* $\tau_r = \tau' x_k$ *with* $\tau' \in N$ ***do***
13: $\quad\quad\quad \phi(\tau', x_k) = \tau_r$
14: $\quad\quad$ ***end for***
15: $\quad$ ***end if***
16: ***end while***

Where the internal functions in the algorithm are

(1) InsertNext[τ, List] Inserts all the products τx in List, where $x \in X$, and keeps List in increasing order w.r.t. the order $\prec_e$.
(2) NextTerm[List] returns the first element from List and deletes it from that set.
(3) Member[*obj*, G] returns the position j of *obj* in G if *obj* $\in G$ and false otherwise.

For the proof of correctness of the algorithm we refer the reader to [5]. Note that by the construction, those representatives of the cosets given in N such that are syndromes corresponding to vectors in $B(\mathcal{C}, t)$ are the smallest terms in $\mathcal{T}$ w.r.t. $\prec_e$, i.e. they are the standard terms whose images by ψ are the error vectors.

An important byproduct of this construction is the following theorem that allows us to compute the error correcting capability of a code (see [5] for a proof)

Theorem 3.1. *Let* List *be the list of words in Step 3 of the previous algorithm and let* τ *be the first element analyzed by* NextTerm[List] *such that* τ *does not belong to* N *and* τ *is in standard representation. Then*

$$t = |\mathrm{ind}(\tau)| - 1. \tag{9}$$

Note that we do not need to run the whole algorithm in order to compute such element τ in the theorem above, we just need to compute the first one.

Example 3.1. Consider the binary linear code $\mathcal{C}$ in $\mathbb{F}_2^6$ with generator matrix:

$$G = \begin{pmatrix} 1\,0\,0\,1\,1\,1 \\ 0\,1\,0\,1\,0\,1 \\ 0\,0\,1\,0\,1\,1 \end{pmatrix}.$$

The set of codewords is
$\mathcal{C} = \{(0, 0, 0, 0, 0, 0), (1, 0, 1, 1, 0, 0), (1, 1, 0, 0, 1, 0), (0, 1, 0, 1, 0, 1)$
$(0, 0, 1, 0, 1, 1), (1, 1, 1, 0, 0, 1), (0, 1, 1, 1, 1, 0), (1, 0, 0, 1, 1, 1)\}$.

Let $\prec_{ad}$ be the degree reverse lexicographical ordering induced by $x_1 \prec x_2 \prec \ldots \prec x_6$. Running Algorithm 3.1 it computes

$$N = \{1, x_1, x_2, x_3, x_4, x_5, x_6, x_1x_6\}$$

and ϕ is represented as a matrix of positions (pointer matrix) as follows

$$\begin{aligned}
&[\,[[0,\ 0,\ 0,\ 0,\ 0,\ 0],\ 1,\ [2,3,4,5,6,7]],[[1,\ 0,\ 0,\ 0,\ 0,\ 0],\ 1,\ [1,6,5,4,3,8]],\\
&[[0,\ 1,\ 0,\ 0,\ 0,\ 0],\ 1,\ [6,1,8,7,2,5]],[[0,\ 0,\ 1,\ 0,\ 0,\ 0],\ 1,\ [5,8,1,2,7,6]],\\
&[[0,\ 0,\ 0,\ 1,\ 0,0],\ 1,\ [4,7,2,1,8,3]],[[0,\ 0,\ 0,\ 0,\ 1,\ 0],\ 1,\ [3,2,7,8,1,4]],\\
&[[0,\ 0,\ 0,\ 0,\ 0,\ 1],\ 1,\ [8,5,6,3,4,1]],[[1,\ 0,\ 0,\ 0,\ 0,\ 1],\ 0,\ [7,4,3,6,5,2]]\,]
\end{aligned}$$

where in each triple the first entry correspond to the elements $\psi(\tau)$ where $\tau \in N$ ($\tau = N[i]$), the second entry is 1 if $\psi(\tau) \in B(\mathcal{C}, t)$ or 0 otherwise, and the third component points to the values $\phi(\tau, x_j)$, for $j = 1, \ldots, 6$.

4. Reduced and border bases

Following the analogy between the Gröbner representation of a code $\mathcal{C}$ and the Gröbner representation of an ideal presented in section 2 we will consider the border basis of the code $\mathcal{C}$ w.r.t. the error vector ordering $\prec_e$ given by the set of binomials

$$\mathrm{Bor}_{\prec_e}(\mathcal{C}) = \{\tau x - \tau' \mid \tau, \tau' \in N, x \in X, \tau x \neq \tau', \xi_{\mathcal{C}}(\tau x) = \xi_{\mathcal{C}}(\tau')\}. \quad (10)$$

Note that this set is closely related with the structure matphi since

$$\mathrm{Bor}_{\prec_e}(\mathcal{C}) = \{\tau x - \phi(\tau, x) \mid \tau \in N, x \in X\} \setminus \{0\}, \quad (11)$$

i.e., the border basis of the code $\mathcal{C}$ w.r.t. $\prec_e$ contains all the binomials corresponding to the non trivial pairs $(\tau x, \phi(\tau, x)) \in R_{\mathcal{C}}$.

As in every Gröbner bases technology, we will define a reduction to the set of canonical forms in N by the following statement

Definition 4.1 (One step reduction). *Let N, ϕ be as in Definition 3.2 for a code $\mathcal{C}$ and $\tau \in N$, $x \in X$, we say that $\phi(\tau, x)$ is the canonical form of τx, i.e. τx reduces in one step to $\phi(\tau, x)$.*

This reduction definition can be extended to the set $\mathcal{T}$ as follows: Let $\mathbf{x}^{\mathbf{a}} = x_{i_1} \ldots x_{i_k} \in \mathcal{T}$, $x_{i_j} \prec_e x_{i_k}$ for all $j \leq k-1$ and consider the recursive function

$$\begin{array}{rcl} \mathrm{Can}_{\prec_e} : \mathcal{T} & \longrightarrow & N \\ 1 & \mapsto & 1 \\ \mathbf{x}^{\mathbf{a}} & \mapsto & \phi(\mathrm{Can}_{\prec_e}(x_{i_1} \ldots x_{i_{k-1}}), x_{i_k}). \end{array} \quad (12)$$

where the initial case is the empty word represented by 1. The element $\mathrm{Can}_{\prec_e}(\mathbf{x}^{\mathbf{a}}) \in N$ with the same syndrome as $\mathbf{x}^{\mathbf{a}}$ since

$$\begin{aligned} \xi_{\mathcal{C}}\left(\mathrm{Can}_{\prec_e}(x_{i_1} \ldots x_{i_k})\right) &= \xi_{\mathcal{C}}\left(\phi\left(\mathrm{Can}_{\prec_e}(x_{i_1} x_{i_2} \ldots x_{i_{k-1}}), x_{i_k}\right)\right) \\ &= \xi_{\mathcal{C}}\left(\mathrm{Can}_{\prec_e}(x_{i_1} x_{i_2} \ldots x_{i_{k-1}}) x_{i_k}\right) \\ &= \xi_{\mathcal{C}}\left(\mathrm{Can}_{\prec_e}(x_{i_1} x_{i_2} \ldots x_{i_{k-1}})\right) + \xi_{\mathcal{C}}(x_{i_k}) \end{aligned} \quad (13)$$

where the second equality in (13) holds by the definition of ϕ and the third one due to the additivity of $\xi_{\mathcal{C}}$, and we now compute by recursion

$$\xi_{\mathcal{C}}\left(\mathrm{Can}_{\prec_e}(x_{i_1} \ldots x_{i_k})\right) = \xi_{\mathcal{C}}\left(\mathrm{Can}_{\prec_e}(1) x_{i_1} x_{i_2} \ldots x_{i_k}\right) = \xi_{\mathcal{C}}\left(x_{i_1} x_{i_2} \ldots x_{i_k}\right)$$

thus both syndromes are equal. It remains to prove that $\mathrm{Can}_{\prec_e}$ is well defined for all the elements on $\mathcal{T}$ by the recurrence in (12), but this follows from steps 10 and 11 in Algorithm 3.1. Note that the recurrence procedure we just have described is just recursive applications of border basis reduction.

Finally we introduce the notion of reduced basis for the code $\mathcal{C}$ as follows

Definition 4.2 (Reduced basis of a code). *The reduced basis in $\mathbb{F}[X]$ for the code $\mathcal{C}$ w.r.t the ordering $\prec_e$ is a set $\mathrm{Red}_{\prec_e}(\mathcal{C}) \subseteq \mathrm{Bor}_{\prec_e}(\mathcal{C})$ such that*

(1) For all $(\tau, x) \in N \times X$ *such that* $\tau x \in \mathrm{T}(\mathrm{Bor}_{\prec_e})$*, there exists* $\tau_1 \in \mathrm{T}(\mathrm{Red}_{\prec_e}(\mathcal{C}))$ *such that* $\tau_1 \mid \tau x$.

(2) Given $\tau_1, \tau_2 \in \mathrm{T}(\mathrm{Red}_{\prec_e}(\mathcal{C}))$ *then* $\tau_1 \nmid \tau_2$ *and* $\tau_2 \nmid \tau_1$.

Note that in the first case in definition above we have that $\tau x \neq \phi(\tau, x)$, i.e. $\tau x \notin N$. Note also that although the definitions of reduced Gröbner basis and reduced basis of a code are very similar in general the reduced basis of a code can not be used for an effective reduction process due to the non admissibility of the ordering $\prec_e$ (see Example 3.2 in [5]). However, the structure of Gröbner representation always works and by this way we have an effective reduction process for any code. In the binary case, the reduced basis can be used as well.

4.1. *Binary codes*

We will make the assumption that we are working with a code $\mathcal{C}$ defined over the field with two elements $\mathbb{F}_2$ during the rest of this section. Consider the binomial ideal

$$\mathrm{I}(\mathcal{C}) := \langle \{\tau_1 - \tau_2 \mid (\psi(\tau_1), \psi(\tau_2)) \in R_{\mathcal{C}}\} \rangle \subset \mathbb{F}[X] \tag{14}$$

where $\mathbb{F}$ is an arbitrary field. In the binary case we have that $x_i^2 - 1 \in \mathrm{I}(\mathcal{C})$, for all $x_i \in X$ and it follows from Theorem 3.1 that if the code corrects at least one error we have $x_i^2 - 1 \in \mathrm{Red}_{\prec_e}(\mathcal{C})$, i.e. all the variables x_i correspond to canonical forms. If the code has 0 correcting capability there exists at least one x_i such that it is not a canonical form, and $x_i^2 - 1 \in \mathrm{Red}_{\prec_e}(\mathcal{C})$ or $x_i \in \mathrm{T}(\mathrm{Red}_{\prec_e}(\mathcal{C}))$, for each x_i thus all the other elements of $\mathrm{T}(\mathrm{Red}_{\prec_e}(\mathcal{C}))$ will be standard words (i.e. the exponent of each variable is at most one).

By the above discussion in the case of standard words, the order $\prec_e$ and the total degree term ordering are exactly the same. So, the reduced basis of the code w.r.t. $\prec_e$ will be exactly the reduced Gröbner basis of $\mathrm{I}(\mathcal{C})$ w.r.t. the total degree term ordering related to the same admissible ordering used for defining $\prec_e$, thus in this case (binary case) the reduced basis of a code can be used for a effective (Notherian) reduction process.

Example 4.1. If we consider the same code as in Example 3.1 then we have that the reduced basis is

$$\begin{aligned}\mathrm{Red}_{\prec_e}(\mathcal{C}) = \{&x_1^2 - 1, x_2^2 - 1, x_3^2 - 1, x_4^2 - 1, x_5^2 - 1, x_6^2 - 1,\\ &x_1x_2 - x_5, x_1x_3 - x_4, x_1x_4 - x_3, x_1x_5 - x_2,\\ &x_2x_3 - x_1x_6, x_2x_4 - x_6, x_2x_5 - x_1, x_2x_6 - x_4,\\ &x_3x_4 - x_1, x_3x_5 - x_6, x_3x_6 - x_5,\\ &x_4x_5 - x_1x_6, x_4x_6 - x_2, x_5x_6 - x_3\}.\end{aligned}$$

5. Applications

Although it is not our purpose in this paper to fully describe the applications of the Gröbner representation of a linear code we present here several essential facts. All this applications are implemented in GAP [8] using the package GBLA-LC [6].

5.1. *Gradient decoding*

Complete decoding [12] for a linear block code has proved to be an NP-hard computational problem [2], i.e. it is unlikely that a polynomial time (space) complete decoding algorithm can be found. In the literature several attempts have been made to improve the syndrome decoding idea for a general linear code. Usually they look for a smaller structure than the syndrome table to perform the decoding, the main idea is finding for each coset the smaller weight of the words in that coset instead of storing the candidate error vector (see for example the Step-by-Step algorithm in [17] or the test set decoding in [1], in particular those based on zero-neighbors and zero-guards [9–11]). Following the notation in [1] we will call these procedures gradient decoding algorithms.

In the same fashion we use the reduction given by the structures computed above matphi or the border basis to give a procedure to decode any arbitrary linear code. Also we give a step further for binary codes where the reduction given by the reduced basis is Notherian (i.e. it can be used for decoding) and the reduced basis is often smaller than matphi.

The theorem below, is independent of whether we used matphi or border basis for reduction in any linear code or the reduced basis in a binary code.

Theorem 5.1 (See [5]). *Let $\mathcal{C}$ be a linear code. Let $\tau \in \mathcal{T}$ and $\tau' \in N$ its corresponding canonical form. If $\mathrm{w}_h(\psi(\tau')) \leq t$ then $\psi(\tau')$ is the error vector corresponding to $\psi(\tau)$. Otherwise, if $\mathrm{w}_h(\psi(\tau')) > t$, $\psi(\tau)$ contains more than t errors. (t is the error correcting capability)*

Proof. Note that each element has one and only one canonical form. If $\psi(\tau) \in B(C,t)$ then it follows that $\mathrm{w}_h(\psi(\tau')) \leq t$, that is, $\psi(\tau')$ is the error vector, and $\psi(\tau) - \psi(\tau')$ is the codeword corresponding to $\psi(\tau)$. If $\psi(\tau) \notin B(C,t)$ it is clear that $\psi(\tau') \notin B(C,t)$ (they both have the same syndrome), therefore if $\mathrm{w}_h(\psi(\tau')) > t$ means that we had more than t errors. □

Note that the decoding procedure derived from Theorem 5.1 is a complete decoding procedure, that is it always finds the codeword that is closest to the received vector. The procedure can be modified to an incomplete decoding (bounded-distance decoding) procedure in order to further reduce the decoding computation needed.

Example 5.1. We consider the code defined in Example 3.1 and its matphi, and the reduced basis showed in Example 4.1.

Decoding process using matphi.

(1) If $y \in B(\mathcal{C}, t)$
$y = (1,1,0,1,1,0)$; $w_y := x_1x_2x_4x_5$; $\phi(1, x_1) = x_1$; $\phi(x_1, x_2) = x_5$; $\phi(x_5, x_4) = x_2x_3$; $\phi(x_2x_3, x_5) = x_4$, this means $\mathrm{w}_h(\psi(x_4)) = 1$, then the codeword corresponding to y is $c = y - \psi(x_4) = (1,1,0,0,1,0)$.

(2) If $y \notin B(\mathcal{C}, t)$
$y = (0,1,0,0,1,1)$; $w_y := x_2x_5x_6$; $\phi(1, x_2) = x_2$; $\phi(x_2, x_5) = x_1$; $\phi(x_1, x_6) = x_2x_3$, thus, $\mathrm{w}_h(\psi(x_2x_3)) > 1$; consequently, we report an error in the transmission process, in this case the reader can check that the vector y is outside the set $B(\mathcal{C}, 1)$ for the set $\mathcal{C}$ given in Example 3.1. Note that we could also give the value $y - \psi(x_2x_3)$ as a result; this could be useful for applications of codes when it is necessary to always give a result.

Using the reduced basis for decoding Let us work with the same two cases above. By $w \xrightarrow{g} v$ we mean w is reduced to v modulo the binomial g of the reduced basis.

(1) $x_1x_2x_4x_5 \xrightarrow{x_1x_2 - x_5} x_4x_5^2$, $x_4x_5^2 \xrightarrow{x_5^2 - 1} x_4$.

(2) $x_2x_5x_6 \xrightarrow{x_2x_5 - x_1} x_1x_6$.

The following result gives us the "worst case" complexity of our decoding procedure

Proposition 5.1.

Preprocessing *(Möller's algorithm for codes) Algorithm 3.1 performs $\mathcal{O}(mnq^{n-k})$ iterations.*

Decoding *For any linear code the reduction to the candidate error vector is performed in $\mathcal{O}(mn(p-1))$ applications of the matrix matphi or border basis reduction.*

Computing the error correction capability *Algorithm 3.1 computes the error correcting capability of a linear code after at most $m \cdot n \cdot S(t+1)$*

iterations where

$$S(l) = \sum_{i=0}^{l} \binom{n}{i} (q-1)^i.$$

We refer the reader to [5] for a proof of this proposition. Note that the algorithm we refer for computing the error correction capability is the one derived from Theorem 3.1 ,i.e. run Möller's algorithm until one element in the theorem is found.

5.2. *Permutation equivalent codes*

Let $\mathcal{C}$ be a code of length n over $\mathbb{F}_q$ and let $\sigma \in S_n$, where S_n denotes the symmetric group of degree n, we define:

$$\sigma(\mathcal{C}) = \{(y_{\sigma^{-1}(i)})_{i=1}^n \mid (y_i)_{i=1}^n \in \mathcal{C}\},$$

and we say that $\mathcal{C}$ and $\sigma(\mathcal{C})$ are permutation-equivalent or σ-equivalent and we denote it by $\mathcal{C} \sim \sigma(\mathcal{C})$.

The problem of finding whether two codes are permutation equivalent or not is studied in several places in the literature (see [19] and the references therein). In [18] the authors proved that the *Code Equivalence Problem* is not an NP-complete problem, but it is at least as hard as the *Graph Isomorphism Problem*. We transform the problem using a combinatorial definition of permutation equivalent matphi as

Definition 5.1 (Permutation equivalent matphi). *Let $\phi : N \times X \longrightarrow N$ and $\phi^\star : N^\star \times X \longrightarrow N^\star$ be two matphi functions. Then $\phi \sim \phi^\star$ if and only if the following two conditions hold:*

(1) There exists a $\sigma \in S_n$ such that $N^\star = \sigma(N)$, and
(2) For all $v \in N$ and $i = 1, \ldots, mn$ we have $\phi^\star(\sigma(v), \sigma(x_i)) = \sigma(\phi(v, x_i))$.

Our contribution to determine if two codes are permutation equivalent or not is stated in the following theorem

Theorem 5.2. *Let ϕ be a matphi function for the code $\mathcal{C}$, and $\phi^\star$ a matphi for a code $\mathcal{C}^\star$. Then $\mathcal{C} \sim \mathcal{C}^\star \iff \phi \sim \phi^\star$.*

See [5] for a proof. In that paper several heuristic and incremental procedures are shown for dealing with the *Code Equivalence Problem* (some of them are implemented in the package GBLA-LC [6]).

In the binary case we can make use of the reduced basis. The main idea is the following, if two codes are equivalent then, under the appropriate permutation, words of the same weight must be sent to each other. Note also, that it will be used only the level $t+1$ of the reduced bases, which is the first interesting level, from level 1 to t all the elements are canonical forms (we define level l of a reduced basis as the set of binomials of the reduced basis which their maximal terms have cardinal of the set of indices equal to l). The number of elements at this level can be large for big codes but it is considerable smaller than the whole basis. Note that the same reasoning by levels could be used for checking the permutation equivalence of two matphis, thus, it is possible to use a part of a big structure and not the whole object.

5.3. *Gröbner codewords for binary codes*

During this section all codes $\mathcal{C}$ are binary, i.e. defined over the field with two elements $\mathbb{F}_2$ and we will work with an error term ordering such that it is a degree compatible monomial ordering $\prec_{dc}$ and $x_1 \prec_{dc} x_2 \prec_{dc} \ldots$. Let $\mathrm{Td}\,(f)$ denote the total degree of the polynomial f and let $G = \mathrm{Red}_{\prec_{dc}}(\mathcal{C})$ be the reduced basis of the binomial ideal associated to the code in (14) w.r.t. $\prec_{dc}$. For each element in $g = \tau_1 - \tau_2 \in \mathrm{I}(\mathcal{C})$ we define $\mathbf{c}_g$ as the codeword associated to the binomial, i.e. $\mathbf{c}_g = \psi(\tau_1) + \psi(\tau_2)$. We define the set of Gröbner words of the code $\mathcal{C}$ w.r.t. $\prec_{dc}$ as the set

$$\mathcal{C}_G = \left\{\mathbf{c}_g \in \mathcal{C} \mid g \in G \setminus \left\{x_i^2 - 1\right\}_{i=1}^{n}\right\}. \tag{15}$$

From Section 5.1 we know that this set can be used to perform a gradient decoding procedure in the code, we will show two further combinatorial properties of this set (See [3] for the proofs).

Proposition 5.2 (Codewords of minimal weight). *Let* $\mathbf{c}$ *be a codeword of minimal weight* d.

(1) If d *is odd then there exists* $g \in G$ *such that* $\mathbf{c} = \mathbf{c}_g$ *and* $\mathrm{Td}\,(g) = t+1$.
(2) If d *is even then either there exists* $g \in G$ *such that* $\mathbf{c} = \mathbf{c}_g$ *and* $\mathrm{Td}\,(g) = t+1$ *or there exist* $g_1, g_2 \in G$ *such that* $\mathbf{c} = \mathbf{c}_{g_1} + \mathbf{c}_{g_2} = \psi(\tau_1) + \psi(\tau_2)$, *where* $g_1 = \tau_1 - \tau$, $g_2 = \tau_2 - \tau$ $(\tau_1 = \mathrm{T}\,(g_1), \tau_2 = \mathrm{T}\,(g_2), \tau = \mathrm{Can}(g_1, G) = \mathrm{Can}(g_2, G))$, *with* $t+1 = \mathrm{Td}\,(g_1) = \mathrm{Td}\,(g_2)$.

A codeword $\mathbf{c}$ is called minimal if does not exist $\mathbf{c}_1 \in \mathcal{C} \setminus \{\mathbf{c}\}$ such that $\mathrm{supp}(\mathbf{x}^{\mathbf{c}_1}) \subset \mathrm{supp}(\mathbf{x}^{\mathbf{c}})$. Then we have the following result for a set of Gröbner codewords.

Proposition 5.3. *The elements of the set* $\mathcal{C}_G$ *of Gröbner codewords are minimal codewords of the code* $\mathcal{C}$.

Proposition 5.4 (Decomposition of a codeword). *Any codeword* $\mathbf{c} \in \mathcal{C}$ *can be decomposed as a sum of the form* $\mathbf{c} = \sum_{i=1}^{l} \mathbf{c}_{g_i}$*, where* $\mathbf{c}_{g_i} \in \mathcal{C}_G$*,* $\mathrm{w}_h(\mathbf{c}_{g_i}) \leq \mathrm{w}_h(\mathbf{c})$*, and*

$$\mathrm{Td}\,(g_i) \leq \left[\frac{(\mathrm{w}_h(\mathbf{c}) - 1)}{2}\right] + 1, \text{ for all } i = 1, \ldots, l.$$

Using the connection between the set of cycles in graph and binary codes [3, 17] the propositions above enable us to compute all the minimal cycles of a graph according to their lengths and a minimal cycle basis (see [3] for further details).

Example 5.2. The set of Gröbner codewords for the code of the Example 3.1 and the reduced basis of Example 4.1 is

$$\mathcal{C}_G = \left\{ \begin{array}{l} (1,1,0,0,1,0),(1,0,1,1,0,0),(0,1,0,1,0,1), \\ (1,1,1,0,0,1),(1,0,0,1,1,1) \end{array} \right\}.$$

By Proposition 5.2 and taking into account that $d = 3$ and the codewords of minimal weight of $\mathcal{C}$ are $(1,1,0,0,1,0)$,$(1,0,1,1,0,0)$,$(0,1,0,1,0,1)$. Let $\mathbf{c} = (0,1,1,1,1,0) \notin \mathcal{C}_G$. Applying Proposition 5.4 we get

$$\mathbf{c} = \mathbf{c}_{g_1} + \mathbf{c}_{g_2} = (1,1,0,0,1,0) + (1,0,1,1,0,0).$$

Acknowledgments

The authors wish to express their gratitude to Teo Mora for many helpful suggestions. They also want to thank David Joyner for his active interest in the publication of this survey. This work has been partially conducted during the Special Semester on Gröbner Bases, February 1— July 31, 2006 organized by RICAM, Austrian Academy of Sciences, and RISC, Johannes Kepler University, Linz, Austria.

References

[1] A. Barg. Complexity issues in coding theory. In *Handbook of Coding Theory, Elsevier Science*, vol. 1, 1998.

[2] E. Berlekamp, R. McEliece, H. van Tilborg. On the inherent intractability of certain coding problems. *IEEE Trans. Inform. Theory*, IT-24, 384–386, 1978.

[3] M. Borges-Quintana, M. A. Borges-Trenard, P. Fitzpatrick and E. Martínez-Moro. On Gröbner basis and combinatorics for binary codes. Appl. Algebra Engrg. Comm. Comput., 1–13 (Submitted, 2006).

[4] M. Borges-Quintana, M. A. Borges-Trenard and E. Martínez-Moro. A general framework for applying FGLM techniques to linear codes. In AAAECC 16, *Lecture Notes in Comput. Sci.*, Springer, Berlin, vol. 3857, 76–86, 2006

[5] M. Borges-Quintana, M. A. Borges-Trenard and E. Martínez-Moro. On a Gröbner bases structure associated to linear codes. *J. Discrete Math. Sci. Cryptogr.*, 1–41 (To appear, 2007).

[6] M. Borges-Quintana, M. A. Borges-Trenard and E. Martínez-Moro. GBLA-LC: Gröbner basis by linear algebra and codes. International Congress of Mathematicians 2006 (Madrid), Mathematical Software, *EMS (Ed)*, 604–605, 2006. Avaliable at `http://www.math.arq.uva.es/~edgar/GBLAweb/`.

[7] J.C. Faugère, P. Gianni, D. Lazard, T. Mora. Efficient Computation of Zero-Dimensional Grobner Bases by Change of Ordering. *J. Symbolic Comput.*, vol. 16(4), 329–344, 1993.

[8] The GAP Group. GAP – Groups, Algorithms, and Programming, Version 4.4.9, 2006. `http://www.gap-system.org`.

[9] Y. Han. A New Decoding Algorithm for Complete Decoding of Linear Block Codes. *SIAM J. Discrete Math.*, vol. 11(4), 664–671, 1998.

[10] Y. Han, C. Hartmann. The zero-guards algorithm for general minimum-distance decoding problems. *IEEE Trans. Inform. Theory*, vol. 43, 1655–1658, 1997.

[11] L. Levitin, C. Hartmann. A new approach to the general minimum distance decoding problem: the zero-neighbors algorithm. *IEEE Trans. Inform. Theory*, vol. 31, 378–384, 1985.

[12] F.J. MacWilliams, N.J.A. Sloane. The theory of error-correcting codes. Parts I, II. (3rd repr.). North-Holland Mathematical Library, *North- Holland (Elsevier)*, vol. 16, 1985.

[13] M.G. Marinari, H.M. Möller. Gröbner Bases of Ideals Defined by Functionals with an Application to Ideals of Projective Points.Appl. Algebra Engrg. Comm. Comput., vol. 4, 103–145, 1993.

[14] H.M. Möller, B. Buchberger. The construction of multivariate polynomials with preassigned zeros. *Lecture Notes Comp. Sci.*, Springer-Verlag, vol. 144, 24–31, 1982.

[15] T. Mora. *Solving polynomial equation systems II. Macaulay's paradigm and Gröbner technology.* Encyclopedia of Mathematics and its Applications, Cambridge University Press, vol. 99, 2005.

[16] T. Mora. A survey on Combinatorial Duality Approach to Zero-dimensional Ideals 1: Möller Algorithm and the FGLM problem. Submitted to the special volumen *Gröbner Bases, Coding, and Cryptography RISC Book Series (Springer, Heidelberg).*

[17] W. W. Peterson, E. J. Jr. Weldon. *Error-Correcting Codes (2nd ed.).* MIT Press, Cambridge, Massachusetts, London, England, 1972.

[18] E. Petrank, R. M. Roth. Is code equivalence easy to decide? *IEEE Trans. Inform. Theory*, vol. 43(5), 1602–1604, 1997.

[19] N. Sendrier. Finding the permutation between equivalent linear codes: the support splitting algorithm. *IEEE Trans. Inform. Theory*, vol. 46(4), 1193–1203, 2000.

Arcs, minihypers, and the classification of three-dimensional Griesmer codes

Harold N. Ward

Department of Mathematics,
University of Virginia
Charlottesville, VA 22904, USA
E-mail: hnw@virginia.edu

We survey background material involved in the geometric description of codes. Arcs and minihypers figure prominently, appearing here as multisets. We reprove several results, but our main goal is setting the stage for a recent minihyper approach to the classification of three-dimensional codes meeting the Griesmer bound.

1. Introduction

Ray Hill and the author [20] recently began a systematic classification of certain three-dimensional codes meeting the Griesmer bound. We employed minihypers as the basis for the classification, mainly because of a natural inductive process inherent within that framework. But we were pleasantly surprised by the variety of geometric structures that arise in the description of some of the key minihypers. The present paper outlines the minihyper framework, presenting background, concepts, and vocabulary. It also gives proofs of several related geometric and coding results, most of which are known. The final section contains examples of the classification that was carried out in the cited paper. The references given are not meant to be exhaustive, but they are intended to provide access to an extensive literature.

2. Codes and the Griesmer bound

The subject of this paper is linear codes and developments centering on the Griesmer bound. The alphabet for the codes is the finite field $\mathbb{F}_q$ of prime-power size q. Traditionally, an $[n,k]_q$ code is a k-dimensional subspace of the ambient space $\mathbb{F}_q^n$ of words of length n. When the minimum weight d

of the code is specified, the code parameters are displayed as $[n,k,d]_q$. A linear code is usually presented as the row space of an n-columned matrix of rank k, a generator matrix for the code. We shall shortly give a geometric presentation for codes that will be a major theme of the paper.

Given the field and two of the code parameters, one can seek to maximize or minimize the third, as the case may be. Thus a *distance-optimal* code is one for which d is as large as possible for given n and k, while a *length-optimal* code is one with smallest n, given k and d. Finding codes displaying the extremes has been a major activity in coding theory. We shall be most interested in length-optimal codes, usually simply called optimal codes. The guiding bound for them is the Griesmer bound, proved for $q = 2$ by Griesmer [14] and for general q by Solomon and Stiffler [29].

Theorem 2.1. *(Griesmer bound) For an $[n,k,d]_q$ code,*

$$n \geq g_q(k,d) = d + \left\lceil \frac{d}{q} \right\rceil + \ldots + \left\lceil \frac{d}{q^{k-1}} \right\rceil .$$

A code meeting the Griesmer bound is called a *Griesmer code.*

There is a formula for $g_q(k,d)$ that will be useful later (Hill [18]): for a vector space V over $\mathbb{F}_q$, let $\mathrm{PG}(V)$ be the projective space based on V, whose points are the 1-dimensional subspaces of V. We denote $\mathrm{PG}(\mathbb{F}_q^{t+1})$ by $\mathrm{PG}(t,q)$, or simply Π_t if q is understood. The number of points in $\mathrm{PG}(t,q)$ is $\theta_t := (q^{t+1}-1)/(q-1) = q^t + \ldots + 1$. Note that $\theta_0 = 1$ and $\theta_{i+1} = q\theta_i + 1$; we set $\theta_{-1} = 0$ (some authors write v_{t+1} for θ_t). Now let $\lceil d/q^{k-1} \rceil = \delta$, and expand $\delta q^{k-1} - d$ base q as $\sum_{i=0}^{k-2} \delta_i q^i$, with $0 \leq \delta_i \leq q-1$. Then $d = \delta q^{k-1} - \sum_{i=0}^{k-2} \delta_i q^i$, and

$$g_q(k,d) = \delta\theta_{k-1} - \sum_{i=0}^{k-2} \delta_i \theta_i. \qquad (1)$$

3. Codes and multisets

The alternative description of codes we shall use was presented by Assmus and Mattson [1] at the beginning of the development of algebraic coding theory. For a recent exposition of this idea aimed at codes over rings, see Wood [34]. Let $\mathbb{F}_q^k$ be construed as the *message space* for a code. Then let $\lambda_1, \ldots, \lambda_n$, the *coding functionals*, be n members of the vector-space dual $(\mathbb{F}_q^k)^*$ of $\mathbb{F}_q^k$; we shall identify $(\mathbb{F}_q^k)^*$ with $\mathbb{F}_q^k$ itself. Message $\mathbf{v}$ is encoded as $\lambda(\mathbf{v}) = (\lambda_1(\mathbf{v}), \ldots, \lambda_n(\mathbf{v}))$, and the image $\lambda(\mathbb{F}_q^k)$ in $\mathbb{F}_q^n$ is the corresponding code. If $\mathbf{v}_1, \ldots, \mathbf{v}_k$ is a basis of $\mathbb{F}_q^k$, then the $k \times n$ matrix $[\lambda_j(\mathbf{v}_i)]$ is a generator matrix of the code. The λ_i must satisfy the *coding axiom*: λ is to be one-to-one. A code is called *full length* if none of the λ_i is the 0-functional. Permuting and scaling the λ_i replaces $\lambda(\mathbb{F}_q^k)$ with a monomially equivalent code. Thus as far as the weight structure of the code is concerned, only the points $\langle\lambda_i\rangle$ in $\Pi_{k-1} = \mathrm{PG}(k-1, q)$ ($= \mathrm{PG}((\mathbb{F}_q^k)^*)$) are of significance.

Let $\mathcal{L}$ be the *multiset* in Π_{k-1} comprising the $\langle\lambda_i\rangle$: the members of $\mathcal{L}$ are the $\langle\lambda_i\rangle$, but with multiplicities. More formally, $\mathcal{L}$ is the mapping $\Pi_{k-1} \to \mathbb{N}$ for which $\mathcal{L}(P)$ is the number of times point P appears in the list $\langle\lambda_1\rangle, \ldots, \langle\lambda_n\rangle$. Two such multisets $\mathcal{L}$ and $\mathcal{L}'$ correspond to (monomially) equivalent codes exactly when there is a projectivity τ of Π_{k-1} (induced by a linear transformation) with $\mathcal{L}' = \mathcal{L} \circ \tau$, making $\mathcal{L}$ and $\mathcal{L}'$ *projectively equivalent.* We can use the notation $C(\mathcal{L})$ for the code $\lambda(\mathbb{F}_q^k)$ if we bear in mind that $\mathcal{L}$ determines a code only to equivalence. As we wish to concentrate on the connection between multisets and codes, we shall assume that all codes are full-length.

Several authors have investigated this use of multisets in coding theory, among them Dodunekov and Simonis [11], Hamada (cited later), Landjev [25], and Storme [30]. The classic paper by Calderbank and Kantor [9] explores connections between codes with just two non-zero weights and various geometric structures.

3.1. *Arcs*

Let $\mathcal{A}$ be any multiset in Π_t for some t. The *multiplicity* of point P is $\mathcal{A}(P)$, and we let $\max(\mathcal{A})$ and $\min(\mathcal{A})$ denote the maximum and minimum of the $\mathcal{A}(P)$. If $\max(\mathcal{A}) = 1$, we shall refer to $\mathcal{A}$ as a set. (The code $C(\mathcal{L})$ is sometimes called *projective* if $\mathcal{L}$ is a set.) For a subset X of Π_t, let $\mathcal{A}(X) = \sum_{P:P\in X} \mathcal{A}(P)$, the *strength* of X. Thus for the multiset $\mathcal{L}$ of an $[n,k]_q$ code, $n = \mathcal{L}(\Pi_{k-1})$. Moreover, if $\mathbf{v} \in \mathbb{F}_q^k$, $\mathbf{v} \neq 0$, then the weight of the corresponding codeword $\mathbf{c} = \lambda(\mathbf{v})$ is $\mathrm{wt}(\mathbf{c}) = n - \mathcal{L}(\mathbf{c}^\perp)$, where $\mathbf{c}^\perp$ is the hyperplane in Π_{k-1} comprising the points $\langle\lambda\rangle$ for which $\lambda(\mathbf{v}) = 0$. We can think of the message space in projective terms, too: the nonzero vectors in the projective point $\langle\mathbf{v}\rangle$, $\mathbf{v} \neq 0$, give codewords in the corresponding point $\langle\lambda(\mathbf{v})\rangle$ all having the same weight. If the code is an $[n,k,d]_q$ code, then for all hyperplanes H, $\mathcal{L}(H) \leq n-d$, with equality for some H (every H has the form $\mathbf{c}^\perp$ for some $\mathbf{c}$). Such a multiset $\mathcal{L}$ is called an $(n, n-d)$-*arc* (Landjev [25]). There is a one-to-one correspondence between classes of projectively equivalent $(n, n-d)$-arcs and classes of equivalent codes.

Here is the Griesmer bound in the language of arcs:

Proposition 3.1. *Let $\mathcal{K}$ be an (n, r)-arc in $\Pi_t = \mathrm{PG}(t, q)$: $\mathcal{K}(\Pi_t) = n$ and $\mathcal{K}(H) \leq r$ for all hyperplanes H, with equality for some H. Suppose that $d \leq n - r$. Then*

$$n \geq d + \left\lceil \frac{d}{q} \right\rceil + \ldots + \left\lceil \frac{d}{q^t} \right\rceil. \tag{2}$$

Proof. Let J be a $t-2$ subspace in Π_t (J is the empty set if $t = 1$). Then

$$n = \mathcal{K}(\Pi_t) = \sum_{H:J\subset H} \mathcal{K}(H) - q\mathcal{K}(J) \leq (q+1)r - q\mathcal{K}(J),$$

the sum over the $q+1$ hyperplanes H containing J. Thus $\mathcal{K}(J) \leq r - (n-r)/q \leq r - d/q$. If H is a hyperplane with $\mathcal{K}(H) = r$, this means that $\mathcal{K}|H$ is an (r, r')-arc in Π_{t-1} for some $r' \leq r - d/q$; so $r - r' \geq \lceil d/q \rceil$. As $\lceil a/(bc) \rceil = \lceil \lceil a/b \rceil /c \rceil$ in general, we obtain $r \geq \lceil d/q \rceil + \ldots + \lceil d/q^t \rceil$ by induction (at $t = 1$, $n \leq (q+1)r$), which gives the stated inequality. □

Of course, for arcs one expects an *upper* bound on n, and that is what (2) really is: with $d = n - r$, the inequality becomes

$$\left\lceil \frac{n-r}{q} \right\rceil + \ldots + \left\lceil \frac{n-r}{q^t} \right\rceil \leq r.$$

If $n > r$, this gives the elementary lower bound $(n-r)/q + (t-1) \leq r$, or $n \leq (q+1)r - (t-1)q$. If $\mathcal{K}$ is a set, then $r \leq \theta_{t-1}$. Thus at $t = 2$, arcs with $r = q+2$ *must* be multisets. For them, the bound is $n \leq q^2 + q + 2$, since $n = q^2+q+2$ gives $q+1$ on the left and $n = q^2+q+3$ gives $q+3$. The paper by Ball et al. [3] contains an extensive study of $(q^2+q+2, q+2)$-arcs in Π_2. Bounds for arcs that are sets have been widely investigated, and Hirschfeld and Storme [22] provide a recent survey.

3.2. *Combinations*

If $\mathcal{A}$ and $\mathcal{B}$ are multisets in Π_t, the function combination $a\mathcal{A} + b\mathcal{B}$ ($a, b \in \mathbb{Q}$) makes sense as a multiset if its values are always in $\mathbb{N}$. However, as Wood [34] observes, one can usefully define "virtual codes" whose multisets are unrestricted functions to $\mathbb{Q}$. The empty multiset $\mathcal{N}$ has $\mathcal{N}(P) = 0$ for all P, and the set Π_t itself, denoted by $\mathcal{P}$, has $\mathcal{P}(P) = 1$ for all P. If $\mathcal{L}$ and $\mathcal{L}'$ are multisets for two codes C and C' of the same dimension, then $\mathcal{L} + \mathcal{L}'$ is the multiset for a code denoted $C|C'$ that has as generator matrix the juxtaposition of those for C and C', the matrices being set up relative to the same basis of the message space. For $m > 0$, $m\mathcal{L}$ corresponds to the

m-fold *replication* $m \times C = C|\dots|C$ of C. In particular, $m\mathcal{P}$ represents an m-fold *replicated simplex code*, all of whose nonzero words have the same weight. Linear codes with this property are called *constant weight* linear codes. As will be shown in Corollary 4.1, these replicated simplex codes are the only full-length constant weight linear codes (Assmus and Mattson [1]; Bonisoli [6]).

4. Minihypers

An (f, h)-*minihyper* in Π_t is a multiset $\mathcal{M}$ for which $\mathcal{M}(\Pi_t) = f$ and $\mathcal{M}(H) \geq h$ for all hyperplanes H, with equality for some H. Minihypers were defined by Hamada and Tamari [17] and have been used extensively by Hamada and others for classifying Griesmer codes (see, for example, the surveys by Hamada [15, 16] and the one by Storme [30]). If $\mathcal{K}$ is an (n, r)-arc in Π_t with $\max(\mathcal{K}) \leq m$, then $\mathcal{M} = m\mathcal{P} - \mathcal{K}$ is an $(m\theta_t - n, m\theta_{t-1} - r)$-minihyper. When $\mathcal{L}$ is the arc for an $[n, k]_q$ code and $b = \max(\mathcal{L})$, $\mathcal{M} = b\mathcal{P} - \mathcal{L}$ presents the functionals (as projective points) that must be omitted from b copies of $(\mathbb{F}_q^k)^*$ to define the code. Such a structure is closely related to the concept of an *anticode* introduced by Farrell (see, for example, Farrell [12] and MacWilliams and Sloane [27], Chapter 17, Section 6). Conversely, if $\mathcal{M}$ is an (f, h)-minihyper in Π_{k-1} and $m \geq \max(\mathcal{M})$, then $\mathcal{L} = m\mathcal{P} - \mathcal{M}$ defines an $(n, n-d)$-arc with $n = m\theta_{k-1} - f$ and $d = n - (m\theta_{k-2} - h) = mq^{k-1} - f + h$. The shortest corresponding code C_0 comes from taking m to be $\max(\mathcal{M})$, and the other codes have the form $C_0|S$, where S is a replicated simplex code.

Here are some results for codes proved in the language of minihypers, based on this preliminary observation: let $\mathcal{M}$ be an (f, h)-minihyper in Π_t, $t \geq 1$, and let $P \in \Pi_t$. Then

$$\sum_{H:P\in H} \mathcal{M}(H) = \theta_{t-1}\mathcal{M}(P) + \theta_{t-2} \sum_{Q \neq P} \mathcal{M}(Q) = q^{t-1}\mathcal{M}(P) + \theta_{t-2}f, \quad (3)$$

H in the left sum running through the hyperplanes containing P.

Proposition 4.1. *Suppose that the weights of all nonzero words in an* $[n, k]_q$ *code* C *are congruent modulo* Δ, *for some* Δ *relatively prime to* q. *Then* C *is equivalent to* $(\Delta \times C_0)|S$ *for a code* C_0 *and a replicated simplex code* S.

Proof. Let $\mathrm{wt}(\mathbf{c}) \equiv w(\mathrm{mod}\ \Delta)$ for all nonzero $\mathbf{c}$ in C. With $\mathcal{M} = b\mathcal{P} - \mathcal{L}$, $b = \max(\mathcal{L})$ as above, $\mathcal{M}(\mathbf{c}^{\perp}) = b\theta_{k-2} - n + \mathrm{wt}(\mathbf{c})$, so that $\mathcal{M}(H) \equiv b\theta_{k-2} - n + w(\mathrm{mod}\,\Delta)$ for all hyperplanes H. Then (3) implies that $q^{t-1}\mathcal{M}(P)$

is constant mod Δ, and as Δ and q are relatively prime, $\mathcal{M}(P)$ itself is constant mod Δ. The same is then true for $\mathcal{L}(P)$: $\mathcal{L}(P) = x_P\Delta + y$, y independent of P ($y < \Delta$). Defining $\mathcal{L}_0$ by $P \to x_P$, we have $\mathcal{L} = \Delta\mathcal{L}_0 + y\mathcal{P}$ and we take $C_0 = C(\mathcal{L}_0)$, $S = C(y\mathcal{P})$. (If all x_P are 0, then there is no code C_0.) □

The special case that $w = 0$, when all word weights are divisible by Δ (and $y = 0$), was proved in Ward [31] and reproved in Dodunekov and Simonis [11] in the arc framework. As a corollary of Proposition 4.1 we obtain the Assmus-Mattson-Bonisoli theorem:

Corollary 4.1. *A full-length constant-weight linear code is equivalent to a replicated simplex code.*

Proof. The hypothesis of the proposition is satisfied for any Δ relatively prime to q, the field size. For $\Delta > n$, there can be no term $\Delta \times C_0$, and C must be equivalent to S. □

We also obtain a slight generalization of Corollary 2 in Delsarte [10]:

Corollary 4.2. *Let a code C over $\mathbb{F}_q$ have exactly two nonzero codeword weights. Suppose that $\min(\mathcal{L}) = 0$ for the arc $\mathcal{L}$ of C, and that C is not a replicated code (these conditions hold if C is projective). Then the two weights differ by a power of the prime dividing q.*

Proof. If the difference between the weights had a factor Δ relatively prime to q, the two weights would be congruent mod Δ. Then the hypothesis implies that $y = 0$ in Proposition 4.1. But now the non-replication assumption requires $\Delta = 1$. □

4.1. *The Hamada bound*

Hamada [15] presented a bound for minihypers that are sets and described the generalization to multisets. We shall prove this generalization in detail and mention some connections. We need the following numerical arrangement: given q, let $e \geq 0$ be fixed. For any $a \geq 0$, write $a = \sum_{i=0}^{e} a_i\theta_i$, where $a_e = \lfloor a/\theta_e \rfloor$ and for $j < e$, $a_j = \left\lfloor (a - \sum_{i=j+1}^{e} a_i\theta_i)/\theta_j \right\rfloor$. We express this $(e+1)$-*term θ-expansion* as $a = [a_e, a_{e-1}, \ldots, a_0]$. The expansion has these properties: $a_e \geq 0$; $0 \leq a_j \leq q$ for $j < e$; and if $a_j = q$ for some $j < e$, then $a_{j-1} = \ldots = a_0 = 0$. The mapping $a \to [a_e, a_{e-1}, \ldots, a_0]$ is one-to-one from

$\mathbb{N}$ onto the set of lists of length $e+1$ having these three properties. Numerical order corresponds to lexicographic order (denoted $\prec$) of the $(e+1)$-term θ-expansions. If $a = [a_e, \ldots, a_0]$, then $a+1 = [a_e, \ldots, a_0 + 1]$ if none of $a_{e-1}, \ldots, a_0$ is q; but if $a = [a_e, \ldots, a_i, q, 0, \ldots 0]$ (possibly with no zeros after the q), then $a + 1 = [a_e, \ldots, a_i + 1, 0, \ldots, 0]$, with one more zero.

Theorem 4.1. *(Hamada bound) Let $\mathcal{M}$ be an (f,h)-minihyper in* $\mathrm{PG}(t,q)$*, and let the t-term θ-expansion of h be $h = [h_{t-1}, \ldots, h_0]$. Then*

$$\begin{aligned} f \geq f(h) &= [h_{t-1}, \ldots, h_0, 0] \quad \textit{(a $(t+1)$ term θ-expansion)} \\ &= qh + \sum_{i=0}^{t-1} h_i. \end{aligned}$$

Proof. That $f(h) = qh + \sum_{i=0}^{t-1} h_i$ follows from the relation $\theta_{i+1} = q\theta_i + 1$. For the proof of the bound, induct on t. At $t = 1$, $h = [h]$. The hyperplanes are just the points, for which $\mathcal{M}(P) \geq h$. Then $f \geq (q+1)h = [h, 0]$.

For $t \geq 2$, let J be a $(t-2)$-subspace in $\mathrm{PG}(t,q)$. Summing over the $q+1$ hyperplanes H containing J, we obtain

$$f = \sum_{H:J \subset H} \mathcal{M}(H) - q\mathcal{M}(J) \geq (q+1)h - q\mathcal{M}(J).$$

The bound will be established if there is a J for which $(q+1)h - q\mathcal{M}(J) \geq [h_{t-1}, \ldots, h_0, 0]$, which simplifies to $[h_{t-1}, \ldots, h_1] \geq \mathcal{M}(J)$ (the left side is a $(t-1)$-term θ-expansion). Thus suppose that $[h_{t-1}, \ldots, h_1] < \mathcal{M}(J)$ for all J. Let H be a hyperplane, and let $[h'_{t-2}, \ldots, h'_0]$ be the minimum of the $\mathcal{M}(J)$ for $J \subset H$; $[h_{t-1}, \ldots, h_1] \prec [h'_{t-2}, \ldots, h'_0]$. By induction, $\mathcal{M}(H) \geq [h'_{t-2}, \ldots, h'_0, 0]$. But $[h_{t-1}, \ldots, h_1] \prec [h'_{t-2}, \ldots, h'_0]$ implies that $[h_{t-1}, \ldots, h_1, h_0] \prec [h'_{t-2}, \ldots, h'_0, 0]$; that is, $h < \mathcal{M}(H)$. As this holds for all hyperplanes H, the fact that $\mathcal{M}(H) = h$ for some H is contradicted. □

This corollary was also given in Hamada [15]:

Corollary 4.3. *Let $f = [h_{t-1}, \ldots, h_0, h_{-1}]$, and let $h = [h_{t-1}, \ldots, h_0]$. Suppose that $\mathcal{M}$ is a multiset in Π_t for which $\mathcal{M}(\Pi_t) = f$ and $\mathcal{M}(H) \geq h$ for all hyperplanes H. Then $\mathcal{M}$ is an (f,h)-minihyper.*

Proof. $\mathcal{M}$ is an (f,h')-minihyper for some $h' \geq h$, and we need $h' = h$. Let $h' = [h'_{t-1}, \ldots, h'_0]$, so that $f(h') = [h'_{t-1}, \ldots, h'_0, 0]$. By the theorem, $f(h') \leq f$, which reads $[h'_{t-1}, \ldots, h'_0, 0] \preceq [h_{t-1}, \ldots, h_0, h_{-1}]$ for the θ-expansions. If $h < h'$, then $[h_{t-1}, \ldots, h_0] \prec [h'_{t-1}, \ldots, h'_0]$; but, as above, that implies the contradiction $[h_{t-1}, \ldots, h_0, h_{-1}] \prec [h'_{t-1}, \ldots, h'_0, 0]$. □

The Hamada bound can be applied to multiarcs: let $\mathcal{K}$ be an (n,r)-arc in $\mathrm{PG}(t,q)$, and let $b = \max(\mathcal{K})$; thus $b \geq \lceil r/\theta_{t-1} \rceil$. Then $b\mathcal{P} - \mathcal{K}$ is a $(b\theta_t - n, b\theta_{t-1} - r)$-minihyper, so that $b\theta_t - n \geq f(b\theta_{t-1} - r)$; that is, $n \leq b\theta_t - f(b\theta_{t-1} - r)$. Write $b\theta_{t-1} - r = (b - \lceil r/\theta_{t-1} \rceil)\theta_{t-1} + (\lceil r/\theta_{t-1} \rceil \theta_{t-1} - r)$. Let the $(t-1)$-term θ-expansion of $\lceil r/\theta_{t-1} \rceil \theta_{t-1} - r$ be $[r_{t-2}, \ldots, r_0]$. Since $\lceil r/\theta_{t-1} \rceil \theta_{t-1} - r < \theta_{t-1}$, we have $r_{t-2} \leq q$. Thus the t-term θ-expansion of $(b - \lceil r/\theta_{t-1} \rceil)\theta_{t-1} + (\lceil r/\theta_{t-1} \rceil \theta_{t-1} - r)$ is $[b - \lceil r/\theta_{t-1} \rceil, r_{t-2}, \ldots, r_0]$. Then

$$
\begin{aligned}
f(b\theta_{t-1} - r) &= q(b\theta_{t-1} - r) + (b - \lceil r/\theta_{t-1} \rceil) + \sum_{i=0}^{t-2} r_i \\
&= b\theta_t - qr - \lceil r/\theta_{t-1} \rceil + \sum_{i=0}^{t-2} r_i.
\end{aligned}
$$

In the resulting inequality, b disappears:

Proposition 4.2. *If* $\mathcal{K}$ *is an* (n,r)*-arc in* $\mathrm{PG}(t,q)$*, then* $n \leq qr + \lceil r/\theta_{t-1} \rceil - \sum_{i=0}^{t-2} r_i$. *Here* $\lceil r/\theta_{t-1} \rceil \theta_{t-1} - r = [r_{t-2}, \ldots, r_0]$.

This bound should be compared with the elementary bound $n \leq (q+1)r - (t-1)q$ obtained after Proposition 3.1.

How does the Hamada bound relate to the Griesmer bound? The arc $\mathcal{L}$ of a full-length $[n,k,d]_q$ code is an $(n, n-d)$-arc in $\mathrm{PG}(k-1,q)$ (we take $k \geq 3$), and with $r = n - d$, the bound inequality in Proposition 4.2 can be written as

$$
0 \leq q(n-d) - n + \lceil (n-d)/\theta_{k-2} \rceil - \sum_{i=0}^{k-3} r_i,
$$

with $\lceil (n-d)/\theta_{k-2} \rceil \theta_{k-2} - (n-d) = [r_{k-3}, \ldots, r_0]$. For fixed q, k, and d, the right-hand side of the inequality is a nondecreasing function $R(n)$ of n, as one verifies by examining the change from n to $n+1$; the delicate point is the effect in the t-term θ-expansion of a when a changes to $a+1$ as covered by the formulas before Theorem 4.1. From (1), $g_q(k,d) = \delta\theta_{k-1} - \sum_{i=0}^{k-2} \delta_i \theta_i$ when $d = \delta q^{k-1} - \sum_{i=0}^{k-2} \delta_i q^i$. Then $g_q(k,d) - d = \delta\theta_{k-2} - \sum_{i=1}^{k-2} \delta_i \theta_{i-1}$ and $\lceil (g_q(k,d) - d)/\theta_{k-2} \rceil = \delta$, so that $[r_{k-3}, \ldots, r_0] = [\delta_{k-2}, \ldots, \delta_1]$. With this, $R(g_q(k,d))$ comes out to be δ_0, correctly nonnegative. But for $n = g_q(k,d) - 1$, $[r_{k-3}, \ldots, r_0] = [\delta_{k-2}, \ldots, \delta_1 + 1]$, since each δ_i is actually at most $q-1$. Thus $\lceil (n-d)/\theta_{k-2} \rceil$ is still δ, and $R(n) = R(g_q(k,d)) - q = \delta_0 - q < 0$. So $g_q(k,d)$ is the smallest value of n for which the inequality holds, in line with the Griesmer bound.

Let $\mathcal{L}$ be the arc for an $[n,k,d]_q$ Griesmer code, C. With $b = \max(\mathcal{L})$, $\mathcal{M} = b\mathcal{P} - \mathcal{L}$ is the (f,h)-minihyper (a *Griesmer minihyper*) for this code.

We give the parameters f and h explicitly. If $d = \delta q^{k-1} - \sum_{i=0}^{k-2} \delta_i q^i$, then $n = g_q(k,d) = \delta\theta_{k-1} - \sum_{i=0}^{k-2} \delta_i \theta_i$, from the preceding paragraph. Since $\sum_{i=0}^{k-2} \delta_i\theta_i < \theta_{k-1}$, we have $\lceil n/\theta_{k-1} \rceil = \delta$; and since $f = \mathcal{M}(\Pi_{k-1}) = b\theta_{k-1} - n \geq 0$, it follows that $b \geq \delta$. If $\mathcal{L}(\langle\lambda\rangle) = b$, let C' be the *shortened* code whose coordinate functionals are the restrictions to $\ker\lambda$ of the coordinate functionals of C that are not in $\langle\lambda\rangle$. (In conventional terms, C' consists of the words of C that have zeros at the positions indexed by members of $\langle\lambda\rangle$, with those positions then deleted.) Then C' is an $[n-b, k-1, d']_q$ code with $d' \geq d$. The Griesmer bound for C' gives $n - b \geq \sum_{i=0}^{k-2} \lceil d/q^i \rceil = n - \lceil d/q^{k-1} \rceil$, so that $b \leq \lceil d/q^{k-1} \rceil = \delta$. Consequently $b = \delta$ (this argument is adapted from Hill [18]). Thus with $h = b\theta_{k-2} - (n-d)$, then since $n - d = \delta\theta_{k-2} - \sum_{i=1}^{k-2} \delta_i\theta_{i-1}$ (again from the preceding paragraph), we have

$$f = \sum_{i=0}^{k-2} \delta_i\theta_i = [\delta_{k-2}, \ldots, \delta_0] \text{ and } h = \sum_{i=1}^{k-2} \delta_i\theta_{i-1} = [\delta_{k-2}, \ldots, \delta_1] .$$

4.2. *Achievement of the Griesmer bound*

The following theorem guides the search for optimal codes.

Theorem 4.2. *Let k and q be fixed. Then for sufficiently large d, $[n,k,d]_q$ Griesmer codes exist.*

Finding the cases for which Griesmer codes do not exist for given q and k thus is a finite problem, upon which much effort has been put. The two surveys Hill [18] and Hill and Kolev [19] cover background and the state of affairs at their publication dates. Maruta [28] illustrates the kind of information sought for particular values of k. A web server maintained by A. E. Brouwer [7] provides up-to-date lower and upper bounds for distance-optimal codes, substantiated by references, from which bounds for length-optimal codes can be inferred. We give a proof of Theorem 4.2 adapted from Baumert and McEliece [5] (which uses ideas in Solomon and Stiffler [29]), which, however, provides only a crude lower bound for d. Hill [18] presents constructions due to Belov and others for Griesmer codes that lower the bound on d; indeed, much of Hamada's work has been in the direction of generalizing these constructions by using minihypers.

Proof. In Π_{k-1}, let T be a subspace of projective dimension $t \leq k-1$ as a minihyper $\mathcal{P}_T$: $\mathcal{P}_T$ is the characteristic function of T. If $T \subseteq H$ for a hyperplane H, then $\mathcal{P}_T(H) = \theta_t$; while if $T \not\subseteq H$, $H \cap T$ is a $(t-1)$-space and $\mathcal{P}_T(H) = \theta_{t-1}$. Thus $\mathcal{P}_T$ is a (θ_t, θ_{t-1})-minihyper (the second entry is 0 when $t = 0$). Now suppose that $d = \delta q^{k-1} - \sum_{i=0}^{k-2} \delta_i q^i$, with the δ_i fixed, but δ not prescribed. Let $\mathcal{M} = \sum_{i=0}^{k-2} \delta_i \mathcal{P}_i$, where $\mathcal{P}_i = \mathcal{P}_{T_i}$, T_i a subspace of dimension i. Then $\mathcal{M}(\Pi_{k-1}) = \sum_{i=0}^{k-2} \delta_i \mathcal{P}_i(\Pi_{k-1}) = \sum_{i=0}^{k-2} \delta_i \theta_i$; and for a hyperplane H, $\mathcal{M}(H) = \sum_{i=0}^{k-2} \delta_i \mathcal{P}_i(H) \geq \sum_{i=0}^{k-2} \delta_i \theta_{i-1} = \sum_{i=1}^{k-2} \delta_i \theta_{i-1}$. By Corollary 4.3, $\mathcal{M}$ is a $(\sum_{i=0}^{k-2} \delta_i \theta_i, \sum_{i=1}^{k-2} \delta_i \theta_{i-1})$-minihyper (which is not hard to prove directly). If $\delta \geq \max(\mathcal{M})$, then $\delta\mathcal{P} - \mathcal{M}$ is an $(n, n-d')$-arc with $n = \delta\theta_{k-1} - \sum_{i=0}^{k-2} \delta_i \theta_i$ and

$$d' = n - (\delta\theta_{k-2} - \sum_{i=1}^{k-2} \delta_i \theta_{i-1}) = \delta q^{k-1} - \sum_{i=0}^{k-2} \delta_i q^i = d.$$

So $C(\delta\mathcal{P} - \mathcal{M})$ is a desired $[g_q(k,d), k, d]_q$ Griesmer code. Now $\mathcal{M}(P) \leq \sum_{i=0}^{k-2} \delta_i \leq (k-1)(q-1)$. Therefore this construction works for $\delta \geq (k-1)(q-1)$ and so for $d \geq (k-1)(q-1)q^{k-1}$. □

When $\delta = 1$, $\mathcal{M}$ is required to be a set, and if it is to be a sum of certain $\mathcal{P}_i$, the corresponding subspaces must be disjoint. A number of authors have sought constructions of such $\mathcal{M}$; see, for example, the paper of Ferret and Storme [13] that surveys and improves earlier work.

5. Divisibility

If all the word weights of a linear code share a common divisor $\Delta > 1$, the code is called *divisible* (by Δ), and Δ is a *divisor* of the code. Generalized Reed-Muller codes and formally self-dual codes covered by the Gleason-Pierce theorem are prominent examples; see the survey by Ward [33]. This theorem from Ward [32] generalizes an earlier theorem of Dodunekov for binary codes:

Theorem 5.1. *If the minimum weight of a Griesmer code over $\mathbb{F}_p$, p a prime, is divisible by p^e, then the code itself is divisible by p^e.*

The following generalizing conjecture appears in Ward [33] (where it is proved for $q = 4$):

Conjecture 5.1. *Let C be a Griesmer code over $\mathbb{F}_q$ whose minimum weight is divisible by $p^e \geq q$, p the prime dividing q. Then C is divisible by p^{e+1}/q.*

If $\mathcal{L}$ is the $(n, n-d)$-arc in Π_{k-1} of a (full-length) $[n, k, d]_q$ code divisible by Δ, then as in Proposition 4.1, the strengths of all hyperplanes will be congruent mod Δ, and the same will be true of the corresponding minihyper. But such congruence properties of geometrical objects are more widespread than just those inherited from codes. Polynomial methods have led to many of these properties, as surveyed, for example, by Ball [2]. For instance, Landjev [25] proves Theorem 5.1 (but not–alas—its generalization) with polynomial methods. Theorem 5.1 of Ball et al. [3], which concerns $(q^2+q+2, q+2)$-arcs in $PG(2, q)$, q a power of the prime p, has a polynomial proof of the fact that if the number of points with multiplicity 2 is more than $(q-1)p^{e-1}$, then the line strengths are all congruent to q^2+q+2 $\bmod p^e$ (2 is the maximum point multiplicity for such an arc). These arcs correspond to $[q^2+q+2, 3, q^2]_q$ codes, which, though distance-optimal, are not Griesmer codes. And perhaps most famously, Ball, Blokhuis, and Mazzocca [4] used the polynomial approach to show that in $PG(2, q)$ with q odd, there are no sets that are (n, r)-arcs, with $r < q$, meeting the bound $n \leq (q+1)r - q$ with equality. (At $r = q$, the complement of a line *is* such an arc.)

6. Three-dimensional Griesmer codes

For $k = 1$, Griesmer codes, like *all* codes, are trivial. At $k = 2$, with $d = \delta q - \delta_0$ and $n = \delta(q+1) - \delta_0$, the corresponding arc is an (n, δ)-arc. Such an arc can be created, for example, by initially assigning multiplicity δ to all $q+1$ points and then lowering the multiplicity by 1 at δ_0 of them–this is actually an example of a Belov construction!

For $k = 3$, the situation is totally different. Here $d = \delta q^2 - \delta_1 q - \delta_0$ and $n = \delta q^2 + (\delta - \delta_1)q + \delta - \delta_1 - \delta_0$. The arc is an $(n, \delta q + \delta - \delta_1)$-arc and the minihyper has parameters $(\delta_1(q+1) + \delta_0, \delta_1)$. As pointed out earlier, arcs in Π_2 have been studied intensively. Our paper, Hill and Ward [20], is meant to initiate a classification of Griesmer codes for $k = 3$ from the minihyper viewpoint. It was inspired in part by the work of Jones et al. [24] on four-dimensional divisible Griesmer codes over $\mathbb{F}_8$. One of the results of that paper was that there is no $[93, 4, 80]_8$ code; it can be shown that such a code would have to be divisible by 4, in line with Conjecture 5.1. Proving nonexistence required some analysis of $[92, 3, 80]_8$ subcodes, but only the weight distributions of these codes were needed. There are four different distribution possibilities, each of which corresponds to at least one code. It is thus of interest to characterize the actual codes, whose minihypers have parameters $(54, 6)$. In our general analysis we focused on $(x(q+1), x)$-minihypers ($\delta_1 = x$, $\delta_0 = 0$), with $x < q$. We did this in part because of the divisibility aspects mentioned below, and in part because in existence questions for higher dimensional Griesmer codes, it is frequently the case that key values of d are multiples of q. This is so because solutions for these cases often lead to solutions for a sequence of codes with minimum distances going down from d. Three-dimensional codes will be involved in an analysis based on induction in dimension.

In our work, two important aspects came into play: minihypers that are "orphans" and divisibility.

6.1. *Orphans*

If $\mathcal{M}_i$ is a (f_i, h_i)-minihyper in Π_t for $i = 1, 2$, then their sum is an $(f_1 + f_2, h)$-minihyper for some $h \geq h_1 + h_2$. Hence in classifying minihypers, one could begin by looking for *indecomposable* minihypers (a term suggested by Ivan Landjev), those not the sum of two "smaller" minihypers. Even with the indecomposable minihypers in hand, to projective equivalence, one would have to deal with possible ways of adding representatives of the equivalence classes. The simplest instance is that one of the minihypers is the characteristic function of a subspace. Thus in Π_2, let l be a line and let $\mathcal{P}_l$ be its characteristic function, a $(q+1, 1)$-minihyper. Then the move from $\mathcal{M}$ to $\mathcal{M} + \mathcal{P}_l$ might be referred to as "adding a line." Coining a phrase, we call $\mathcal{M} + \mathcal{P}_l$ a *child* of its *parent* $\mathcal{M}$, and now the emphasis shifts to classifying *orphans*, minihypers with no parents. The status of orphanage for a minihyper puts extra constraints on its geometric structure.

If $\mathcal{M}$ is an $(x(q+1), x)$-minihyper, then $\min(\mathcal{M}) = 0$, since otherwise $x(q+1) = \mathcal{M}(\Pi_2) \geq q^2 + q + 1$, contradicting the standing assumption that $x < q$. Let P be a point with $\mathcal{M}(P) = 0$. As $\mathcal{M}(l) \geq x$ for each of the $q+1$ lines l on P and $\sum_{l:P\in l} \mathcal{M}(l) = x(q+1)$, it can only be that $\mathcal{M}(l) = x$ for each such line. Thus if $\mathcal{M}(l) > x$ for some line l, then l has no 0-point on it. If, in fact, $\mathcal{M}(l) \geq x + q$, then each point P on l has $\mathcal{M}(P) > 0$. Such an $\mathcal{M}$ will not be an orphan, for the following reason: let $\mathcal{M}' = \mathcal{M} - \mathcal{P}_l$ (so that $\mathcal{M} = \mathcal{M}' + \mathcal{P}_l$). Then $\mathcal{M}'$ is an $((x-1)(q+1), x-1)$-minihyper. This is so because $\mathcal{M}'(l) \geq x-1$; and for any other line l', $\mathcal{M}'(l') = \mathcal{M}(l') - 1 \geq x - 1$. Since $M(l') = x$ for some line $l' \neq l$, we have $\mathcal{M}'(l') = x - 1$. Therefore in seeking orphans, we may assume that $\mathcal{M}(l) < x + q$ for all lines l. We include the empty minihyper $\mathcal{N}$ for convenience; each $\mathcal{P}_l$ is a child of $\mathcal{N}$.

6.2. *Divisibility*

The second aspect also puts extra constraints on the geometric structure of the minihypers, namely divisibility again. The codes corresponding to an

$(x(q+1), x)$-minihyper $\mathcal{M}$ in Π_2 are $[\delta q^2 + (\delta - x)q + \delta - x, 3, (\delta q - x)q]_q$ Griesmer codes. For a nonzero codeword $\mathbf{c}$, $\mathcal{M}(\mathbf{c}^{\perp}) = x(q+1) - \delta q^2 + \mathrm{wt}(\mathbf{c})$. So divisibility of all values $\mathrm{wt}(\mathbf{c})$ by a divisor q_0 of q^2 implies that all line strengths (the hyperplanes are lines) are congruent to $x(q+1) \bmod q_0$.

For example, when q is a prime, the fact that q divides d gives $\mathcal{M}(l) \equiv x(q+1) \equiv x (\mathrm{mod}\, q)$, by Theorem 5.1. If $\mathcal{M}$ were an orphan, then $\mathcal{M}(l) < x + q$ and the congruence would force $\mathcal{M}(l) = x$. In the present case, (3) reads

$$x(q+1) = \sum_{l:P \in l} \mathcal{M}(l) = q\mathcal{M}(P) + x(q+1),$$

making $\mathcal{M}(P) = 0$ for all P and so $\mathcal{M} = \mathcal{N}$. That is:

Theorem 6.1. *If q is a prime, then there are no orphan $(x(q+1), x)$-minihypers $(x < q)$ in Π_2 other than $\mathcal{N}$. In other words, each $(x(q+1), x)$-minihyper is a sum of lines.*

If p is the prime dividing q and $p^e | x$ (so $p^e < q$), then $p^e q$ divides the minimum weight of the corresponding code. Assuming that Conjecture 5.1 is true, we would get that the code is divisible by p^{e+1}, and then $\mathcal{M}(l) \equiv x (\mathrm{mod}\, p^{e+1})$ for all lines l. But in fact this congruence *is* true–never mind the conjecture–because the polynomial proof in Ball et al. [3] mentioned after Conjecture 5.1 can be invoked with appropriate changes. Having the congruence in hand, we can strengthen Theorem 6.1:

Theorem 6.2. *Suppose that $\mathcal{M}$ is a nonempty orphan $(x(q+1), x)$-minihyper in Π_2 and p is the prime dividing q. Then $x > q - q/p$.*

(There is an analogous theorem in the paper by Landjev and Honold [26]; this paper develops multiset ideas for codes defined over chain rings.) We can also bound line strengths and point multiplicities more sharply:

Proposition 6.1. *Let $\mathcal{M}$ be a nonempty orphan $(x(q+1), x)$-minihyper in Π_2 and suppose that $x \le y < q$ with $p^e | y$. Then*
i) $\mathcal{M}(l) \le x + q - p^{e+1}$ for each line l, and
ii) $\max \mathcal{M} \le x - p^{e+1}$.

Both Theorem 6.2 and Proposition 6.1 are proved using an observation like that at the end of Subsection 6.1: suppose that $\mathcal{M}$ is an $(x(q+1), x)$-minihyper, and let l_0 be a line with $\mathcal{M}(l_0) = x$. If $x \le y < q$, let $\mathcal{M}' = \mathcal{M} + (y - x)\mathcal{P}_{l'}$ for a chosen line l' different from l_0. Then $\mathcal{M}'(l) = \mathcal{M}(l) + (y - x)$ for $l \ne l'$, while $\mathcal{M}'(l') = \mathcal{M}(l') + (y-x)(q+1)$. Thus $\mathcal{M}'(l) \ge y$ for all lines

l, and since $\mathcal{M}'(l_0) = y$, $\mathcal{M}'$ is a $(y(q+1), y)$-minihyper. Now if $p^e < q$, then $\mathcal{M}'(l) \equiv y(\bmod p^{e+1})$ if and only if $\mathcal{M}(l) \equiv x(\bmod p^{e+1})$, a transfer of congruences.

6.3. *The* $[92, 3, 80]_8$ *codes*

As pointed out, the $[92, 3, 80]_8$ Griesmer codes correspond to $(54, 6)$-minihypers in PG$(2, 8)$. Since $80 = 2 \times 8^2 - 6 \times 8$, so that $\delta = 2$, $\max \mathcal{M} = 2$ for a corresponding minihyper $\mathcal{M}$. Thus if $\mathcal{M}$ is a sum of lines, no three are concurrent and their configuration is a $(6, 2)$-arc in the dual plane Π_2^* of Π_2. To projective equivalence, there are five such arcs (Hirschfeld [21], Section 14.6): one comprises six lines (as dual points) on a conic in the dual plane and a second five lines on a conic plus its nucleus (line). The remaining three are *complete*–they cannot be augmented to $(7, 2)$-arcs. They are projectively distinct, but they are equivalent under collineations induced by *semilinear* transformations that involve the automorphisms of $\mathbb{F}_8$. So there are five monomially inequivalent corresponding codes.

The "oldest" (smallest x) orphan minihyper is $\mathcal{N}$, and by Theorem 6.2, the next oldest could have $x = 8 - 8/2 + 1 = 5$. For a minihyper, two counts are important: a_i is the number of lines of strength i (i-lines), the sequence of a_i forming the *spectrum* of the minihyper; and p_j is the number of points of multiplicity j (j-points). These numbers are connected by standard equations derived by counting arguments (generally *double*-counting arguments!). Examination of the possibilities for the number of i-lines on a j-point is also an important ingredient in the analysis. There is indeed a $(45, 5)$ orphan minihyper, $\mathcal{H}_1$ (in the notation of Hill and Ward [20]). By Proposition 6.1, $\max(\mathcal{H}_1) \leq 5 - 4 = 1$ (from $y = 6$); so $\mathcal{H}_1$ is a set. It has the spectrum $a_5 = 63$, $a_9 = 10$; and $p_0 = 28$, $p_1 = 45$. It follows quickly that the ten 9-lines form a *dual hyperoval*, a $(10, 2)$-arc in Π_2^*. This is projectively unique (Hirschfeld [21], Section 14.6 again): it consists of the nine lines of a dual conic and the nucleus. However, $\mathcal{H}_1$ has three inequivalent children, according as the added line is a line of the conic, the nucleus, or one of the remaining 63 lines (and all line choices here are projectively equivalent).

The final code corresponds to the unique orphan $(54, 6)$-minihyper, $\mathcal{H}_2$. Its spectrum is $a_6 = 61$, $a_{10} = 12$; and $p_0 = 22$, $p_1 = 48$, $p_2 = 3$. The three 2-points are collinear on a 6-line, l. Each of the them is on four additional 6-lines. Moreover, any point of intersection of two of these twelve lines that go through two different 2-points is on a 6-line through the third 2-point. This means that l and the twelve additional 6-lines form a *dual projective triad* (Hirschfeld [21], p. 335), which is also projectively unique. We shall return to $\mathcal{H}_2$ shortly.

We have now described the minihypers for *nine* distinct $[92, 3, 80]_8$ codes.

6.4. *Duality*

The possibility of introducing configurations in some sense dual to given ones seems implicit from the beginning, where the arc for a code is actually set up in the *dual* of the message space. In their fundamental paper, Brouwer and van Eupen [8] showed how to obtain new codes from old by constructing arcs in the projective space of the message space itself (geometrically dual configurations have been known for centuries, of course). The points of these arcs were certain points $\langle \mathbf{v} \rangle$, with multiplicities, chosen by the weights of their codewords $\lambda(\mathbf{v})$. Dodunekov and Simonis [11] elaborated upon the framework that Brouwer and van Eupen set up, and that elaboration was invoked, for example, by Jaffe and Simonis [23], to produce a number of codes better than any previously known. Duality also plays a role in the paper of Calderbank and Kantor [9] cited. However, as Jaffe and Simonis point out, the choice of which codewords to use in creating a dual is something of an art: "A key problem is to understand *why* the dual transform method ... produces good codes so often."

To illustrate the idea, we show how a natural dual for $\mathcal{H}_2$ produces another orphan minihyper, labeled $\mathcal{H}_7$ in our paper, with parameters $(63, 7)$. In the analysis of a minihyper, one assigns a *complexion* to a line, the description of how many points of given multiplicity lie on the line. A convenient short-hand is a symbol $a^\alpha b^\beta \ldots$ indicating that the line contains α a-points, β b-points, and so on. A similar notation is used for points to describe the strengths of the lines on them. First we give the line complexions for $\mathcal{H}_2$, along with the multiplicities used for the dual structure. (The type names suggest the number of 0-points on the lines.)

line strength	line type	complexion	number	dual multiplicity
6	s	$2^3 0^6$ (the line l)	1	3
6	f	$2^1 1^4 0^4$	12	1
6	t	$1^6 0^3$	48	1
10	z	$2^1 1^8$	12	0

Then here are the point complexions, multiplicities, and strengths as lines in the dual. We have also identified how many lines there are of each type on the point.

point multiplicity	complexion	line names	number	dual strength
2	$10^4 6^5$	$z^4s^1f^4$	3	7
1	$10^2 6^7$	$z^2f^1t^6$	48	7
0	6^9	f^3t^6	16	9
0	6^9	s^1t^8	6	11

The thirteen points s and f in the dual form a projective triad. The type f points are collinear in fours with the type s point, the three lines of collinearity in the dual being the 2-points in the original.

Acknowledgment

My thanks go to Ray Hill for his hospitality and mathematical insights while we completed the paper Hill and Ward [20]. I am also grateful to Ray and to Cary Huffman for reading this manuscript.

References

[1] E. F. Assmus, Jr., and H. F. Mattson, Error-correcting codes: an axiomatic approach. *Information and Control* **6** (1963), 315–330.

[2] Simeon Ball, Polynomials in finite geometries. *Surveys in Combinatorics, 1999 (Canterbury)*, 17–35, London Math. Soc. Lecture Note Ser., **267**, Cambridge Univ. Press, Cambridge, 1999.

[3] Simeon Ball, Ray Hill, Ivan Landjev, and Harold Ward, On $(q^2+q+2, q+2)$-arcs in the projective plane $PG(2,q)$. *Des. Codes Cryptogr.* **24** (2001), no. 2, 205–224.

[4] Simeon Ball, Aart Blokhuis, and Francesco Mazzocca, Maximal arcs in Desarguesian planes of odd order do not exist. *Combinatorica* **17** (1997), no. 1, 31–41.

[5] L. D. Baumert and R. J. McEliece, A note on the Griesmer bound. *IEEE Trans. Information Theory* **IT-19** (1973), no. 1, 134–135.

[6] Arrigo Bonisoli, Every equidistant linear code is a sequence of dual Hamming codes. *Ars Combin.* **18** (1984), 181–186.

[7] A. E. Brouwer, `http://www.win.tue.nl/~aeb/voorlincod.html`

[8] A. E. Brouwer and M. van Eupen, The correspondence between projective codes and 2-weight codes. *Des. Codes Cryptogr.* **11** (1997), no. 3, 261–266.

[9] R. Calderbank and W. M. Kantor, The geometry of two-weight codes. *Bull. London Math. Soc.* **18** (1986), no. 2, 97–122.

[10] Ph. Delsarte, Weights of linear codes and strongly regular normed spaces. *Discrete Math.* **3** (1972), 47–64.

[11] Stefan Dodunekov and Juriaan Simonis, Codes and projective multisets. *Electron. J. Combin.* **5** (1998), Research Paper 37.

[12] P. G. Farrell, An introduction to anticodes. *Algebraic Coding Theory and Applications*, 180–229, Lectures from the Summer School held at the International Centre for Mechanical Sciences (CISM), Udine, July 1978, Edited by Giuseppe Longo, *CISM Courses and Lectures,* **258**, Springer-Verlag, Vienna, 1979.

[13] S. Ferret and Leo Storme, Minihypers and linear codes meeting the Griesmer bound: improvements to results of Hamada, Helleseth and Maekawa. *Des. Codes Cryptogr.* 25 (2002), no. 2, 143–162.

[14] J. H. Griesmer, A bound for error-correcting codes. *IBM J. Res. Develop.* **4** (1960), 532–542.

[15] Noboru Hamada, A characterization of some $[n, k, d; q]$-codes meeting the Griesmer bound using a minihyper in a finite projective geometry. *Discrete Math.* **116** (1993), no. 1-3, 229–268.

[16] Noboru Hamada, A survey of recent work on characterization of minihypers in $\mathrm{PG}(t, q)$ and nonbinary linear codes meeting the Griesmer bound. *J. Combin. Inform. System Sci.* **18** (1993), no. 3-4, 161–191.

[17] Noboru Hamada and Fumikazu Tamari, On a geometrical method of construction of maximal t-linearly independent sets. *J. Combin. Theory Ser. A* **25** (1978), no. 1, 14–28.

[18] Ray Hill, Optimal linear codes. *Cryptography and Coding, II (Cirencester, 1989)*, 75–104, Inst. Math. Appl. Conf. Ser. New Ser., **33**, Oxford Univ. Press, New York, 1992.

[19] Ray Hill and Emil Kolev, A survey of recent results on optimal linear codes. *Combinatorial Designs and Their Applications (Milton Keynes, 1997)*, 127–152, Chapman & Hall/CRC Res. Notes Math., **403**, Chapman & Hall/CRC, Boca Raton, FL, 1999.

[20] Ray Hill and Harold N. Ward, A geometric approach to classifying Griesmer codes, *Des. Codes Cryptogr.*, to appear.

[21] J. W. P. Hirschfeld, *Projective Geometries over Finite Fields.* Second edition. Oxford Mathematical Monographs. The Clarendon Press, Oxford University Press, New York, 1998.

[22] J. W. P. Hirschfeld and Leo Storme, The packing problem in statistics, coding theory and finite projective spaces: update 2001. *Finite Geometries*, 201–246, *Dev. Math.*, **3**, Kluwer Acad. Publ., Dordrecht, 2001.

[23] David B. Jaffe and Juriaan Simonis, New binary linear codes which are dual transforms of good codes. *IEEE Trans. Inform. Theory* **45** (1999), no. 6, 2136–2137.

[24] Chris Jones, Angela Matney, and Harold Ward, Optimal four-dimensional codes over GF(8). *Electron. J. Combin.* **13** (2006), no. 1, Research Paper 43.

[25] I. N. Landjev, The geometric approach to linear codes. *Finite Geometries*, 247–256, *Dev. Math.*, **3**, Kluwer Acad. Publ., Dordrecht, 2001.

[26] I. Landjev and T. Honold, Arcs in projective Hjelmslev planes. *Discrete Math. Appl.* **11** (2001), no.1, 53–70.

[27] F. J. MacWilliams and N. J. A. Sloane, *The Theory of Error-Correcting Codes.* North-Holland Mathematical Library, Vol. 16. North-Holland Publishing Co., Amsterdam-New York-Oxford, 1977.

[28] Tatsuya Maruta, On the minimum length of q-ary linear codes of dimension four. *Discrete Math.* **208/209** (1999), 427–435.

[29] G. Solomon and J. J. Stiffler, Algebraically punctured cyclic codes. *Information and Control* **8** (1965) 170–179.

[30] Leo Storme, Linear codes meeting the Griesmer bound, minihypers and geometric applications. *Matematiche (Catania)* **59** (2004), no. 1-2, 367–392 (2006).

[31] Harold N. Ward, Divisible codes. *Arch. Math. (Basel)* **36** (1981), no. 6, 485–494.

[32] Harold N. Ward, Divisibility of codes meeting the Griesmer bound. *J. Combin. Theory Ser. A* **83** (1998), no. 1, 79–93.

[33] Harold N. Ward, Divisible codes—a survey. *Serdica Math. J.* **27** (2001), no. 4, 263–278.

[34] Jay A. Wood, The structure of linear codes of constant weight. *Trans. Amer. Math. Soc.* **354** (2002), no. 3, 1007–1026.

Optical orthogonal codes from Singer groups

T. L. Alderson

Mathematical Sciences
University of New Brunswick
Saint John, NB.
E2L 4L5, Canada
E-mail: tim@unbsj.ca
www.unbsj.ca

Keith E. Mellinger

Department of Mathematics
University of Mary Washington
Fredericksburg, VA, 22401, USA
E-mail: kmelling@umw.edu
www.umw.edu

We construct some new families of optical orthogonal codes that are asymptotically optimal. In particular, for any prescribed value of λ, we construct infinite families of (n, w, λ)-OOCs that in each case are asymptotically optimal. Our constructions rely on various techniques in finite projective spaces involving normal rational curves and Singer groups. These constructions generalize and improve previous constructions of OOCs, in particular, those from conics [1] and arcs [2].

Keywords: optical orthogonal codes; Singer cycles; cyclically permutable constant weight codes; normal rational curves.

1. Introduction

There is interest in applying code-division multiple-access (CDMA) techniques to optical networks (OCDMA) and the codes used in an OCDMA system are called *optical orthogonal codes.* An $(n, w, \lambda_a, \lambda_c)$-optical orthogonal code (OOC) is a family of binary sequences (codewords) of length n, and constant hamming weight w satisfying the following two conditions:

- (auto-correlation property) for any codeword $c = (c_0, c_1, \ldots, c_{n-1})$ and for any integer $1 \le t \le n-1$, there holds $\sum_{i=1}^{n-1} c_i c_{i+t} \le \lambda_a$
- (cross-correlation property) for any two distinct codewords c, c' and for any integer $0 \le t \le n-1$, there holds $\sum_{i=0}^{n-1} c_i c'_{i+t} \le \lambda_c$

where each subscript is reduced modulo n.

As stated above, an application of optical orthogonal codes is to optical CDMA communication systems where binary codewords with strong correlation properties are required (see Refs. 3–5 for more details). Subsequently, OOCs have been used for multimedia transmissions in networks using fiber-optics [6]. Optical orthogonal codes have also been called cyclically permutable constant weight codes in the construction of protocol sequences for multiuser collision channels without feedback [7]. Mathematically, OOCs have been studied in their own right because of their connection to various problems that arise naturally in combinatorics. For instance, there is a fundamental equivalence between optimal OOCs and maximum cyclic t-difference packings [8].

An $(n, w, \lambda_a, \lambda_c)$-OOC with $\lambda_a = \lambda_c$ is denoted an (n, w, λ)-OOC. The number of codewords is the *size* of the code. For fixed values of n, w, λ_a and λ_c, the largest size of an $(n, w, \lambda_a, \lambda_c)$-OOC is denoted $\Phi(n, w, \lambda_a, \lambda_c)$. An $(n, w, \lambda_a, \lambda_c)$-OOC of size $\Phi(n, w, \lambda_a, \lambda_c)$ is said to be *optimal*. In applications, optimal OOCs facilitate the largest possible number of asynchronous users to transmit information efficiently and reliably. From the Johnson Bound for constant weight codes it follows [4] that

$$\Phi(n, w, \lambda) \le \left\lfloor \frac{1}{w} \left\lfloor \frac{n-1}{w-1} \left\lfloor \frac{n-2}{w-2} \left\lfloor \cdots \left\lfloor \frac{n-\lambda}{w-\lambda} \right\rfloor \right\rfloor \cdots \right\rfloor \right\rfloor \right\rfloor . \tag{1}$$

Much of the literature is restricted to (n, w, λ)-OOCs. If C is an $(n, w, \lambda_a, \lambda_c)$-OOC with $\lambda_a \ne \lambda_c$ then we obtain a bound on the size of C by taking $\lambda = max\{\lambda_a, \lambda_c\}$ in (1). Alternatively, Yang and Fuja [9] discuss OOCs with $\lambda_a > \lambda_c$ and a corresponding bound is established. The codes we construct in Sections 3, 4 and 5 all have $\lambda_a = \lambda_c$ and, as such, (1) seems the only applicable bound.

We now carefully define the concept of an OOC being asymptotically optimal. Let F be an infinite family of OOCs of varying length n with $\lambda_a = \lambda_c$. For any (n, w, λ)-OOC $C \in F$ containing at least one codeword, the number of codewords in C is denoted by $M(n, w, \lambda)$ and the corresponding Johnson bound is denoted by $J(n, w, \lambda)$.

Definition 1.1. The family F is called asymptotically optimal if

$$\lim_{n\to\infty} \frac{M(n,w,\lambda)}{J(n,w,\lambda)} = 1. \tag{2}$$

For $\lambda = 1, 2$ there are many constructions of (asymptotically) optimal families of (n, w, λ)-OOCs. For $\lambda > 2$ however, constructive examples seem relatively scarce. In Ref. 1, 2, 10, 11, methods of projective geometry are successfully employed to provide asymptotically optimal families of OOCs with $\lambda \geq 2$. In the present work we generalize the previous constructions. In particular, for each prescribed $\lambda \geq 2$ we provide several new asymptotically optimal families of OOCs (Theorems 3.3, 5.1 and Corollaries 4.1, 4.2, 5.1, and 5.2). The codes constructed in Theorem 5.1 have the same or similar parameters to those constructed in Ref. 1 yet compare more favorably with the Johnson Bound (JB). For instance, Table 1 shows how the sizes of some of our codes compare to some previously known codes. We remark that the construction given in Ref. 1 is a special case of our Corollary 4.1 by taking $k = 3$. We also mention that the construction provided in Corollary 4.2 is a strict improvement to the main results of Ref. 2.

Table 1. Comparison of constructions of $(n, 9, \lambda)$-OOCs

n	λ	$\|C\|$	JB	$\|C\|/JB$	Reference
585	2	456	673	0.6775631501	1, Proposition 6
511	2	448	510	0.8784313727	Theorem 5.1 ($k = 3$, $q = 8$)
4681	3	14450752	33845825	0.4269581846	2, Theorem 9
4681	3	14479433	33845825	0.4278055860	Corollary 4.2 ($k = 4$, $q = 8$)

2. Preliminaries

As our work relies heavily on the structure of finite projective spaces, we start with a short overview of the fundamental concepts needed. We let $PG(k, q)$ represent the finite projective geometry of dimension k and order q. Due to a result of Veblen and Young in the early 1900s, all finite projective spaces of dimension not equal to two are equivalent up to the order. The space $PG(k, q)$ can be modeled easily with the vector space of dimension $k+1$ over the finite field $GF(q)$. In this model, the one-dimensional subspaces represent the points, two-dimensional subspaces represent lines, etc. Using this model, it is not hard to show by elementary counting that the number of points of $PG(k, q)$ is given by $\theta(k, q) = \frac{q^{k+1}-1}{q-1}$. We will continue to use the symbol $\theta(k, q)$ to represent this number.

The Fundamental Theorem of Projective Geometry states that the full automorphism group of $PG(k,q)$ is the group $P\Gamma L(k+1,q)$ of semilinear transformations acting on the underlying vector space. The subgroup $PGL(k+1,q) \cong GL(k+1,q)/Z_0$ (where Z_0 represents the center of the group $GL(k+1,q)$) of projective linear transformations is easily modeled by matrices and will be referred to in some of our discussions. A Singer group is a cyclic group acting sharply transitively on the points and hyperplanes of $PG(k,q)$, and the generator of such a group is known as a Singer cycle. Singer groups are known to exists in projective spaces of any order and dimension.

Another property that will provide some assistance is the principle of duality. For any result about points of $PG(k,q)$, there is always a corresponding result about hyperplanes (subspaces, or *flats*, of dimension $k-1$). More generally, for any result dealing with flats of $PG(k,q)$, replacing each reference to an m-flat, $m < k$, with a reference to a $(k-m-1)$-flat, yields a corresponding *dual* statement that has the same truth value. For instance, a result about a set of points of $PG(k,q)$, no three of which are collinear, could be rewritten dually about a set of hyperplanes of $PG(k,q)$, no three of which meet in a common $(k-2)$-flat.

Chung, Salehi, and Wei [4] provide a method for constructing $(n,w,1)$-OOCs using lines of the projective geometry $PG(k,q)$. As our methods may be viewed as a generalization of this construction, we describe the technique in detail. The idea makes use of a Singer cycle that is most easily understood by modeling a finite projective space using a finite field. If we let ω be a primitive element of $GF(q^{k+1})$, the points of $\Sigma = PG(k,q)$ can be represented by the field elements $\omega^0 = 1, \omega, \omega^2, \ldots, \omega^{n-1}$ where $n = \frac{q^{k+1}-1}{q-1}$. Hence, in a natural way a point set A of $PG(k,q)$ corresponds to a binary n-tuple (or codeword) $(a_0, a_1, \ldots, a_{n-1})$ where $a_i = 1$ if and only if $\omega^i \in A$.

Recall that the non-zero elements of $GF(q^{k+1})$ form a cyclic group under multiplication. Moreover, it is not hard to show that multiplication by ω induces an automorphism, or collineation, on the associated projective space $PG(k,q)$. Denote by ϕ the collineation of Σ defined by $\omega^i \mapsto \omega^{i+1}$. The map ϕ clearly acts transitively on the points (and dually on the hyperplanes) of Σ. It is important to note that if A is a point set of Σ corresponding to the codeword $c = (a_0, a_1, \ldots, a_{n-1})$, then ϕ induces a cyclic shift on the coordinates of c. Furthermore, ϕ is a Singer cycle for $PG(k,q)$.

For each line ℓ of $\Sigma = PG(k,q)$, consider its orbit $\mathcal{O}_\ell$ under ϕ. We say $\mathcal{O}_\ell$ is a *full orbit* if it has size $n = \theta(k,q)$. Let $\mathcal{L}(k,q)$ denote the number of full line orbits. A variety of techniques for determining $\mathcal{L}(k,q)$ exist in the literature; in sections 4,5 of Ref. 3 Bird and Keedwell employ methods of design theory, whereas in section 5 of Ref. 12, Ebert *et. al.* take a more geometrical approach. If $\mathcal{O}_\ell$ is a full orbit, then a representative line and corresponding codeword is chosen. Short orbits are discarded. Two lines of Σ intersect in at most one point and each line contains $q+1$ points. It follows that the codewords satisfy both $\lambda_a \leq 1$ and $\lambda_c \leq 1$ and the following is obtained.

Theorem 2.1. *For any prime power q and any positive integer k, there exists a $(\theta(k,q), q+1, 1)$-OOC consisting of $\mathcal{L}(k,q) = \left\lfloor \frac{q^k-1}{q^2-1} \right\rfloor$ codewords.*

Our new constructions of asymptotically optimal OOCs will also rely on orbits of Singer groups. However, we consider the orbits of flats of varying dimension. As such, we let $\begin{bmatrix} k+1 \\ d+1 \end{bmatrix}_q$ denote the number of d-flats in $PG(k,q)$. Elementary counting can be used to show that

$$\begin{bmatrix} k+1 \\ d+1 \end{bmatrix}_q = \frac{(q^{k+1}-1)(q^{k+1}-q)\cdots(q^{k+1}-q^d)}{(q^{d+1}-1)(q^{d+1}-q)\cdots(q^{d+1}-q^d)} \approx q^{(k-d)(d+1)}.$$

Moreover, it is well understood that in $PG(k,q)$, not all orbits of d-flats are full orbits (having size $\theta(k,q)$). The number of orbits of d-flats of varying lengths was investigated in Ref. 13. We let $\mathcal{N}_q(d,k)$ be the number of full d-flat orbits in $PG(k,q)$. Hence, using the notation from the construction above, $\mathcal{N}_q(1,k) \equiv \mathcal{L}(k,q)$. The following lemma is a consequence of Theorem 2.1 of Ref. 13 and shall prove useful in our new constructions of asymptotically optimal OOCs. Note that the count in Theorem 2.1 is a special case of the following.

Lemma 2.1. *Using the notation above,*

$$\mathcal{N}_q(d,k) = \left\lfloor \frac{1}{\theta(k,q)} \begin{bmatrix} k+1 \\ d+1 \end{bmatrix}_q \right\rfloor \approx q^{(k-d-1)d}.$$

The final concept from finite projective geometry that we make use of is that of an *arc*. An *m-arc* in $PG(d,q)$ is a collection of $m > d$ points that meets some hyperplane in d points and meets no hyperplane in as many as $d+1$ points. It follows that if $\mathcal{K}$ is an m-arc in $PG(d,q)$ then no $d+1$ points of $\mathcal{K}$ lie on a hyperplane, no d lie on a $(d-2)$-flat,..., no 3 lie on a line. An arc is called *complete* if it is maximal with respect to inclusion. The concept of an arc generalizes naturally. We define an m-arc of *degree* r $(\geq d)$ in $PG(d,q)$ to be a set of m points of $PG(d,q)$ that meets some hyperplane in r points and meets no hyperplane in as many as $r+1$ points. Hence, arcs of degree d are simply arcs. In the plane $PG(2,q)$, for instance, an arc of degree 2 is simply an arc, and an arc of degree 3 (also known as a *cubic arc*) is a set of points that intersects at least one line in 3 points and intersects no line in as many as 4 points. There is a great deal of literature regarding the connection between arcs and other classes of error-correcting codes including low-density parity-check codes [14] and MDS codes [15].

In $PG(2,q)$, a (non-degenerate) conic is a $(q+1)$-arc and elementary counting shows that this arc is complete when q is odd. In fact, a well-known result of Segre says that every complete arc of $PG(2,q)$, q odd, is a conic. The $(q+2)$-arcs (hyperovals) exist in $PG(2,q)$ if q is even and they are necessarily complete. Conics are a special case of the so called normal rational curves. A *rational curve* $\mathcal{C}_n$ of order n in $PG(d,q)$ is a set of points

$$\{P(t) = (g_0(t_0,t_1),\ldots,g_d(t_0,t_1)) \,|\, t_0,t_1 \in GF(q)\}$$

where each g_i is a binary form of degree n and the highest common factor of $g_0, g_1, \ldots, g_d$ is 1. The curve $\mathcal{C}_n$ may also be written

$$\{P(t) = (f_0(t),\ldots,f_d(t)) \,|\, t \in GF(q) \cup \{\infty\}\} \tag{3}$$

where $f_i(t) = g_i(1,t)$.

Definition 2.1. A normal rational curve (NRC) in $PG(d,q)$, $2 \le d \le q-2$ is a rational curve (of order d) projectively equivalent to

$$\{(1,t,\ldots,t^d) \,|\, t \in GF(q)\} \cup \{(0,\ldots,0,1)\}.$$

It is well-known that an NRC is, in fact, a $(q+1)$-arc. If $\mathcal{C}$ is an NRC in $PG(d,q)$ then the subgroup of $PGL(d+1,q)$ leaving $\mathcal{C}$ fixed is (isomorphic to) $PGL(2,q)$ (see Ref. 16 Theorem 27.5.3). It follows that if $\nu(d,q)$ denotes the number of distinct normal rational curves in $PG(d,q)$ then

$$\nu(d,q) = \frac{|PGL(d+1,q)|}{|PGL(2,q)|} = \frac{(q^{d+1}-1)(q^{d+1}-q)\cdots(q^{d+1}-q^d)}{(q^2-1)(q^2-q)} \tag{4}$$

The following is a well known property of NRCs (see Ref. 17).

Theorem 2.2. *For $2 \le d \le q-2$, a $(d+3)$-arc in $PG(d,q)$ is contained in a unique normal rational curve.*

Definition 2.2. Let $\pi = PG(d,q)$. A collection $\mathcal{F}$ of m-arcs (perhaps of varying degrees) in π is said to be a t-family if every pair of distinct members of $\mathcal{F}$ meet in at most t points. By $\mathcal{F}_q^d(m,r,t)$ we denote the maximal size in $PG(d,q)$ of a t-family of m-arcs each having degree at most $r\ (\geq d)$. If $r = d$ (and consequently all arcs are of degree d) we write $\mathcal{F}_q^d(m,t)$.

Remark 2.1. $\mathcal{F}_q^1(q+1,t) = 1$ for all $t \geq 1$ and in light of Theorem 2.2, $\mathcal{F}_q^d(q+1,d+i) \geq \nu(d,q)$ for all $i \geq 2$.

3. A construction from arcs in d-flats

Our first construction relies on arcs lying in d-flats of a large projective space over sufficiently large order q. Using families of arcs as defined in Definition 2.2, we obtain the following.

Theorem 3.1. *Fix k and d with $k > d \geq 1$. For each prime power $q \geq d$ there exists an $(\theta(k,q), m, d)$-OOC with*

$$|C| = \mathcal{F}_q^d(m,d) \cdot \mathcal{N}_q(d,k).$$

Proof. Let $\Sigma = PG(k,q)$, let ω be a primitive element of $GF(q^{k+1})$ with associated Singer cycle ϕ, and let $N = \mathcal{N}_q(d,k)$. Let $\langle \Pi_1 \rangle, \langle \Pi_2 \rangle, \ldots, \langle \Pi_N \rangle$ be the full orbits of d-flats in Σ. Within each Π_i, let $\mathcal{F}_i$ be a d-family of m-arcs with $|\mathcal{F}_i| = \mathcal{F}_q^d(m,d)$, $i = 1,2,\ldots,N$. Let

$$\mathcal{F} = \bigcup_{i=1}^{N} \mathcal{F}_i$$

and identify each member of $\mathcal{F}$ with the corresponding codeword of length $\theta(k,q)$ and weight m.

For the auto-correlation, let $\mathcal{K}$ be a member of $\mathcal{F}$, where say $\mathcal{K}$ is an m-arc in Π_k. For each i, $\mathcal{K} \cap \phi^i(\mathcal{K}) \subset \Pi_k \cap \phi^i(\Pi_k)$. Here, we use $\phi(\mathcal{K})$ to represent the image of $\mathcal{K}$ under the Singer cycle ϕ. Therefore, for all i with $1 \leq i \leq \theta(k,q) - 1$, the number $\left|\mathcal{K} \cap \phi^i(\mathcal{K})\right|$ is bounded above by the maximal intersection of $\mathcal{K}$ with a $(d-1)$-flat contained in Π_k which, by the definition of arc, is d. It follows that $\lambda_a \leq d$.

For the cross-correlation consider two distinct members of $\mathcal{F}$, say $\mathcal{K}$ and $\mathcal{K}'$ where $\mathcal{K}$ and $\mathcal{K}'$ are m-arcs in say Π_s and Π_t respectively (where perhaps $s = t$). We wish to investigate the maximal cardinality:

$$\max_{1 \leq i,j \leq \theta(k,q)} \left\{ |\phi^i(\mathcal{K}) \cap \phi^j(\mathcal{K}')| \right\}.$$

We have that $\phi^i(\mathcal{K}) \cap \phi^j(\mathcal{K}') \subseteq \phi^i(\Pi_s) \cap \phi^j(\Pi_t)$. If $s \neq t$ then $\phi^i(\Pi_s)$ and $\phi^j(\Pi_t)$ are in different orbits of d-flats, implying that $\phi^i(\Pi_s) \cap \phi^j(\Pi_t)$ is contained in a $(d-1)$-flat. If $s = t$ but $i \neq j$, then $\phi^i(\Pi_s) \neq \phi^j(\Pi_s)$, implying that $\phi^i(\Pi_s) \cap \phi^j(\Pi_s)$ is still contained in a $(d-1)$-flat. Therefore, by definition of an arc in $PG(d,q)$, $\phi^i(\mathcal{K}) \cap \phi^j(\mathcal{K}')$ must have cardinality at most d. It follows that $\lambda_c \leq d$. □

The following appears in Ref. 2; for the sake of completeness we include a proof.

Theorem 3.2. *In $\pi = PG(d,q)$, $d \geq 2$, there exists a d-family $\mathcal{F}$ of $(q+1)$-arcs where $|\mathcal{F}| = (q^{d+1} - q^2)(q^{d+1} - q^3) \cdots (q^{d+1} - q^d)$.*

Proof. Consider $\pi = PG(d,q)$ as a (Baer) subspace of $\Pi = PG(d,q^2)$. Let $\Pi^* = \Pi \setminus \pi$. Choose a point $P = (\alpha_0, \alpha_1, \ldots, \alpha_d) \in \Pi^*$.

With reference to Equation (3), consider the collection of NRCs in Π having polynomial coefficients in $GF(q)$. Denote by X_P the number of such NRCs containing P. Note that any such NRC intersects π in an NRC of π. To determine X_P, we count ordered pairs $(\mathcal{N}, Q)$ where $\mathcal{N}$ is an NRC of Π over $GF(q)$ and Q is a point of $\mathcal{N}$ in Π^*. This gives us the following.

$$\frac{|PGL(d+1,q)|}{|PGL(2,q)|}[(q^2+1)-(q+1)] = \left[\left(\frac{(q^2)^{d+1}-1}{q^2-1}\right) - \left(\frac{(q)^{d+1}-1}{q-1}\right)\right] X_P.$$

After some simplification we arrive at

$$X_P = (q^{d+1} - q^2)(q^{d+1} - q^3) \cdots (q^{d+1} - q^d).$$

Let $\mathcal{C}$ be an NRC in Π over $GF(q)$ containing P. Then the point $P^q = (\alpha_0^q, \alpha_1^q, \ldots, \alpha_d^q)$ conjugate to P is also contained in $\mathcal{C}$ (and is also in Π^*). As such, any two of the NRCs counted above have at most d common points in π. Hence, by restricting to the intersection of these NRCs with π we have a d-family of $(q+1)$-arcs in π having size X_P. □

Corollary 3.1. *If q is a prime power, then in $PG(d,q)$, the maximum size of a d-family of $(q+1)$-arcs, denoted by $\mathcal{F}_q^d(q+1,d)$ satisfies*

$$\mathcal{F}_q^d(q+1,d) \geq (q^{d+1} - q^2)(q^{d+1} - q^3) \cdots (q^{d+1} - q^d).$$

Theorem 3.3. *Fix k and d with $k > d \geq 2$. For each prime power $q \geq d$ there exists a $(\theta(k,q), q+1, d)$-OOC C with*

$$|C| \geq (q^{d+1} - q^2)(q^{d+1} - q^3) \cdots (q^{d+1} - q^d) \cdot \left\lfloor \frac{1}{\theta(k,q)} \begin{bmatrix} k+1 \\ d+1 \end{bmatrix}_q \right\rfloor \approx q^{kd-d-1}$$

Proof. Follows from Theorem 3.1, Corollary 3.1 and Lemma 2.1. □

Now fix $k > d \geq 1$ and consider the infinite family of $(\theta(k,q), q+1, d)$-OOCs constructed as in Theorem 3.3. The Johnson Bound for these codes is

$$\begin{aligned} J(\theta(k,q), q+1, d) &= \left\lfloor \tfrac{1}{q+1} \left\lfloor \tfrac{\theta(k,q)-1}{q} \left\lfloor \tfrac{\theta(k,q)-2}{q-1} \left\lfloor \cdots \left\lfloor \tfrac{\theta(k,q)-d}{q+1-d} \right\rfloor \right\rfloor \right\rfloor \right\rfloor \right\rfloor \\ &\approx q^{kd-d-1}. \end{aligned}$$

With reference to Definition 1.1 we see that the codes constructed as in Theorem 3.3 satisfy the following limit:

$$\lim_{n \to \infty} \frac{M(n,w,\lambda)}{J(n,w,\lambda)} = 1.$$

Hence, we obtain the following.

Theorem 3.4. *Each infinite family of OOCs in Theorem 3.3 is asymptotically optimal.*

4. A construction from arcs of higher degree

We now show that for $d > 1$ it is possible to improve the codes constructed above. Again, we rely on families of arcs lying in certain flats of a large projective space with sufficiently large order q. For this construction, however, we vary the dimension of the flats where the arcs lie.

Theorem 4.1. *Fix k and d with $k > d \geq 1$. For each prime power $q \geq d$ and for each $m > d$ there exists a $(\theta(k,q), m, d)$-OOC C with*

$$|C| = \sum_{i=1}^{d} \mathcal{F}_q^i(m,d,d) \cdot \mathcal{N}_q(i,k).$$

Proof. Let $\Sigma = PG(k,q)$. For fixed s, $1 \leq s \leq d$, let $N_s = \mathcal{N}_q(s,k)$, the number of full orbits of s-flats in $PG(k,q)$. For each s, $1 \leq s \leq d$ let $\Pi_{s,1}, \Pi_{s,2}, \ldots, \Pi_{s,N_s}$ be s-flats chosen one from each of the full s-flat orbits

under ϕ. In each $\Pi_{s,t}$ $1 \le s \le d$, $1 \le t \le N_s$ let $\mathcal{F}_{s,t}$ be a d-family of m-arcs each of degree at most d with $|\mathcal{F}_{s,t}| = \mathcal{F}_q^s(m,d,d)$.
Let

$$\mathcal{F} = \bigcup_{s,t} \mathcal{F}_{s,t}.$$

Identify each member of $\mathcal{F}$ with the corresponding codeword of length $\theta(k,q)$ and weight m. We claim that the code C comprised of all such codewords is a $(\theta(k,q),m,d)$-OOC. That C is of constant weight m is clear.

The auto-correlation, $\lambda_a = d$:
Let $\mathcal{K}$ be a member of $\mathcal{F}$, say $\mathcal{K} \in \mathcal{F}_{s,t}$ is an m-arc of degree r ($\le d$) in the s-flat Π, $1 \le s \le d$, and $\langle \Pi \rangle$ is a full orbit under ϕ. It suffices to show

$$|\phi^i(\mathcal{K}) \cap \phi^j(\mathcal{K})| \le d, \text{ for all } i \ne j, 1 \le i,j \le \theta(k,q).$$

For any i,j, $i \ne j, 1 \le i,j \le \theta(k,q)$, since $\langle \Pi \rangle$ is a full orbit, $\phi^i(\Pi) \ne \phi^j(\Pi)$ which implies that

$$dim\left(\phi^i(\Pi) \cap \phi^j(\Pi)\right) \le s-1.$$

Therefore, since $\phi^i(\mathcal{K}) \cap \phi^j(\mathcal{K}) \subset \phi^i(\mathcal{K}) \cap \left(\phi^i(\Pi) \cap \phi^j(\Pi)\right)$ and since $\phi^i(\Pi) \cap \phi^j(\Pi)$ is at most an $(s-1)$-flat, we are computing the maximum size of the intersection of an m-arc of degree r lying in an s-flat with an $(s-1)$-flat. It follows that

$$|\phi^i(\mathcal{K}) \cap \phi^j(\mathcal{K})| \le |\phi^i(\mathcal{K}) \cap \left(\phi^i(\Pi) \cap \phi^j(\Pi)\right)| \le r$$

by the definition of an arc of degree r. Since $r \le d$ we have $\lambda_a \le d$.

The cross-correlation, $\lambda_c = d$:
Let $\mathcal{K} \ne \mathcal{K}' \in \mathcal{F}$ where $\mathcal{K} \in \mathcal{F}_{s,t}$ is an m-arc of degree $r \le d$ in the s-flat Π, $\langle \Pi \rangle$ a full orbit and $\mathcal{K}' \in \mathcal{F}_{s',t'}$ is an m-arc of degree $r' \le d$ in the s'-flat Π', $\langle \Pi' \rangle$ a full orbit. It suffices to show

$$|\phi^i(\mathcal{K}) \cap \phi^j(\mathcal{K}')| \le d, \text{ for all } i,j,\ 1 \le i,j \le \theta(k,q).$$

For any i,j, $1 \le i,j \le \theta(k,q)$, either $s = s'$ or, without loss of generality, $s' < s$. If $s' < s$ then $dim(\phi^i(\Pi) \cap \phi^j(\Pi')) \le s' < s$ and therefore (as in the first part of the proof)

$$|\phi^i(\mathcal{K}) \cap \phi^j(\mathcal{K}')| \le r \le d.$$

If $s = s'$ we consider two cases:
Case 1: $\Pi = \Pi'$. In this case $\Pi, \Pi' \in \mathcal{F}_{s,t}$. Therefore if $i = j$, then (by definition of a d-family) $|\phi^i(\mathcal{K}) \cap \phi^j(\mathcal{K}')| \leq d$. If $i \neq j$ then $dim\left(\phi^i(\Pi) \cap \phi^j(\Pi)\right) \leq s - 1$ whence

$$|\phi^i(\mathcal{K}) \cap \phi^j(\mathcal{K}')| \leq r \leq d.$$

Case 2: $\Pi \neq \Pi'$. In this case $\langle \Pi \rangle \neq \langle \Pi' \rangle$, so $\phi^i(\Pi) \neq \phi^j(\Pi')$ and again $dim\left(\phi^i(\Pi) \cap \phi^j(\Pi)\right) \leq s - 1$ whence

$$|\phi^i(\mathcal{K}) \cap \phi^j(\mathcal{K}')| \leq r \leq d.$$

It follows that $\lambda_c \leq d$.

□

If we use the 2-family of arcs (in this case, conics) in the plane as in Theorem 3.2, and embed into the ambient space $PG(k, q)$, we obtain the following asymptotically optimal class of OOCs.

Corollary 4.1. *For $k > 2$ and for each prime power $q \geq 2$ there exists a $(\theta(k,q), q+1, 2)$-OOC C with*

$$\begin{aligned} |C| &= (q^3 - q^2) \cdot \mathcal{N}_q(2,k) + \mathcal{N}_q(1,k) \\ &= (q^3 - q^2) \cdot \left\lfloor \frac{1}{\theta(k,q)} \begin{bmatrix} k+1 \\ 3 \end{bmatrix}_q \right\rfloor + \left\lfloor \frac{1}{\theta(k,q)} \begin{bmatrix} k+1 \\ 2 \end{bmatrix}_q \right\rfloor \end{aligned} \tag{5}$$

codewords.

Remark 4.1. For $k = 3$ above, we get the main result of Ref. 1.

Table 2 compares some of the classes of codes constructed as in Corollary 4.1 with the number of codes given by the Johnson Bound.

Table 2. Values of $\frac{M(n,w,\lambda)}{J(n,w,\lambda)}$, $n = \theta(k,q)$, $w = q+1$, $\lambda = 2$

q	$k=3$	$k=4$	$k=5$
7	0.6404255318	0.6330472103	0.6318161869
11	0.7546353523	0.7521739130	0.7519103045
121	0.9754142500	0.9754115290	0.9754115020
343	0.9912792665	0.9912791440	0.9912791434
1721	0.9982578413	0.9982575034	0.9982578401

Teaming the result of Theorem 4.1 with the construction in Theorem 3.2 for large families of arcs we can improve upon the codes constructed as in Theorem 3.3. That is, codes of the same parameters and of larger size result. Indeed, fix d, let our ambient space be $PG(k,q)$, $k > d$, and consider the full Singer orbits of flats of dimension d or less. For our first class of codewords, we take a d-family of arcs in a representative d-flat from each full d-flat orbit. As in Corollary 3.1 we have

$$\mathcal{F}_q^d(q+1,d,d) \geq (q^{d+1}-q^2)(q^{d+1}-q^3)\cdots(q^{d+1}-q^d).$$

For our second class of codewords, we look at the $(d-1)$-flats. In the construction outlined in Theorem 4.1 a d-family of arcs of degree d in a representative $(d-1)$-flat from each full orbit is used. For such families, a general construction yielding a family of significant size appears difficult. However, a $(d-1)$-family of arcs in $PG(d-1,q)$ is easily constructed (as in Theorem 3.2) and may be considered (perhaps rather trivially) as a d-family of arcs of degree at most d. That is,

$$\mathcal{F}_q^{d-1}(q+1,d,d) \geq \mathcal{F}_q^{d-1}(q+1,d-1,d-1) \geq (q^d-q^2)(q^d-q^3)\cdots(q^d-q^{d-1}).$$

For subsequent classes of codewords, we consider in turn the $(d-i)$-flats, for each $i \geq 2$. By Theorem 2.2 the collection of all NRCs in a $(d-i)$-flat $i \geq 2$ is a $(d-i+2)$-family of arcs (of degree $(d-i)$). Hence again we arrive at an ostensibly loose lower bound:

$$\mathcal{F}_q^{d-i}(q+1,d,d) \geq \nu(d-i,q) \text{ for each } i \geq 2.$$

Putting all of these classes of codewords together establishes the following.

Corollary 4.2. *For $k > d \geq 3$ and for each prime power $q \geq d$ there exists a $(\theta(k,q), q+1, d)$-OOC C consisting of*

$$\mathcal{N}_q(d,k)\cdot\prod_{i=2}^{d}(q^{d+1}-q^i)+\mathcal{N}_q(d-1,k)\cdot\prod_{i=2}^{d-1}(q^d-q^i)+\sum_{i=1}^{d-2}(\nu(i,q)\cdot\mathcal{N}_q(i,k))$$
$$=\left\lfloor\frac{1}{\theta(k,q)}\begin{bmatrix}k+1\\d+1\end{bmatrix}_q\right\rfloor\cdot\prod_{i=2}^{d}(q^{d+1}-q^i)+\left\lfloor\frac{1}{\theta(k,q)}\begin{bmatrix}k+1\\d\end{bmatrix}_q\right\rfloor\cdot\prod_{i=2}^{d-1}(q^d-q^i)$$
$$+\textstyle\sum_{i=1}^{d-2}\left(\nu(i,q)\cdot\left\lfloor\begin{bmatrix}k+1\\i+1\end{bmatrix}_q\right\rfloor\right) \qquad (6)$$

codewords.

Remark 4.2. Taking $k = d+1$ in the above yields codes of the same parameters as those constructed in Ref. 2. Moreover, the size of the codes constructed in Ref. 2 correspond to the first and last terms in the expansion (6). Consequently, for $\lambda > 2$ we obtain a strict improvement to the main construction of (n,w,λ)-OOCs in Ref. 2.

Tables 3 and 4 compare some of the classes of codes constructed as in Corollary 4.2 with the number of codes given by the Johnson Bound.

Table 3. Values of $\frac{M(n,w,\lambda)}{J(n,w,\lambda)}$, $n = \theta(k,q)$, $w = q+1$, $\lambda = 3$

q	$k=4$	$k=5$	$k=6$
7	0.3723672313	0.3778141740	0.3788019688
11	0.5503002252	0.5542495934	0.5546309684
121	0.9512311850	0.9512954758	0.9512960102
343	0.9826092131	0.9826175386	0.9826175623
1721	0.9965177060	0.9965180418	0.9965180423

Table 4. Values of $\frac{M(n,w,\lambda)}{J(n,w,\lambda)}$, $n = \theta(k,q)$, $w = q+1$, $\lambda = 5$

q	$k=6$	$k=7$	$k=8$
7	0.0663583530	0.0677297426	0.0679268867
11	0.2100588301	0.2118051740	0.2119642548
121	0.8822149212	0.8822751817	0.8822756800
343	0.9570511656	0.9570593005	0.9570593242
1721	0.9913154765	0.9913158107	0.9913158117

5. Affine constructions

For our final construction, we will work in the finite affine space $AG(k,q)$. Our basic technique follows the work of Ref. 18 where the authors use d-flats of $AG(k,q)$ to construct some OOCs, some of which are optimal. One way to model $AG(d,k)$ is to simply start in the projective space $PG(d,k)$ and delete any hyperplane Σ. The remaining points form the points of $AG(d,k)$ and the flats of $AG(d,q)$ are simply the flats of $PG(d,k)$ with any points of Σ deleted.

It is well-known that $AG(d,q)$ does not admit a Singer group in the same fashion as $PG(d,q)$. However, we can still apply the same general techniques as above. One way to model $AG(k,q)$ is with a k-dimensional vector space over $GF(q)$. In this model, the vectors represent the affine points. The finite field $GF(q^k)$ is one example of such a vector space. As the non-zero field elements of $GF(q^k)$ form a cyclic group under multiplication, we can obtain a similar group (to that of a Singer group of $PG(d,q)$) by simply removing the point corresponding to the zero element of $GF(q^k)$.

Briefly, let $\Sigma = AG(k,q)$ and denote by 0 the zero vector in Σ. Take α to be a primitive element of $GF(q^k)$. Just as in the projective case, each nonzero vector in Σ corresponds in the natural way to α^j for some j, $0 \le j \le q^{k-2}$. Denote by $\hat{\phi}$ the (Singer-like) mapping of Σ defined by $\hat{\phi}(\alpha^j) = \alpha^{j+1}$ and $\hat{\phi}(0) = 0$. Hence, for all of our constructions below, our code lengths will be of the form $q^k - 1$ where the coordinates of the codewords correspond to the non-zero elements of the finite field $GF(q^k)$ (see e.g. Ref. 19). Just as in the previous sections, we will make use of certain families of arcs lying in $AG(k,q)$.

Definition 5.1. Let $\pi = AG(d,q)$. A collection $\mathcal{F}$ of m-arcs (perhaps of varying degrees) in π is said to be a t-family if every pair of distinct members of $\mathcal{F}$ meet in at most t points. By $\mathcal{E}_q^d(m,r,t)$ we denote maximal size in $AG(d,q)$ of a t-family of m-arcs each having degree at most r $(\ge d)$. If $r = d$ (and consequently all arcs are of degree d) we write $\mathcal{E}_q^d(m,t)$.

Consider the space $AG(k,q)$ with the origin removed, and consider the d-flats that do not contain the origin as a point. We wish to count the number of full orbits of these d-flats under the action of the group described above on the points of $AG(d,q)$ minus the origin. We let $\mathcal{M}_q(d,k)$ be the number of such full d-flat orbits in $AG(k,q)$. It follows from Theorem 8 of Ref. 19 that

$$\mathcal{M}_q(d,k) = \frac{q^{k-d}-1}{q^d-1} \cdot \begin{bmatrix} k \\ d \end{bmatrix}_q = \frac{(q^{k-1}-1)(q^{k-2}-1)\cdots(q^{k-d}-1)}{(q^d-1)(q^{d-1}-1)\cdots(q-1)}.$$

Theorem 5.1. *For each prime power $q \ge 2$ there exists a $(q^k-1, q+1, 2)$-OOC C with*

$$|C| = (q^3 - q^2)M_q(2,k).$$

Proof. Our technique is exactly as in Theorem 3.1. We consider a family of $(q+1)$-arcs lying in a plane π of $AG(k,q)$ not containing the origin. We only need to show that the 2-family of $(q+1)$-arcs of $PG(2,q)$ constructed in Theorem 3.2 can still be constructed in $AG(2,q)$.

Let $\Pi = PG(2,q^2)$ and let $\pi \cong PG(2,q)$ be the natural Baer subplane of Π consisting of the set of points whose homogeneous coordinates lie in the subfield $GF(q)$ of the field $GF(q^2)$. Let P be any point of $\Pi \setminus \pi$. As in Theorem 3.2, there are $q^3 - q^2$ arcs of Π, the family $\mathcal{F}$, that meet the Baer subplane π in a sub-arc of size $q+1$. We refer to these sub-arcs as $GF(q)$-arcs. Now, consider the line PP^q, that is, the line joining P with is conjugate point P^q. It's a simple consequence of the classical theory that this line meets the subplane π in a Baer subline. Since the points P and P^q both lie on each of the arcs of $\mathcal{F}$, it follows that no other points of the line PP^q lie on any of the $GF(q)$-arcs. Hence, if we remove the Baer subline of PP^q lying in π from the Baer subplane π, we are left with an isomorphic copy of $AG(2,q)$ containing a set of $q^3 - q^2$ arcs, pairwise meeting in at most two points.

We now embed the affine plane $AG(2,q)$ in $AG(k,q)$ and associate with each arc of the family a codeword. The results on auto and cross correlation now follow as in Theorem 3.1. □

We can increase the number of codewords in the code above by adding the lines of $AG(2,q)$ as additional codewords. In $AG(2,q)$, however, lines contain q points. Hence, in order to keep our codewords of constant weight, we start by removing one point (randomly) from each arc of the family $\mathcal{F}$. Using these q-arcs together with the lines of $AG(2,q)$ gives us the following.

Corollary 5.1. *For $k > 2$ and for each prime power $q \geq 2$ there exists a $(q^k - 1, q, 2)$-OOC C with*

$$|C| = (q^3 - q^2)\mathcal{M}_q(2,k) + \mathcal{M}_q(1,k).$$

Just as with the projective case, the construction above generalizes naturally. The proof of the following is entire similar to that of Theorem 4.1.

Theorem 5.2. *Fix k and d with $k > d \geq 1$. For each prime power $q \geq d$ and for each $m > d$ there exists a $(q^k - 1, m, d)$-OOC C with*

$$|C| = \sum_{i=1}^{d} \mathcal{E}_q^i(m,d,d) \cdot \mathcal{M}_q(i,k).$$

We now establish some lower bounds on $\mathcal{E}_q^i(m,d,d)$, $i \leq d$.

Lemma 5.1. *In $AG(d,q)$, $d \geq 2$, there exists a d-family $\mathcal{F}_0$ of $(q-d+3)$-arcs with $|\mathcal{F}_0| = (q^{d+1} - q^2)(q^{d+1} - q^3)\cdots(q^{d+1} - q^d)$.*

Proof. As in Theorem 5.1, consider $\pi = PG(d,q)$ as a (Baer) subspace of $\Pi = PG(d,q^2)$ and choose a point P of Π outside of π. As discussed in Theorem 3.2, there are $(q^{d+1} - q^2)(q^{d+1} - q^3)\cdots(q^{d+1} - q^d)$ NRCs (the family $\mathcal{F}$) passing through P and meeting π in a sub-arc, and this collection of $(q+1)$-arcs forms a d-family of $GF(q)$-arcs in π. The line PP^q of Π meets π in a Baer subline l_0. Now consider *any* $(d-1)$-flat, say π_0, of π that contains the line l_0. The hyperplane π_0 extends to a hyperplane of the entire space Π that contains the points P and P^q. By the definition of arc, any of the arcs in our family $\mathcal{F}$ meet this $(d-1)$-flat of Π in at most d points, two of which are P and P^q. Hence, if we delete the hyperplane π_0 from π, we delete at most $d-2$ points from any arc of the family $\mathcal{F}$. This gives us a family of arcs we call $\mathcal{F}_0$. For any arc of $\mathcal{F}$ not meeting π_0 in $d-2$ points, we (randomly) remove points so that each arc of $\mathcal{F}_0$ has size $(q+1)-(d-2) = q-d+3$. Hence, every member of $\mathcal{F}_0$ is a $(q-d+3)$-arc. □

From Lemma 5.1 we have

$$\mathcal{E}_q^d(q-d+3, d, d) \geq (q^{d+1} - q^2)(q^{d+1} - q^3)\cdots(q^{d+1} - q^d).$$

Moreover, an analysis similar to that preceding Corollary 4.2 yields

$$\begin{aligned}\mathcal{E}_q^{d-1}(q-d+3, d, d) &\geq \mathcal{E}_q^{d-1}(q-d+3, d-1, d-1)\\ &\geq (q^d - q^2)(q^d - q^3)\cdots(q^d - q^{d-1}),\end{aligned}$$

and

$$\mathcal{E}_q^{d-i}(q-d+3, d, d) \geq \nu(d-i, q) \text{ for each } i \geq 2$$

which, with Theorem 5.2, establishes the following.

Corollary 5.2. *For $k > d \geq 3$ and for each prime power $q \geq d$ there exists a $(q^k - 1, q-d+3, d)$-OOC C consisting of*

$$\mathcal{M}_q(d,k)\cdot\prod_{i=2}^{d}(q^{d+1}-q^i)+\mathcal{M}_q(d-1,k)\cdot\prod_{i=2}^{d-1}(q^d-q^i)+\sum_{i=1}^{d-2}(\nu(i,q)\cdot\mathcal{M}_q(i,k))$$

codewords.

As discussed before Corollary 4.2, we can potentially increase the number of codewords by using arcs of higher degree. In particular, if there exists a t-family $\mathcal{F}$ of m-arcs of degree r in $PG(d,q)$, then there would exist a t-family $\mathcal{F}_0$ of $(m-r)$-arcs of degree r in $AG(d,q)$. Such a family $\mathcal{F}_0$ could potentially be larger than the family described in Lemma 5.1 which would lead to larger codes. In addition, notice in the constructions above that we were forced to remove some points from our arcs for the sole purpose of maintaining a constant codeword weight. Avoiding this might improve the parameters of our codes.

Tables 5, 6, and 7 compare some of the classes of codes constructed as in Corollary 5.2 with the number of codes given by the Johnson Bound. Of particular note are the codes for $\lambda = 2$ (Table 5) whose ratio with the Johnson bound is extremely close to 1.

Table 5. Values of $\frac{M(n,w,\lambda)}{J(n,w,\lambda)}$, $n = q^k - 1$, $w = q + 1$, $\lambda = 2$

q	$k = 3$	$k = 4$	$k = 5$
7	0.8621700881	0.9804586940	0.9972032137
11	0.9104589917	0.9918766332	0.9992610944
121	0.9917366568	0.9999317079	0.9999994356
343	0.9970845975	0.9999915002	0.9999999753
1721	0.9994189427	0.9999996625	0.9999999997

Table 6. Values of $\frac{M(n,w,\lambda)}{J(n,w,\lambda)}$, $n = q^k - 1$, $w = q$, $\lambda = 3$

q	$k = 4$	$k = 5$	$k = 6$
7	0.2960072911	0.3428831335	0.3497386757
11	0.4885791465	0.5365032592	0.5409140118
121	0.9432368020	0.9510961860	0.9511611465
343	0.9797276160	0.9825922863	0.9826006382
1721	0.9959379975	0.9965170310	0.9965173675

Table 7. Values of $\frac{M(n,w,\lambda)}{J(n,w,\lambda)}$, $n = q^k - 1$, $w = q - 2$, $\lambda = 5$

q	$k = 6$	$k = 7$	$k = 8$
8	0.0023607860	0.0026977534	0.0027405412
11	0.0307547138	0.0338295282	0.0341113883
121	0.7894584074	0.7960372276	0.7960916016
343	0.9210483659	0.9237414898	0.9237493411
1721	0.9838383895	0.9844103884	0.9844107212

Note: Our code construction for the tables above involves d-families of $(q - d + 3)$-arcs. To avoid trivial arcs we considered only values of q for which $q - d + 3 > d$.

6. Conclusion

We have exhibited a very general construction of optical orthogonal codes that gives rise to a robust class of asymptotically optimal codes. Our codes generalize and improve the prior constructions involving conics [1] and arcs [2] by expanding the families of intersecting arcs and by working in higher dimensional projective spaces. One next step might be to consider subgeometries $PG(k,q)$ embedded in $PG(k,q^n)$ and use large families of arcs in these subgeometries to find other classes of OOCs whose size approaches that given by the Johnson Bound.

In the last section of Ref. 18 the authors discuss the possibility of OOCs with different weight classes. In the constructions of Section 5, points were arbitrarily removed from certain arcs for the sole purpose of maintaining a constant codeword weight. Hence, the methods of Section 5 provide a construction for large non-constant weight codes with strong auto and cross correlations properties. The investigation into bounds on the size of such OOCs with different weight classes seems an interesting problem as well.

Acknowledgments

The first author acknowledges support from the N.S.E.R.C. of Canada. The second author acknowledges support by a Jepson Fellowship from the University of Mary Washington and National Security Agency grant #H98230-06-1-0080

References

[1] Miyamoto, Nobuko and Mizuno, Hirobumi and Shinohara, Satoshi, Optical orthogonal codes obtained from conics on finite projective planes, Finite Fields Appl., 10, 2004, no. 3, 405–411.

[2] Alderson, T.L., Optical Orthogonal Codes and Arcs in $PG(d,q)$, Finite Fields Appl., *to appear.*

[3] Bird, C. M. and Keedwell, A. D., Design and applications of optical orthogonal codes—a survey, Bull. Inst. Combin. Appl., 11, 1994, 21–44.

[4] Chung, Fan R. K. and Salehi, Jawad A. and Wei, Victor K., Optical orthogonal codes: design, analysis, and applications, IEEE Trans. Inform. Theory, 35, 1989, 3, 595–604.

[5] Healy, Timothy J., Coding and decoding for code division multiple user communication systems, IEEE Trans. Comm., Institute of Electrical and Electronics Engineers. Transactions on Communications, 33, 1985, 4, 310–316.

[6] Maric, S. V and Moreno, O. and Corrada, C., Multimedia transmission in fiber-optic LANs using optical CDMA, J. Lightwave Technol., 14, 1996, 2149–2153.

[7] Nguyen, Q. A and Györfi, László and Massey, James L., Constructions of binary constant-weight cyclic codes and cyclically permutable codes, IEEE Trans. Inform. Theory, 38, 1992, 3, 940–949.

[8] Fuji-Hara, Ryoh and Miao, Ying, Optical orthogonal codes: their bounds and new optimal constructions, IEEE Trans. Inform. Theory, 46, 2000, 7, 2396–2406.

[9] Yang, Guu-chang and Fuja, Thomas E., Optical orthogonal codes with unequal auto- and cross-correlation constraints, IEEE Trans. Inform. Theory, 41, 1, 1995, 96-106.

[10] Alderson, T.L. and Mellinger, Keith E., Constructions of optical orthogonal codes from finite geometry, *to appear*.

[11] Alderson, T.L. and Mellinger, Keith E., Optical orthogonal codes from arcs in root subspaces, *to appear*.

[12] Ebert, G. L. and Metsch, K. and Szönyi, T., Caps embedded in Grassmannians, Geom. Dedicata, 70, 1998, 2, 181–196.

[13] Drudge, Keldon, On the orbits of Singer groups and their subgroups, Electron. J. Combin., 9, 2002, 1, 10 pp.

[14] Droms, Sean V. and Mellinger, Keith E. and Meyer, Chris, LDPC codes generated by conics in the classical projective plane, Des. Codes Cryptogr., 40, 2006, 3, 343–356.

[15] Alderson, T.L. and Bruen, A. A. and Silverman, R., Maximum distance separable codes and arcs in projective spaces, J. Combin. Theory Ser. A, *to appear*.

[16] Hirschfeld, J. W. P. and Thas, J. A., General Galois geometries, Oxford Mathematical Monographs, The Clarendon Press Oxford University Press, New York, 1991, xiv+407.

[17] Thas, Joseph A., Projective geometry over a finite field, Handbook of incidence geometry, 295–347, North-Holland, Amsterdam, 1995.

[18] Omrani, R. and Moreno, O. and Kumar, P.V., Improved Johnson bounds for optical orthogonal codes with $\lambda > 1$ and some optimal constructions, Proc. Int. Symposium on Information Theory, 2005, 259–263.

[19] Rao, C. Radhakrishna, Cyclical generation of linear subspaces in finite geometries, Combinatorial Mathematics and its Applications (Proc. Conf., Univ. North Carolina, Chapel Hill, N.C., 1967), 515–535, Univ. North Carolina Press, 1969.

Codes over F_{p^2} and $F_p \times F_p$, lattices, and theta functions

T. Shaska *

367 Science and Engineering Building,
Department of Mathematics and Statistics,
Oakland University,
Rochester, MI, 48309.
Email: shaska@oakland.edu

C. Shor

Department of Mathematics,
Bates College,
3 Andrews Road,
Lewiston, ME, 04240.
Email: cshor@bates.edu

Let $\ell > 0$ be a square free integer and $\mathcal{O}_K$ the ring of integers of the imaginary quadratic field $K = Q(\sqrt{-\ell})$. Codes C over K determine lattices $\Lambda_\ell(C)$ over rings $\mathcal{O}_K/p\mathcal{O}_K$. The theta functions $\theta_{\Lambda_\ell}(C)$ of such lattices are known to determine the symmetrized weight enumerator $swe(C)$ for small primes $p = 2, 3$; see [1, 10].

In this paper we explore such constructions for any p. If $p \nmid \ell$ then the ring $\mathcal{R} := \mathcal{O}_K/p\mathcal{O}_K$ is isomorphic to $\mathbb{F}_{p^2}$ or $\mathbb{F}_p \times \mathbb{F}_p$. Given a code C over $\mathcal{R}$ we define new theta functions on the corresponding lattices. We prove that the theta series $\theta_{\Lambda_\ell}(C)$ can be written in terms of the complete weight enumerator of C and that $\theta_{\Lambda_\ell}(C)$ is the same for almost all ℓ. Furthermore, for large enough ℓ, there is a unique complete weight enumerator polynomial which corresponds to $\theta_{\Lambda_\ell}(C)$.

Keywords: codes, lattices, theta functions

1. Introduction

Let $\ell > 0$ be a square free integer, $K = Q(\sqrt{-\ell})$ be the imaginary quadratic field, and $\mathcal{O}_K$ its ring of integers. Codes, Hermitian lattices, and their theta-functions over rings $\mathcal{R} := \mathcal{O}_K/p\mathcal{O}_K$, for small primes p, have been studied by many authors, see [1, 7, 8] among others. In [1], an explicit description

*Partially supported by a NATO grant

of theta functions and MacWilliams identities are given for $p = 2, 3$. For a general reference of the topic, see [6].

In this paper we aim to explore such constructions, under certain restrictions, for any p. Further, we study the weight enumerators of such codes in terms of the theta functions of the corresponding lattices. We aim to find MacWilliams-like identities in such cases and explore to what extent the theta functions of these lattices determine the codes. The last question was studied in [2] and [10] for $p = 2$.

This paper is organized as follows. In section 2 we give a brief overview of the basic definitions for codes and lattices and define theta functions over $\mathbb{F}_p$. In section 3 we define theta-functions on the lattice defined over $\mathcal{R} := \mathcal{O}_K/p\mathcal{O}_K$. For general odd p, among the p^2 lattices, there are $\frac{(p+1)^2}{4}$ associated theta series.

In section 4, we address a special case of a general problem of the construction of lattices: the injectivity of Construction A. For codes defined over an alphabet of size four (regarded as a quotient of the ring of integers of an imaginary quadratic field), the problem is solved completely in [10]. The analogous questions are asked for codes defined over $\mathbb{F}_{p^2}$ or $\mathbb{F}_p \times \mathbb{F}_p$. The main obstacle seems to express the theta function in terms of the symmetric weight enumerator of the code. However, the theta function $\theta_{\Lambda_\ell}(C)$ can be expressed in terms of the complete weight enumerator of the code (cf. section 4). Using such an expression we prove the following two facts:

Theorem: Let p be a fixed prime and ℓ any square free integer such that $K = \mathbb{Q}(\sqrt{-\ell})$ and $\mathcal{R} := \mathcal{O}_K/p\mathcal{O}_K$ is isomorphic to $\mathbb{F}_{p^2}$ or $\mathbb{F}_p \times \mathbb{F}_p$. For a given code C defined over $\mathcal{R}$, the theta series $\theta_{\Lambda_\ell}(C)$ is the same for almost all ℓ.

Theorem: Let C be a code defined over $\mathcal{R}$ and $\theta_{\Lambda_\ell}(C)$ be its corresponding theta function for level ℓ. Then, for large enough ℓ, there is a unique complete weight enumerator polynomial which corresponds to $\theta_{\Lambda_\ell}(C)$.

In contrary to results in [10] we did not attempt to find explicit bounds for ℓ. However, for a given small p it is possible such bounds can be determined using similar techniques as in [10]. This is intended to be completed in further work; see [11].

2. Preliminaries

Let $\ell > 0$ be a square free integer and $K = Q(\sqrt{-\ell})$ be the imaginary quadratic field with discriminant d_K. Recall that

$$d_K = \begin{cases} -\ell & \text{if } \ell \equiv 3 \mod 4, \\ -4\ell & \text{otherwise.} \end{cases}$$

Let $\mathcal{O}_K$ be the ring of integers of K. A lattice Λ over K is an $\mathcal{O}_K$-submodule of K^n of full rank. The Hermitian dual is defined by

$$\Lambda^* = \{x \in K^n \mid x \cdot \bar{y} \in \mathcal{O}_K, for\ all\ y \in \Lambda\}, \tag{1}$$

where $x \cdot y := \sum_{i=1}^n x_i y_i$ and $\bar{y}$ denotes component-wise complex conjugation. In the case that Λ is a free $\mathcal{O}_K$ - module, for every $\mathcal{O}_K$ basis $\{v_1, v_2, ..., v_n\}$ we can associate a Gram matrix G(Λ) given by $G(\Lambda) = (v_i \cdot v_j)_{i,j=1}^n$ and the determinant $\det \Lambda := \det(G)$ defined up to squares of units in $\mathcal{O}_K$. If $\Lambda = \Lambda^*$ then Λ is Hermitian self-dual (or unimodular) and integral if and only if $\Lambda \subset \Lambda^*$. An integral lattice has the property $\Lambda \subset \Lambda^* \subset \frac{1}{det\Lambda}\Lambda$. An integral lattice is called even if $x \cdot x \equiv 0 \mod 2$ for all $x \in \Lambda$, and otherwise it is odd. An odd unimodular lattice is called a Type 1 lattice and even unimodular lattice is called a Type 2 lattice.

The theta series of a lattice Λ in K^n is given by

$$\theta_\Lambda(\tau) = \sum_{z \in \Lambda} e^{\pi i \tau z \cdot \bar{z}},$$

where $\tau \in H = \{z \in \mathbb{C} : Im(z) > 0\}$. Usually we let $q = e^{\pi i \tau}$. Then, $\theta_\Lambda(q) = \sum_{z \in \Lambda} q^{z \cdot \bar{z}}$. The one dimensional theta series (or Jacobi's theta series) and its shadow are given by

$$\theta_3(q) = \sum_{n \in \mathbb{Z}} q^{n^2}, \quad \theta_2(q) = \sum_{n \in \frac{1}{2} + \mathbb{Z}} q^{n^2}.$$

Let $\ell \equiv 3 \mod 4$ and d be a positive number such that $\ell = 4d - 1$. Then, $-\ell \equiv 1 \mod 4$. This implies that the ring of integers is $\mathcal{O}_K = \mathbb{Z}[\omega_\ell]$, where $\omega_\ell = \frac{-1+\sqrt{-\ell}}{2}$ and $\omega_\ell^2 + \omega_\ell + d = 0$. The principal norm form of K is given by

$$Q_d(x, y) = |x - y\omega_\ell|^2 = x^2 + xy + dy^2. \tag{2}$$

The structure of $\mathcal{O}_K/p\mathcal{O}_K$ depends on the value of ℓ modulo p. For $(\frac{a}{p})$ the Legendre symbol,

$$\mathcal{O}_K/p\mathcal{O}_K = \begin{cases} \mathbb{F}_p \times \mathbb{F}_p & \text{if } (\frac{-\ell}{p}) = 1, \\ \mathbb{F}_{p^2} & \text{if } (\frac{-\ell}{p}) = -1, \\ \mathbb{F}_p + u\mathbb{F}_p \text{ with } u^2 = 0 & \text{if } p \mid \ell. \end{cases} \tag{3}$$

We will concern ourselves with the cases where $p \nmid \ell$.

2.1. *Theta functions over* $\mathbb{F}_p$

Let $q = e^{\pi i \tau}$. For integers a and b and a prime p, let $\Lambda_{a,b}$ denote the lattice $a - b\omega_\ell + p\mathcal{O}_K$. The theta series associated to this lattice is

$$\begin{aligned} \theta_{\Lambda_{a,b}}(q) &= \sum_{m,n\in\mathbb{Z}} q^{|a+mp-(b+np)\omega_\ell|^2} \\ &= \sum_{m,n\in\mathbb{Z}} q^{Q_d(mp+a,np+b)} \\ &= \sum_{m,n\in\mathbb{Z}} q^{p^2 Q_d(m+a/p,n+b/p)}. \end{aligned} \tag{4}$$

For a prime p and an integer j, consider the theta series

$$\theta_{p,j}(q) := \sum_{n\in\frac{j}{2p}+\mathbb{Z}} q^{n^2}. \tag{5}$$

Note that $\theta_{p,j}(q) = \theta_{p,k}(q)$ if and only if $j \equiv \pm k \mod 2p$.

The theta series of $\Lambda_{a,b}$ can be written in terms of these series. In particular,

$$\theta_{\Lambda_{a,b}}(q) = \theta_{p,b}(q^{p^2\ell})\theta_{p,2a+b}(q^{p^2}) + \theta_{p,b+p}(q^{p^2\ell})\theta_{p,2a+b+p}(q^{p^2}). \tag{6}$$

The proof of this fact is similar to the proof of Lemma 2.1 in [5].

Lemma 2.1. *For any integers a, b, m, n, if the ordered pair (m, n) is component-wise congruent modulo p to one of*

$$(a,b), (-a,-b), (a+b,-b), (-a-b,b),$$

then

$$\theta_{\Lambda_{m,n}}(q) = \theta_{\Lambda_{a,b}}(q)$$

Proof. We prove this by supposing either

$$\theta_{p,n}(q) = \theta_{p,b}(q) \text{ and } \theta_{p,2m+n}(q) = \theta_{p,2a+b}(q) \tag{7}$$

or

$$\theta_{p,n}(q) = \theta_{p,b+p}(q) \text{ and } \theta_{p,2m+n}(q) = \theta_{p,2a+b+p}(q). \tag{8}$$

From Eq. 7, we have four subcases corresponding to $n \equiv \pm b \mod 2p$ and $2m + n \equiv \pm(2a + b) \mod 2p$. If $n \equiv b \mod 2p$, one finds that $m \equiv a$

mod p or $m \equiv -a-b \mod p$. If $n \equiv -b \mod 2p$, one finds that $m \equiv a+b \mod p$ or $m \equiv -a \mod p$.

From Eq. 8, we have four subcases as well, corresponding to $n \equiv \pm(b+p) \mod 2p$ and $2m+n \equiv \pm(2a+b+p) \mod 2p$. If $n \equiv b+p \mod 2p$, then either $m \equiv a \mod p$ or $m \equiv -a-b \mod p$. And if $n \equiv -b-p \mod 2p$, then either $m \equiv a+b \mod p$ or $m \equiv -a \mod p$.

Therefore, if $n \equiv b \mod p$, then $m \equiv a \mod p$ or $m \equiv -a-b \mod p$. If $n \equiv -b \mod p$, then $m \equiv a+b \mod p$ or $m \equiv -a \mod p$. □

Remark 2.1. Notice that in the case of $p=2$, there are 4 lattices $\Lambda_{a,b}$ corresponding to choices of a and b modulo 2. One finds that $\theta_{\Lambda_{0,1}}(q) = \theta_{\Lambda_{1,1}}(q)$ (which is given as Eq. (3.9) in Lemma 3.1 of [2]), so there are 3 associated theta series.

Remark 2.2. In the case of $p=3$, among the 9 lattices, one finds that

$$\begin{aligned}
&\theta_{\Lambda_{0,1}}(q) = \theta_{\Lambda_{2,1}}(q) = \theta_{\Lambda_{1,2}}(q) = \theta_{\Lambda_{0,2}}(q), \\
&\theta_{\Lambda_{1,1}}(q) = \theta_{\Lambda_{2,2}}(q), \text{ and} \\
&\theta_{\Lambda_{1,0}}(q) = \theta_{\Lambda_{2,0}}(q),
\end{aligned}$$

giving a total of 4 associated theta series.

For general odd p, among the p^2 lattices, there are $\frac{(p+1)^2}{4}$ associated theta series.

3. Theta functions of codes over $\mathcal{R}$

Let $p \nmid \ell$ and

$$\mathcal{R} := \mathcal{O}_K/p\mathcal{O}_K = \{a+b\omega : a,b \in \mathbb{F}_p, \omega^2+\omega+d=0\}.$$

A linear code C of length n over $\mathcal{R}$ is an $\mathcal{R}$-submodule of $\mathcal{R}^n$. The dual is defined as $C^\perp = \{u \in \mathcal{R}^n : u \cdot \bar{v} = 0 \text{ for all } v \in C\}$. If $C = C^\perp$ then C is self-dual. We define

$$\Lambda_\ell(C) := \{x \in \mathcal{O}_K^n : \rho_\ell(x) \in C\},$$

where $\rho_\ell : \mathcal{O}_K \to \mathcal{O}_K/p\mathcal{O}_K \to \mathcal{R}$. In other words, $\Lambda_\ell(C)$ consists of all vectors in $\mathcal{O}_K^n$ which when taken mod $p\mathcal{O}_K$ componentwise are in $\rho_\ell^{-1}(C)$. This method of lattice construction is known as Construction A.

For $0 \le a,b \le p-1$, let $r_{a+pb} = a - b\omega$, so $\mathcal{R} = \{r_0, \ldots, r_{p^2-1}\}$. For a codeword $u = (u_1, \ldots, u_n) \in \mathcal{R}^n$ and $r_i \in \mathcal{R}$, we define the counting function

$$n_i(u) := \#\{i : u_i = r_i\}.$$

The complete weight enumerator of the $\mathcal{R}$ code C is the polynomial

$$cwe_C(z_0, z_1, \ldots, z_{p^2-1}) = \sum_{u \in C} z_0^{n_0(u)} z_1^{n_1(u)} \ldots z_{p^2-1}^{n_{p^2-1}(u)}. \tag{9}$$

We can use this polynomial to find the theta function of the lattice $\Lambda_\ell(C)$.

Lemma 3.1. *Let C be a code defined over $\mathcal{R}$ and cwe_C its complete weight enumerator as above. Then,*

$$\theta_{\Lambda_\ell(C)}(q) = cwe_C(\theta_{\Lambda_{0,0}}(q), \theta_{\Lambda_{1,0}}(q), \ldots, \theta_{\Lambda_{p-1,p-1}}(q))$$

Proof. Since

$$\theta_{\Lambda_\ell(C)}(q) = \sum_{z \in \Lambda_\ell(C)} q^{z \cdot \bar{z}},$$

one has

$$\begin{aligned}
\theta_{\Lambda_\ell(C)}(q) &= \sum_{u \in C} \theta_{\Lambda_\ell(u)}(q), \\
&= \sum_{u \in C} \sum_{x \in u + p\mathcal{O}_K^n} q^{x \cdot \bar{x}}, \\
&= \sum_{u \in C} \prod_{j=1}^{n} \sum_{x \in u_j + p\mathcal{O}_K} q^{x \cdot \bar{x}} \text{ (for } u = (u_1, \ldots, u_n)), \\
&= \sum_{u \in C} \prod_{j=1}^{n} \theta_{u_j + p\mathcal{O}_K}(q), \\
&= \sum_{u \in C} \prod_{i=0}^{p^2-1} (\theta_{\tilde{r}_i + p\mathcal{O}_K}(q))^{n_i(u)} \text{ (where } \tilde{r}_{a+pb} = a - b\omega_\ell \in \mathcal{O}_K), \\
&= cwe_C(\theta_{\tilde{r}_0 + p\mathcal{O}_K}(q), \theta_{\tilde{r}_1 + p\mathcal{O}_K}(q), \ldots, \theta_{\tilde{r}_{p^2-1} + p\mathcal{O}_K}(q)), \\
&= cwe_C(\theta_{\Lambda_{0,0}}(q), \theta_{\Lambda_{1,0}}(q), \ldots, \theta_{\Lambda_{p-1,p-1}}(q)),
\end{aligned}$$

which completes the proof. □

3.1. *A MacWilliams identity*

Let $\mathcal{C}^\perp$ be the dual code to $\mathcal{C}$. From Theorem 4.1 of [1] one has the following MacWilliams identity:

Theorem 3.1. *Let $\chi : (\mathcal{R}, +) \to (\mathbb{C}^*, \times)$ be a character of the additive group of $\mathcal{R}$ whose restriction to any nonzero left ideal of $\mathcal{R}$ is nontrivial.*

Then

$$cwe_{\mathcal{C}^\perp}(z_0,\ldots,z_{p^2-1}) = \frac{1}{p^2}cwe_{\mathcal{C}}(M(z_0,\ldots,z_{p^2-1})),$$

where M is the matrix defined by

$$M = (\chi(r_i\overline{r_j}))_{0\leq i\leq p-1, 0\leq j\leq p-1}.$$

To apply this theorem, we need an appropriate character. Define χ by $\chi(a+b\omega) = e^{2\pi ib/p}$. Any non-zero ideal $I \subset \mathcal{R}$ contains an element of $\mathcal{R} - \{0,1,\ldots,p-1\}$, so there is some $a+b\omega \in I$ with $b \neq 0$, meaning χ acts non-trivially on I. A calculation shows that

$$(a+b\omega)(\overline{s+t\omega}) = (as - at + btd) + (bt - as)\omega,$$

so $\chi((a+b\omega)(\overline{s+t\omega})) = e^{(bs-at)2\pi i/p}$. This is independent of d, so we obtain the same MacWilliams identity for codes over $\mathbb{F}_{p^2}$ and $\mathbb{F}_p \times \mathbb{F}_p$.

In the case of $p = 2$, for example, such identities can be made explicit; see [2] and [1] among others.

3.2. *A generalization of the symmetric weight enumerator polynomial*

In [2], for $p = 2$, the symmetric weight enumerator polynomial $swe_{\mathcal{C}}$ of a code $\mathcal{C}$ over a ring or field of cardinality 4 is defined to be

$$swe_{\mathcal{C}}(X,Y,Z) = cwe_{\mathcal{C}}(X,Y,Z,Z).$$

For $\Lambda_{\mathcal{C}}(q)$ the lattice obtained from $\mathcal{C}$ by Construction A, by Theorem 5.2 of [2], one can then write

$$\theta_{\Lambda_\ell(\mathcal{C})}(q) = swe_{\mathcal{C}}(\theta_{\Lambda_{0,0}}(q), \theta_{\Lambda_{1,0}}(q), \theta_{\Lambda_{0,1}}(q)).$$

These theta functions are referred to as $A_d(q), C_d(q)$, and $G_d(q)$ in [2] and [10].

For $p > 2$, however, there are $\frac{(p+1)^2}{4}$ (which is larger than 3) theta functions associated to the various lattices, so our analog of the symmetric weight enumerator polynomial has more than 3 variables.

Example 3.1. For $p = 3$, from Remark 2.2, we have four theta functions corresponding to the lattices $\Lambda_{a,b}$, namely

$$\theta_{\Lambda_{0,0}}(q), \theta_{\Lambda_{1,0}}(q), \theta_{\Lambda_{1,1}}(q), \theta_{\Lambda_{0,1}}(q).$$

If we define the "symmetric weight enumerator for $p = 3$" to be

$$swe_{\mathcal{C}}(X,Y,Z,W) = cwe_{\mathcal{C}}(X,Y,Y,Z,W,Z,Z,Z,W),$$

then one finds that

$$\theta_{\Lambda_\ell C}(q) = cwe_C(\theta_{\Lambda_{0,0}}(q), \theta_{\Lambda_{1,0}}(q), \ldots, \theta_{\Lambda_{2,2}}(q)), \tag{10}$$

$$= swe_C(\theta_{\Lambda_{0,0}}(q), \theta_{\Lambda_{1,0}}(q), \theta_{\Lambda_{1,1}}(q), \theta_{\Lambda_{0,1}}(q)). \tag{11}$$

Finding such an explicit relation between the theta function and the symmetric weight enumerator polynomial for larger p seems difficult. We suggest the following problem:

Problem 1. *Define an symmetric weight enumerator, analogous to the* $p = 2$ *case, for codes defined over* $\mathcal{R}$ *for* $p > 3$. *Write a MacWilliams identity for the symmetric weight enumerator and determine an explicit relation between the symmetric weight enumerator and theta functions.*

4. The injectivity of construction A

For a fixed prime p, let $\mathcal{R} = \mathcal{O}_K/p\mathcal{O}_K$ and C be a linear code over $\mathcal{R}$ of length n and dimension k. An admissible level ℓ is an integer ℓ such that $\mathcal{R}$ is isomorphic to $\mathbb{F}_{p^2}$ or $\mathbb{F}_p \times \mathbb{F}_p$. For an admissible ℓ, let $\Lambda_\ell(C)$ be the corresponding lattice as in the previous section. Then, the **level ℓ theta function** $\theta_{\Lambda_\ell(C)}(\tau)$ of the lattice $\Lambda_\ell(C)$ is determined by the complete weight enumerator cwe_C of C, evaluated on the theta functions defined on cosets of $\mathcal{O}_K/p\mathcal{O}_K$. We consider the following questions:

i) How do the theta functions $\theta_{\Lambda_\ell(C)}(\tau)$ of the same code C differ for different levels ℓ?

ii) Can non-equivalent codes give the same theta functions for all levels ℓ?

Next we see how this can be made explicit for the case $p = 2$.

4.1. *The case* $p = 2$

For $p = 2$ case these questions are fully answered in [10]. We have the following:

Theorem 4.1 (Thm. 1, [10]). *Let* $p = 2$ *and* C *be a code defined over* $\mathcal{R}$. *For all admissible* ℓ, ℓ' *such that* $\ell > \ell'$, *the following holds*

$$\theta_{\Lambda_\ell}(C) = \theta_{\Lambda_{\ell'}}(C) + \mathcal{O}(q^{\frac{\ell'+1}{4}}).$$

Let C be a code of length n defined over $\mathcal{R}$ and $\theta_{\Lambda_\ell}(C)$ be its corresponding theta function for level ℓ. Let $f(x,y,z) \in F[x,y,z]$ where F is a field of transcendental degree δ. We say that $f(x,y,z)$ is in a family of polynomials of dimension δ.

Theorem 4.2 (Thm. 2, [10]). *Let $p = 2$ and C be a code of length n defined over $\mathcal{R}$ and $\theta_{\Lambda_\ell}(C)$ be its corresponding theta function for level ℓ. Then the following hold:*

i) *For $\ell < \frac{2(n+1)(n+2)}{n} - 1$ there is a δ-dimensional family of symmetrized weight enumerator polynomials corresponding to $\theta_{\Lambda_\ell}(C)$, where $\delta \geq \frac{(n+1)(n+2)}{2} - \frac{n(\ell+1)}{4} - 1$.*

ii) *For $\ell \geq \frac{2(n+1)(n+2)}{n} - 1$ and $n < \frac{\ell+1}{4}$ there is a unique symmetrized weight enumerator polynomial which corresponds to $\theta_{\Lambda_\ell}(C)$.*

Example 4.1. There are two non isomorphic codes

$$C_{3,2} = \omega < [0,1,1] > +(\omega+1) < [0,1,1] >^{\perp}$$
$$C_{3,3} = \omega < [0,0,1] > +(\omega+1) < [0,0,1] >^{\perp} .$$

with symmetrized weight enumerator polynomials

$$swe_{C_{3,2}}(X,Y,Z) = X^3 + X^2Z + XY^2 + 2XZ^2 + Y^2Z + 2Z^3$$
$$swe_{C_{3,3}}(X,Y,Z) = X^3 + 3X^2Z + 3XZ^2 + Z^3$$

Both these codes give the following theta function for level $\ell = 7$:

$$\theta = 1 + 6q^2 + 24q^4 + 56q^6 + 114q^8 + 168q^{10} + 280q^{12} + 294q^{14} + \cdots$$

However, when $\ell = 15$, we are in the second case of the above theorem. Two non equivalent codes cannot give the same theta function for $\ell = 15$ and $n = 3$. Explicit details are given in [10].

The above results were obtained by using the explicit expression of theta in terms of the symmetric weight enumerator valuated on the theta functions of the cosets. Hence, a solution to Problem 1 most likely would lead to obtaining such results for all $p > 2$ and admissible ℓ. In this paper we use the complete weight enumerator polynomial to get similar results.

4.2. *The case $p > 2$*

Let C be a code defined over $\mathcal{R}$ for a fixed $p > 2$. Let the complete weight enumerator of C be the degree n polynomial

$$cwe_C = f(x_0, \ldots, x_r)$$

for $r = p^2 - 1$. Then from Lemma 3.1 we have that

$$\theta_{\Lambda_\ell(C)}(\tau) = f(\theta_{\Lambda_{0,0}}(\tau), \dots, \theta_{\Lambda_{p-1,p-1}}(\tau))$$

for a given ℓ. First we want to address how $\theta_{\Lambda_\ell(C)}(\tau)$ and $\theta_{\Lambda_{\ell'}(C)}(\tau)$ differ for different ℓ and ℓ'. We have the following:

Theorem 4.3. *Let C be a code defined over $\mathcal{R}$. For all admissible ℓ, ℓ' the following holds*

$$\theta_{\Lambda_\ell}(C) - \theta_{\Lambda_{\ell'}}(C) = \sum_{i=0}^{s} a_i q^s$$

for some $a_i \in \mathbb{Z}$ and $s \in \mathbb{Z}^+$.

Corollary 4.1. *Let p be a fixed prime and ℓ any square free integer such that $K = \mathbb{Q}(\sqrt{-\ell})$ and $\mathcal{R} := \mathcal{O}_K / p\mathcal{O}_K$ is isomorphic to $\mathbb{F}_{p^2}$ or $\mathbb{F}_p \times \mathbb{F}_p$. For a given code C defined over $\mathcal{R}$, the theta series $\theta_{\Lambda_\ell}(C)$ is the same for almost all ℓ.*

Theorem 4.4. *Let C be a code defined over $\mathcal{R}$ and $\theta_{\Lambda_\ell}(C)$ be its corresponding theta function for level ℓ. Then, for large enough ℓ, there is a unique complete weight enumerator polynomial which corresponds to $\theta_{\Lambda_\ell}(C)$.*

The proofs of Theorems 4.3 and 4.4 are provided in [11] where explicit bounds for ℓ are provided for small p.

Acknowledgment

Some of the ideas of this paper originated during visits of the first author at The University of Maria Curie Sklodowska, Lublin, Poland. The first author wants to thank the faculty of the Department of Computer Science in Lublin for their hospitality. The paper was presented at the Vlora Conference on Algebra, Coding Theory, and Cryptography.

References

[1] C. Bachoc, Applications of coding theory to the construction of modular lattices. J. Combin. Theory Ser. A 78 (1997), no. 1, 92–119.

[2] K. S. Chua, Codes over GF(4) and $\mathbf{F}_2 \times \mathbf{F}_2$ and Hermitian lattices over imaginary quadratic fields. Proc. Amer. Math. Soc. 133 (2005), no. 3, 661–670 (electronic).

[3] J. H. Conway, N. J. A. Sloane, Sphere packings, lattices and groups. Second edition. Grundlehren der Mathematischen Wissenschaften [Fundamental Principles of Mathematical Sciences], 290. Springer-Verlag, New York, 1993. xliv+679 pp. ISBN: 0-387-97912-3

[4] H. H. Chan, K. S. Chua and P. Solé, Seven-modular lattices and a septic base Jacobi identity, J. Number Theory, volume 99, 2003, 2, pg. 361–372,

[5] H. H. Chan, K. S. Chua and P. Solé, Quadratic iterations to π associated with elliptic functions to the cubic and septic base, Trans. AMS 355, 2003, pg. 1505-1520.

[6] W. Ebeling, Lattices and codes, a course partially based on lectures by F. Hirzenbruch, Vieweg (Braunschweig) 1994.

[7] F. J. MacWilliams, N. J. A. Sloane, The theory of error-correcting codes. II. North-Holland Mathematical Library, Vol. 16. North-Holland Publishing Co., Amsterdam-New York-Oxford, 1977. pp. i–ix and 370–762.

[8] F. J. MacWilliams, N. J. A. Sloane, The theory of error-correcting codes. I. North-Holland Mathematical Library, Vol. 16. North-Holland Publishing Co., Amsterdam-New York-Oxford, 1977. pp. i–xv and 1–369.

[9] N. J. A. Sloane, Codes over GF(4) and complex lattices. Information theory (Proc. Internat. CNRS Colloq., Cachan, 1977) (French), pp. 273–283, Colloq. Internat. CNRS, 276, CNRS, Paris, 1978.

[10] T. Shaska and S. Wijesiri, Codes over rings of size four, Hermitian lattices, and corresponding theta functions, *Proc. Amer. Math. Soc.*, (2007), to appear.

[11] T. Shaska, C. Shor, and S. Wijesiri, Codes, modular lattices, and corresponding theta functions, work in progress.

Goppa codes and Tschirnhausen modules

Drue Coles
Department of Mathematics, Computer Science, and Statistics,
Bloomsburg University,
Bloomsburg PA 17815, USA
E-mail: dcoles@bloomu.edu

Emma Previato
Institut Mittag-Leffler
S-18262 Djursholm, Sweden
Permanent:
Department of Mathematics and Statistics,
Boston University, Boston MA 02215-2411
E-mail: ep@bu.edu

We review the use of rank-2 vector bundles in error-correcting coding theory, introduce the issue of maximal subbundles in this context and give an explicit example of rank-2 bundles naturally associated to an elliptic subcover of the Klein curve. We also describe how codes on curves (and therefore certain associated rank-2 bundles and their maximal subbundles) can be formulated in terms of adeles.

Keywords: Goppa codes; Vector bundles; Klein curve.

Introduction

Goppa codes (more properly, *geometric* Goppa codes, for the earliest codes introduced by Goppa were still associated with rational functions on the line) provide a fertile area of interaction between coding theory and algebraic geometry, specifically algebraic curves over finite fields. Goppa's original idea is based on the explicit representation of the space of sections of a line bundle over the curve, and deep issues regarding 'curves with many points' and asymptotic bounds on the genus and ramification of towers of curves have been brought up in view of this application, cf. [9] for a brief survey. More recently, rank-2 vector bundles over the curve have been interpreted as error-correcting devices [4–6, 12] but not so explicitly. Their line subbundles of highest possible degree are of particular interest for decoding, and our goal in this small note is to initiate a study of these objects in the finite field setting.

Higher-rank vector bundles (meaning higher than 1, for line bundles are quite different and better-known objects) come with a concept of "maximal subbundle" for which we refer to the paper [15] although it made earlier appearances (Corrado Segre 1889), since degrees of subbundles can be be related to the self-intersection numbers of sections of the bundle projectivized fiberwise into a ruled surface. We restrict attention to rank-2 bundles, and for these, a maximal subbundle is a line subbundle of largest possible degree. There has been enormous activity on the topic of maximal subbundles in algebraic geometry, which we do not reference here, and this prompts our proposed line of research. On one hand, the results of [15] are given over an algebraically closed field of characteristic zero. Even from the pure viewpoint of algebraic geometry, it would be worth extending the study to any characteristic, and in addition, restricting the analysis to finite fields. In the same vein as counting (rational) points on curves and points of Brill-Noether loci, we propose to count the number of maximal subbundles. Here we give but one example. We decided to use the Klein curve X as a test case, in part because it is so full of beautiful unique properties among curves of genus 3 (small enough yet highly non-trivial), and partly because its large number of automorphisms has already made it popular in coding theory. Over the finite field $\mathbb{F}_8$ the Klein quartic has 24 points, hence it attains Serre's improvement of the Hasse-Weil bound, $|\#X(\mathbb{F}_q) - (q+1)| \leq g[2\sqrt{q}]$.

As regards the link with error-correcting, a weakness might be that the bundles which correspond to correctable messages are unstable, hence their maximal subbundles have very large degree, too large, roughly speaking, to be interesting in algebraic geometry (except perhaps for the suggestions of [12], to the effect of blowing up unstable strata). Our present result concerns bundles whose maximal subbundles have degree zero, yet we regard it as work towards a potential link with coding theory, for example pursuing the suggestion in [12], that is to look at stable points whose lack of correctability (exceeding the distance from a unique codeword) is not too large, so that error-correction is possible in practice ("For practical purposes this would be almost as good as unique decoding (...) one is then interested in maximal sublinebundles"). Other potential uses of stable bundles are discussed in Section 1.

We adopt three approaches which we believe to be new. The first uses the ideas of [15] to construct all rank-2 bundles with largest-dimensional varieties of subbundles; part of this approach is the study of quotients of the curve by an automorphism, which was done relatively recently [21]. The second approach pertains to one of the constructions of [15], and it consists in determining the rank-2 bundle that presents the curve as a triple cover; this approach has the advantage of bringing in another higher-rank bundle, very natural to the situation and proposed by Miranda in [17], the Tschirnhausen module. In the third approach, we formulate Goppa codes in terms of adeles and pseudo-differentials. Adeles provide another way of looking at the rank-2 bundles that appear in connection to codes on curves, a fact used in [6] to investigate an aspect of code construction. For practical implementation of a (de-)coding algorithm, which is one goal of our program, the first step will necessitate an explicit criterion for (maximal) subbundles in terms of adeles. Then, turning to varieties of maximal subbundles, the Tschirnhausen module will provide the multiplicative structure of the covering curve, thus we believe that determining this bundle is the next step in the direction of the ultimate goal.

1. Goppa Codes and rank-2 Vector Bundles

In this section we review the role of vector bundles in error-correction for Goppa codes.

Let X be a smooth projective curve of genus g defined over a finite field k, with a set of k-rational points denoted $Q, P_1, P_2, \dots, P_n$. Define the divisor $D = P_1 + \cdots + P_n$ and choose an integer m so that $n > m > 2g-2$.

The one-point Goppa code

$$C_L(D, mQ) = \{(f(P_1), \dots, f(P_n)) : f \in \mathcal{L}(mQ)\}$$

has dimension $l(mQ) = m - g + 1$ by the Riemann-Roch theorem. Its minimum distance is at least $n - m$, since any non-zero $f \in \mathcal{L}(mQ)$ can vanish at no more than m of the points P_i.

The space of message functions can be taken more generally as $\mathcal{L}(G)$ for an arbitrary divisor G of degree m supported by k-rational points outside the support of D. However, one-point codes (i.e., G a multiple of a single point) are used in practice to maximize the length n of the code and to simplify the construction of a basis for $\mathcal{L}(G)$.

The dual code to $C_L(D, mQ)$ is also a Goppa code, often described in a more convenient form by defining

$$C_\Omega(D, mQ) = \{(Res_{P_1}(\omega), \dots, Res_{P_n}(\omega)) : \omega \in \Omega_X(mQ - D)\}.$$

The fact that $C_L(D, mQ)$ and $C_\Omega(D, mQ)$ are dual codes is a consequence of the residue theorem, which states the sum of residues of a differential over all points is zero.

Requiring $m > 2g - 2$ makes computing the dimension of $\mathcal{L}(mQ)$ and hence of the code $C_L(D, mQ)$ a simple application of the Riemann-Roch theorem. We actually want $m > 2g$ so that the rational map $\varphi : X \to \mathbb{P}^{m-g}$ determined by the complete linear system $|mQ|$ is guaranteed to be an embedding.

Since the rows of a generator matrix for $C_L(D, mQ)$ are obtained by evaluating the functions of a basis for $\mathcal{L}(mQ)$ at $P_1, \ldots, P_n$, we can view the columns as points $\varphi(P_i)$ on the curve in $\mathbb{P}^{m-g}$. These columns are parity checks for the dual code $C_\Omega(D, mQ)$, so a corrupted codeword of the dual is in effect a linear combination of some of the points $\varphi(P_i)$, namely those points at which errors occurred. More explicitly, if H denotes the parity check matrix and $y = (c+e)$ a received word, with codeword $c \in C_\Omega(D, mQ)$ and error vector e, then $Hy = H(c + e) = He$, and the received word $y = e_1 \cdot \varphi(P_1) + \cdots + e_n \cdot \varphi(P_n)$ can be viewed as a point in the j-secant variety of the curve in $\mathbb{P}^{m-g}$, where $j = |\{i : e_i \neq 0\}|$.

We call $A = \sum_{e_i \neq 0} P_i$ the error divisor. The received word $y = c + e$ is said to be correctable if $\deg A < (d-1)/2$, where $d = m - 2g + 2$ is a lower bound on the mimimum distance, since in this case the received word is closer to the transmitted codeword c than to any other codeword.

We also consider an error vector (e) as a point in $H^0(X, \Omega_C(mQ - D))^*$, and then identify it with the isomorphism class of a rank-2 extension E of the form

$$0 \to \mathcal{O}_X \to E \to \mathcal{O}_X(D - mQ) \to 0$$

in a standard way through

$$\begin{aligned} H^0(X, \Omega(mQ - D))^* &\cong H^1(X, \mathcal{O}_X(mQ - D)) \\ &\cong Ext_{\mathcal{O}_X}(\mathcal{O}_X, \mathcal{O}_X(mQ - D)) \\ &\cong Ext_{\mathcal{O}_X}(\mathcal{O}_X(D - mQ), \mathcal{O}_X). \end{aligned}$$

Lange and Narasimhan [15] showed that $s(E)$(:=deg$E$ − 2max(deg$L$), where L is a subbundle of E) is determined by the smallest integer j such that (e) is contained in the j-secant variety of the curve. Applying their results to our situation and with our notation, we get that A is the error divisor for a correctable word if and only if $\mathcal{O}_X(D - mQ - A)$ is the unique maximal subbundle of E. This abstract connection between decoding and maximal subbundles of rank-2 extensions was first noticed by Johnsen [12].

A decoding algorithm based on this idea would determine the rank-2 bundle E corresponding to the syndrome $He = Hy$ of the received word $y = c + e$, in concrete form for instance as a transition matrix, and then compute its unique maximal subbundle $\mathcal{O}_X(D - mQ - A)$. One might then expect to extract the error divisor A and so obtain the error positions (and then the actual error values via simple linear algebra), but with a caveat: we cannot distinguish $\mathcal{O}_X(D - mQ - A)$ from $\mathcal{O}_X(D - mQ - A')$ when $A \sim A'$, so the most that can be guaranteed about the error divisor computed by such an algorithm without additional assumptions (such as the number of errors being less than the gonality of the curve) is that it is linearly equivalent to the true error divisor.

We note that for correctable words, the associated bundle E is necessarily unstable [12]. Still, computing maximal subbundles of *stable* E's in our extension space may be useful for decoding. If the number of errors in a word y exceeds the error correction capacity of the code, it may happen that there are several codewords of precisely equal Hamming distance from y. In that case, finding maximal subbundles amounts to producing a small list of candidate error divisors, though the issue of linear equivalence discussed above applies here as well. There is a vast coding theory literature on *list decoding*, as it is called.

The study of stable rank-2 bundles on curves with many maximal subbundles defined over a finite field may, in addition to its inherent interest, have a coding theory application, since for particular code parameters the maximum possible number of closest codewords to a given (uncorrectable) word may not be known. This point was discussed in [4], where it was also observed that the recent discovery of families of Goppa codes with exponentially many minimum weight codewords [1] is somewhat related: this result says that for a certain code there is a Hamming sphere of radius d centered at 0 with a huge number of codewords on its boundary; a stable bundle with many maximal subbundles (over the base field) would describe a Hamming sphere of radius greater than d centered at an uncorrectable word with a huge number of codewords on its boundary. One possible way to find rank-2 bundles with lots of maximal subbundles over a finite field is to construct examples with infinitely many in the algebraic closure and then count the ones defined over the base field. We turn to this construction next.

2. The Klein Curve as Cover

In this section, which is composed of old-and-new facts about the Klein curve, we recall some results that were given in characteristic zero in the original references; however, they hold in our more general situation provided the characteristic of the base field k is not 2, 3 or 7 (the divisors of 168 which is the order of Aut$X$ in any other characteristic), and provided k contains a seventh root of unity, as noted in the text, because the results we use are obtained by algebraic operations defined over the integers.

The two most familiar ways (for a third one cf. 2.4) to write an algebraic equation for Klein's curve X are:

$$s^7 = t(1-t)^2,$$

$$x_1^3 x_2 + x_2^3 x_0 + x_0^3 x_1 = 0.$$

Klein, already in his original definition [14] of the unique curve of genus 3 that has the maximal number of automorphisms, presented it at first as a modular curve, then as a (canonical) plane quartic. This double feature already exhibits the curve as a cover, on one hand, a $(7:1)$ cover of $\mathbb{P}^1$, on the other, true of every non-hyperelliptic curve of genus 3, as a $(3:1)$ trigonal cover in a 1-dimensional manifold way. More surprisingly, [3, VIII.75] shows that the Jacobian of the curve is isomorphic as a complex manifold (without principal polarization) to the product of three elliptic curves; more precisely, using the $(7:1)$ cover, Baker computes the period matrix

$$Z = \begin{bmatrix} -\frac{1}{8} + \frac{3\sqrt{7}i}{8} & -\frac{1}{4} - \frac{\sqrt{7}i}{4} & -\frac{3}{8} + \frac{\sqrt{7}i}{8} \\ -\frac{1}{4} - \frac{\sqrt{7}i}{4} & \frac{1}{2} + \frac{\sqrt{7}i}{2} & -\frac{1}{4} - \frac{\sqrt{7}i}{4} \\ -\frac{3}{8} + \frac{\sqrt{7}i}{8} & -\frac{1}{4} - \frac{\sqrt{7}i}{4} & \frac{7}{8} + \frac{3\sqrt{7}i}{8} \end{bmatrix}.$$

As observed in [22], all entries lie in the field generated (over the field k of definition of the curve, $k = \mathbb{Q}$, e.g.) by the character of the representation induced on the differentials of the first kind by the automorphism group of the curve. But another interesting phenomenon occurs: $\mathrm{Jac}(X) = \mathbb{C}^3/\Lambda$, where Λ is the lattice corresponding to $[I\ Z]$, is actually isomorphic to the product of 3 elliptic curves. Indeed, Baker shows that it can be brought by an integral (but not unimodular) transformation into diagonal form:

$$\begin{bmatrix} 1\,0\,0 & \frac{1+i\sqrt{7}}{4} & 0 & 0 \\ 0\,1\,0 & 0 & 2\ \frac{1+i\sqrt{7}}{4} & 0 \\ 0\,0\,1 & 0 & 0 & 2\frac{1+i\sqrt{7}}{4} \end{bmatrix}.$$

He also remarks that this transformation does not give us an algebraic map from X to an elliptic curve; for that we use recent work [21], which gives a bit more: the three elliptic curves are isomorphic as opposed to 2-isogenous as in Baker's decomposition.

We recall some notation and standard facts from [21]. The following three elements generate the automorphism group of X, which is isomorphic to $\mathbb{PSL}_2(\mathbb{F}_7)$: $\sigma(x_0, x_1, x_2) = (x_1, x_2, x_0)$ of order 3, $\tau(x_0, x_1, x_2) = \left(x_1 + \mu_1 x_2 + \frac{1}{\mu_3} x_0, \mu_1 x_1 + \frac{1}{\mu_3} x_2 + x_0, \frac{1}{\mu_3} x_1 + x_2 + \mu_1 x_0\right)$ of order 2 and $\epsilon(x_0, x_1, x_2) = (x_0, \zeta x_1, \zeta^5 x_2)$ of order 7, where ζ is a primitive 7th root of 1 and we let $\mu_i = \zeta^i + \zeta^{-i}$.

Proposition 2.1. *[21] The quotient of X by σ^i, $i = 0, 1, 2$ gives three (canonically isomorphic) elliptic curves T_i with Weierstrass equations:*

$$T_i : y^2 + 3\zeta^{4i} xy + \zeta^{5i} y = x^3 - 2\zeta^{2i} x - 3\zeta^{3i}, \ i = 0, 1, 2,$$

with the $(3:1)$-morphisms $X \to T_i$ given by $\phi_i(x_1, x_2) = (-w_i, v_i)$ where

$$w_i = x + \zeta^{6i}\frac{1}{y} + \zeta^{4i}\frac{y}{x}, \quad v_i = y + \zeta^{6i}\frac{1}{x} + \zeta^{2i}\frac{x}{y}.$$

Given that the above result is algebraic, we can simply replace $\mathbb{Q}[\zeta]$ by a finite field that contains a seventh root of unity, and keep the notation ζ for a primitive one. In fact, it is quite interesting and non-tivial to find Aut$X$ over an algebraically closed field of any characteristic. This was accomplished in [30–32]: if the characteristic is $p \neq 3, 7$ the group is again GL$(3, 2)$. For $p = 3$ (resp. $p = 7$), the group properly contains GL$(3, 2)$ and is of order 6048 (resp. 672). It is thus not true (as had also been observed earlier) that the Hurwitz bound $84(g - 1)$ holds for the number of automorphisms of a curve of genus g (> 1), if the characteristic is not zero; a bound does exist, modified by the contribution of wild ramification in the Riemann-Hurwitz formula, has degree 4 in g, and it is known which curves attain it.

Our program is now the study of maximal subbundles in positive characteristic. Following the seminal article [15], for a rank-2 (algebraic) vector bundle over a curve X of genus g, we define the numerical invariant:

$$s(E) = \text{deg}E - 2\text{max}(\text{deg}L),$$

where L is a line-subbundle of E. By definition, the degree of E and $s(E)$ have the same parity. It is known that $s(E) \leq g$, and the study in [15] addresses the case $s(E) > 0$ (equivalent to E being a stable bundle) or $s(E) \geq 0$ (semi-stable). The relevant geometric object then is $M(E)$, the subvariety of maximal subbundles. This variety can be identified canonically with the space of minimal sections of the ruled surface $\mathbb{P}(E)$, minimal in the sense of having smallest self-intersection number. Let us also denote by $M(d)$ the moduli space of stable bundles of rank 2 and degree d over a curve X of genus $g \geq 2$, and by $M(d, s)$ its stratification into locally closed subsets according to the value of the invariant $s(E)$. For generic E, $M(E)$ is smooth and projective and its dimension is described in terms of the rank and degree of E and the genus of X. It has exactly the Chern numbers of an étale cover of the symmetric product $S^n X$, where $n = \dim M(E)$ [20]. In particular, for the general bundle, $s(E) = g$ if the degree of E has the same parity as the genus, and $s(E) = g - 1$ otherwise. When $s(E) = g$, the variety of maximal subbundles of E is a curve, but when $s(E) = g - 1$, it is generically a finite number of points. It is this number that in the case of positive characteristic could conceivably be smaller, in the case the field is not algebraically closed and the subbundle as a variety is not rational over the field of definition, or perhaps larger, as is the case for the number of automorphisms, due to the wild-ramification contribution in the Riemann-Hurwitz formula, in view of the fact that in [15] a manifold of maximal line subbundles are identified by using covering maps. The number of subbundles does have a topological-degree significance, because of the cited result [20] which computes it as a Chern number, 2^g times a Castelnuovo number, but so does the number of inflections of a plane curve; in point of fact, the Klein curve is the "funny curve" in characteristic 3, and all of its points are inflections [11, Exercise IV.2.4]. It is also interesting to note that the dimension of $M(E)$ can jump, as in the following example [20, Remark 1.5]: the general bundle E with trivial determinant on a curve of genus 3 has a finite number of maximal subbundles, $2^3 = 8$, since $s(E) = g-1$ as we recalled. But $M(E)$ is isomorphic to the curve for the 64 bundles $E = \kappa^{-1} \otimes V$, where κ is a theta characteristic and V is the unique stable rank-2 bundle whose determinant is the canonical bundle, and whose space of sections has the maximal possible $\dim H^0(X, V) = 3$. In fact, in this programmatic note we focus on such 'richest' case only, namely $s(E) = g - 1 (= 2$ in our case) and $\dim M(E)$=1, strictly larger than for general E. In [15] it is determined exactly which E have this property, providing a negative answer to a conjecture of M. Maruyama, to the effect of $\dim M(E)$ being zero for all, not merely general, bundles that have $s(E) \leq g - 1$.

Proposition 2.2 (after 15, Theorem 5.1). *Every degree-2 cover $X \to T$ of an elliptic curve gives a g-dimensional subvariety of $M(d, 2)$, where d is an even number, for all of whose points E, $\dim M(E) = 1$. If X is of genus 3, any trigonality of X gives a 3-dimensional subvariety of $M(d, 2)$*

for all of whose points E, $\dim M(E) = 1$. *For any other* $E \in M(d,2)$, $\dim M(E) = 0$.

We also record the construction of the rank-2 bundles that have a non-generic $\dim M(E)$:

Lemma 2.1 (after 15, Section 5). *(i) If $\pi : X \to T$ is a $(2:1)$ elliptic cover and $g(X) \geq 3$ then to every $L \in \mathrm{Pic}^g T$ where $g = g(X)$ there is associated a vector bundle $E \in M(2,2)$ on X with* $\dim M(E) = 1$. *Varying $L \in \mathrm{Pic}^g T$ and twisting the associated E by a line bundle of degree $\frac{d-2}{2}$ on X yields other elements of $M(d,2)$, while 'factoring' by the one-dimensional families of their maximal subbundles finally gives a g-dimensional algebraic family in $M(d,2)$. (ii) To any trigonality $\pi : X \to \mathbb{P}^1$ of a curve of genus 3 there is associated in a canonical way a vector bundle $E \in M(2,2)$ on X.*

Proof. (i) Pulling back any rank-2 bundle F on the elliptic curve with $s(F) = 1$ as well as the family of line subbundles of appropriate degree gives the examples. They can be described geometrically: the embedding $H^0(T,L) \to H^0(X, \pi^*L)$ (which is of codimension 1) defines a point in $\mathbb{P}^g = \mathbb{P}(H^0(X,\pi^*L)$ which is not on the image of X. This point can be interpreted as a non-split exact sequence on X whose central element is a vector bundle of rank 2 with $s(E) = 2$ and $\det E = \pi^*L \otimes K_X^{-1}$, where K_X is the canonical divisor of X. Projection from the point has degree 2 on the image of X and represents the 2-secants of X through that point, so the maximal subbundles are represented by the points of the elliptic curve embedded in the hyperplane covered by the projection, except possibly the projection of the singular point of the image of X. (ii) Here the bundle E is the middle term of the extension given by the embedding $H^0(\mathbb{P}^1, \mathcal{O}_{\mathbb{P}^1}(2)) \to H^0(X, \pi^*\mathcal{O}_{\mathbb{P}^1}(2))$ so $\det E = \pi^*\mathcal{O}_{\mathbb{P}^1}(2) \otimes K_X^{-1}$ and again the 3-dimensional family of bundles is parametrized by $\mathrm{Pic}^{d-2/2}X$ plus the trigonalities minus 1 for the maximal subbundles, which correspond to the trisecant lines of the embedded curve in $\mathbb{P}^3$ which go through the extension point. □

This lemma together with the proof (which we do not produce) that no other bundle exhibits the jump phenomenon, proves Proposition 2.2.

We are next faced with the task of giving (in an algebraic and explicit way) a $(2:1)$ elliptic subcover of X or a trigonal rationality. We begin with the latter. Rather than take the approach of [7] and determine the quotient of the Klein curve under all cyclic subgroups of automorphisms, we use the interesting analysis proposed in [18], by addressing the additional question: given a trigonality obtained by projecting a smooth plane quartic to a line from a point on it, when is this cover Galois? We take this point of view because we find it potentially interesting to give an addendum to Kowalevski's early result: she proved that a plane quartic is a $(2:1)$ cover of an elliptic curve if and only if four of its 28 bitangents are concurrent, as we recall in Prop. 2.4.

The gonality of a curve is the smallest possible degree of the function field of the curve over a rational field of one variable. We now adapt statements from [18], which assumes the field of definition k to be algebraically closed of characteristic zero. For our purposes we assume that all maps are defined over k in case k isn't algebraically closed (such as a finite field). The Klein curve is not hyperelliptic, hence it is trigonal. For a plane smooth m-gonal curve of degree d the gonality is $d-1$ and any extension $K/k(t)$, where K is the function field of the curve and $k(t)$ is any rational field of degree 1, corresponds to an $(m:1)$ projection from a point of the curve onto a line [18]. In [18], the authors determine the following objects pertaining to a smooth quartic (such as our Klein curve – in fact, their worked-out example is the Fermat curve, whose automorphism group [33] has order 96): for $P \in X$, the projection of X from P to a line is a degree-3 cover, and the Galois group as well as the genus of the corresponding cover are calculated, together with the (finite) number of points P for which the cover is Galois.

Proposition 2.3 (after 18, Theorem 2.1). *For any smooth plane quartic X and any point $P \in X$, the projection from P to a line corresponds to a field extension that does not depend on the line, and if we call $g(P)$ the genus of the smooth curve whose function field is the Galois closure of the field extension corresponding to the projection and P a Galois point when the extension is Galois, then: $g(P)$=3,6,7,8,9, or 10, with $g(P)$ =10 for the general point, with Galois group isomorphic to S_3. The number of Galois points can be 0,1, or 4, and it is zero for a general quartic.*

In [18], part of the criterion for P to be a Galois point is that P be a 2-inflection point. In particular, for the Klein curve, none exists, since the inflections are all distinct and comprise the 24 Weierstrass points, so none of the trigonal covers is Galois.

Similar issues are treated in [18] for the case $P \notin X$, there being more cases to analyze and slightly less complete results. The Klein curve does admit a double cover to an elliptic curve. Indeed, as noted in [14], there are 21 subgroups of order 2 of Aut$X$, each corresponding to a collineation; the centers of projection give $(4:1)$ maps of X to a line which factor through an elliptic curve, the ramification given by the four bitangents to X through the center (each bitangent contains three centers so that there are $\frac{21\times 4}{3}=28$ bitangents). We note however that none of the 4-gonal covers given by projection from $P \notin X$ of the Klein (unlike the Fermat!) quartic are Galois either; the 21 elliptic subfields of $K(X)$ fixed by involutions are one orbit under Aut$X$ [16].

It seems worth recalling Kowalevski's criterion for a smooth plane quartic to be a $(2:1)$ elliptic cover, which again is proved in characteristic zero. Her proof was analytic, a contribution to the theory of reduction, part of her dissertation supervised by Weierstrass. An algebraic proof is given in [8], as part of the properties of Weierstrass points of curves with involution.

Proposition 2.4 (Chap. III, Art. 71, 72, 76 in [3]). *A canonically embedded plane curve of genus 3 admits a $(2:1)$ cover to an elliptic curve if and only if four of its bitangents are concurrent, equivalently in suitable coordinates it has an equation:*

$$(z^2-\phi_2)^2 = 4xy(ax+by)(cx+dy),$$

with ϕ_2 a homogeneous for of degree 2 in x, y.

Here the bitangents are patently represented by the linear forms x, y, $ax+by$ and $cx+dy$, whose cross-ratio is an invariant of the elliptic curve. Note the analogy with genus one: an elliptic curve is the Fermat curve if and only if it can be represented as a plane cubic with three concurrent bitangents, the projection from their common point being Galois. As recalled above, Klein's curve can be written in this way by virtue of its automorphisms of order two. An actual geometric model of the elliptic curve together with the $(2:1)$ projection can be found by embedding X in $\mathbb{P}^3$ via the divisor of degree 6 that pulls back an $L \in \mathrm{Pic}^3 T$, precisely as in Lemma 2.1, obtaining an extension E to be viewed as a point in $\mathbb{P}^3$ and projecting the image of X from that point to a plane; Baker (*loc. cit.* in Prop. 2.4) states this fact concretely presenting the image of X as a space sextic with equations:

$$z^2-\phi_2 = xt, \quad xt^2 = 4y(ax+by)(cx+dy),$$

as obtained by sending $[x,y,z] \mapsto [x,y,z,t] \sim [1, y/x, z/x, (z^2-\phi_2)/x^2]$ by the pole-divisor map of $3P_1+3P_2$, P_1 and P_2 being the points of contact of the bitangent $x=0$.

Remark. One subtle issue that we do not address in this note is the following. A classical result reprised and refined in [13] says that if an abelian surface has more than two elliptic subgroups, then it has infinitely many;

[13] shows also that it has finitely many ones for each bounded degree (the degree can be taken to be the intersection number with any fixed ample divisor). In our case, we would ask how many genuinely distinct (elliptic) subcovers the Klein curve has, in particular over each finite field. We note that much current work is devoted to classifying subcovers of Hermitian curves (of key interest in the area of Goppa codes), for example in [7] a classification is given of the quotients of Hermitian curves by all prime-order automorphisms. For the genus-2 case, an explicit detection of isogenous/isomorphic degree-2 and degree-3 subcovers, as well as partial results for higher degree, is given in [24–27].

Summary. Let X be the Klein curve. For each fixed determinant, the rank-2 bundles $E \in M(2,2)$ with $\dim M(E) = 1$ correspond to a given elliptic-hyperelliptic map or trigonality. The 64 points E mentioned above that exhibit the jump phenomenon as regards $\dim M(E)$ [20] have fixed (even-degree) determinant. It follows from the above construction that each map gives rise to one bundle; the 21 subgroups of order 2 of $\mathrm{Aut}X$ come with three maps each (each group of 4 concurrent bitangents gives an elliptic curve and each bitangent contains three centers), so we recover the $64 = 21 \times 3 +$ [one trigonality] bundles of [20], on which $\mathrm{Aut}X$ acts by permutations. To compute the number of these bundles over a finite field $\mathbb{F}_q$, one of our goals, first we fix a determinant of degree d that is an element of $\mathrm{Pic}^d X(\mathbb{F}_q)$ (there exists one for each degree, and the number of distinct ones is independent of the degree [19, Chap. 3]), then there are as many bundles (semistable and with that determinant), with 'too many subbundles', as there are points of order 2 in $\mathrm{Pic}^0 X(\mathbb{F}_q)$, found [19] (since the Jacobian splits) by splitting the characteristic p in $\mathbb{Z}[\sqrt{-7}]$.

Example. Consider the Klein curve X defined by $x_1^3x_2 + x_2^3x_0 + x_0^3x_1 = 0$ over $\mathbb{F}_8 = \mathbb{F}_2[\beta]/(\beta^3+\beta+1)$. Since the characteristic is 2, we cannot expect the same situation as in characteristic zero, in fact there are no odd theta-characteristics since the tangent line at any point is an inflectionary tangent. However, the maximal-subbundle geometry survives. Fix coordinates so that on the line at infinity $z = 0$, parametrized as $[a, b]$, $P_\infty = [1, 0]$ and let $\pi : X \to \mathbb{P}^1$ be the projection from $Q_3 = [0, 0, 1]$ to the line at infinity, so that $2P_\infty$ pulls back to $6Q_1$, where $Q_1 = [1, 0, 0]$.

Let $\varphi : \mathbb{P}^1 \to \mathbb{P}^2$ denote the embedding $[a, b] \mapsto [1, a/b, a^2/b^2]$. The divisor map $\varphi_{6Q_1} : X \to \mathbb{P}^3$ that makes the following diagram commute is given by $[a, b, c] \mapsto [1, a/b, a^2/b^2, ab/c^2]$.

The injection $H^0(\mathbb{P}^1, 2P_\infty) \overset{\pi^*}{\hookrightarrow} H^0(X, 6Q_1)$ corresponds to the point

$(e) = [0, 0, 0, 1] \in \mathbb{P}^3$. The projection p in the commutative diagram

$$\begin{array}{ccc} \mathbb{P}^3 - \{e\} & \xrightarrow{p} & \mathbb{P}^2 \\ \varphi_{6Q_1} \uparrow & & \uparrow \varphi \\ X & \xrightarrow{\pi} & \mathbb{P}^1 \end{array}$$

is $[a, b, c, d] \overset{p}{\mapsto} [a, b, c]$. The points $\varphi(\mathbb{P}^1)$ parametrize the trisecant lines of $\varphi_{6Q_1}(X)$ containing (e).

Choose a point $Q = [a, 1]$ on the projective line, $a \in \mathbb{F}_8{}^*$. Then the three points $[a, 1, *] \in \pi^{-1}(Q)$ are mapped by φ^* to a trisecant line containing (e). Any two of these points determine a maximal subbundle of E, the rank-2 bundle corresponding to (e). We can compute

$$\pi^{-1}(Q) = \{[a, 1, a^3\beta], [a, 1, a^3\beta^2], [a, 1, a^3\beta^4]\}$$

and it follows that E has $7 \cdot \binom{3}{2} = 21$ maximal subbundles that are rational over $\mathbb{F}_8$, namely those of the form

$$\mathcal{O}_X\left([a, 1, a^3\beta^i] \;+\; [a, 1, a^3\beta^j]\right)$$

where $a \in \mathbb{F}_8{}^*$ and $(i, j) \in \{(1, 2),\ (1, 4),\ (2, 4)\}$.

3. The Tschirnhausen Module of the Cover

In [17], the author sets out to "develop the foundations of the theory of triple coverings in algebraic geometry", working on an algebraically closed field of characteristic unequal to 2 or 3; his result in summary:

A triple cover of an irreducible variety Y is determined by a locally free rank-2 $\mathcal{O}_Y$-module E and a map $\Phi : S^3E \to \wedge^2 E$, and conversely.

It may be worthwhile to determine this rank-2 bundle in our situation, in view of what we described above, even when the cover does not pertain to one of the exceptional rank-2 bundles over the Klein curve. We believe that the object introduced by Miranda has not yet been widely used while being potentially useful in coding theory. We restrict attention to one of the above triple covers $X \to T$, where X is the Klein curve, or one of the trigonalities $X \to \mathbb{P}^1$; we denote the target by Y in either case.

Definition 3.1. E is the Tschirnhausen module of $\mathcal{O}_X$ over $\mathcal{O}_Y$, namely the direct summand in $\mathcal{O}_X = \mathcal{O}_Y \oplus E$ consisting of the functions $a \in \mathcal{O}_X \backslash \mathcal{O}_Y$ whose minimal polynomial over $\mathcal{O}_Y$ has trace zero.

The name given by Miranda to the module refers to the Tschirnhausen transformation [29], used in several instances of reduction of degree of algebraic equations; another important example, the quintic equation, is also related to curves [10]. The conventional way to perform a Tschirnhausen transformation is to allow a substitution $y = x^m + r_{m-1}x^{m-1} + \ldots + r_1 x + r_0$, in order to simultaneously eliminate (by using the r's as free parameters) intermediate terms of any nth- (say) degree equation[a]. In the case of a quintic, to bring it to Bring-Jerrard form: $x^5 + ax + b$, with $y = \sum_{j=0}^{4} a_j x^j$ one has to solve three equations of degrees 1, 2, and 3 in the coefficients of the original equation. In this case [10] it is possible to intersect suitable hypersurfaces in $\mathbb{P}^4$ and find solutions by solving equations of degree at most four. Bring's curve is then of genus four and can be explicitly uniformized as it possesses sufficiently many automorphisms, in particular a $(12 : 1)$ (Galois) cover to an elliptic curve. This provides a solution to the general quintic in terms of modular forms of weight -2.

With this motivation, Miranda defines the Tschirnhausen module of the triple cover $X \to Y$ to be the submodule E in the decomposition of local k-algebras (where k is an algebraically closed field of characteristic unequal to 2 or 3), or sheaves, $\mathcal{O}_X = \mathcal{O}_Y \oplus E$ consisting of the elements $a \in \mathcal{O}_X \backslash \mathcal{O}_Y$ whose minimal polynomial is trace free.

In our situation, for the map in Prop. 2.1 given explicitly as above, the module consists of the elements $\frac{2}{3}a - a^{\sigma} - a^{\sigma^2}$, for all a in the function field of X that are not σ-invariant; is is enough to take $a = x,\ y$ to span the module and the map σ is given explicitly: $x \mapsto y \mapsto z \mapsto x$ so x projects to $\frac{2}{3}x - y - z$ and y to $\frac{2}{3}y - z - x$. This would provide actual equations for the corresponding divisor; however, we give a more theoretic way to identify it.

Miranda computes the ramification and branch locus of the triple cover: the branch locus in Y is a divisor whose associated line bundle is $(\wedge^2 E)^{-2}$ so by the Riemann-Hurwitz formula (which has no inertia components under the assumptions we made on the characteristic), $2g(X) - 2 = 3(2g(Y) - 2) + \text{degree}(\wedge^2 E)$. In conclusion, in our case E has degree 4. Atiyah [2] gave a description of all the semistable bundles over an elliptic curve, but we are further restricted in our situation: the cover is by construction a Galois cover, and Miranda shows that E splits into the sum of two eigenline bundles: $f_*\mathcal{O}_X = \mathcal{O}_Y \oplus L^{-1} \oplus M^{-1}$, $E = L^{-1} \oplus M^{-1}$, where $L^{-1},\ M^{-1}$ are the eigenspaces for σ, σ^2. Since there are exactly two σ-fixed points on X, namely $p_1 = [1, \epsilon, \epsilon^2]$ and $p_2 = [1, \epsilon^2, \epsilon]$ where ϵ is a primitive third root of 1, the bundles L and M are $\mathcal{O}(-2p_i)$.

The trigonality, however, is never Galois as we saw. To compute the

[a]We acknowledge this clear and clever exemplification due to Titus Pierzas III posted on the web: A New Way To Derive The Bring-Jerrard Quintic in Radicals, `www.geocities.com/titus_piezas/Tschirnhausen.pdf`.

Tschirnhausen module which, being a rank-2 bundle over $\mathbb{P}^1$, decomposes into $\mathcal{O}(n) \oplus \mathcal{O}(m)$, we refer to [17, Section 9] for an argument, essentially based on the Riemann-Hurwitz formula, yielding $n = -2$ and $m = -3$.

Summary. The Tschirnhausen module for the possible triple covers of the Klein curve to the elliptic curve T that admits multiplication by a primitive root of 7 as an endomorphism, or to the projective line, are respectively

$$\mathcal{O}_E(-2p_1) \oplus \mathcal{O}_E(-2p_2), \quad \mathcal{O}_{\mathbb{P}^1}(-2) \oplus \mathcal{O}_{\mathbb{P}^1}(-3).$$

4. Goppa Codes and Adeles

We observe in this section that Goppa codes can also be formulated in terms of adeles and pseudo-differentials, and in this setting the duality between $C_L(D, mQ)$ and $C_\Omega(D, mQ)$ can be established without direct appeal to the residue theorem or the analogous result for pseudo-differentials.

An introduction to adeles and pseudo-differentials can be found in the chapters on the Riemann-Roch theorem in the books by Moreno [19, Chap. 2] and Stichtenoth [28, Chap. I.5]. Basic definitions and results needed for our purposes are reviewed below.

4.1. *Adeles and pseudo-differentials*

Let K denote the function field of the curve X, and k the field of constants. In this subsection, D denotes an arbitrary divisor. As usual, $l(D) = \dim_k \mathcal{L}(D)$, where $\mathcal{L}(D)$ is the Riemann-Roch space of D. By Riemann's theorem, $l(D) \geq \deg D - g + 1$, and the *index of specialty* is $i(D) = l(D) - \deg D + g - 1$.

An *adele*[b] is a mapping $\alpha : X \to K$ that associates a function α_P to every point $P \in X$ in such a way that $\alpha_P \in \mathcal{O}_P$ for all but finitely many points P. It is convenient to define the order of an adele α at a point P by $\operatorname{ord}_P(\alpha) = \operatorname{ord}_P(\alpha_P)$.

The set A of all adeles is called the *adele space*. We can add adeles componentwise: the P-component of $\alpha + \alpha'$ is $(\alpha + \alpha')_P = \alpha_P + \alpha'_P$, which is again an adele. Componentwise multiplication also makes sense, turning A into a ring. More to the point for our purposes, it is a vector space over k, and the k-subspace $A(D)$ for a divisor D is defined in analogy to $\mathcal{L}(D)$,

$$A(D) = \{\alpha \in A : \operatorname{ord}_P(\alpha) + \operatorname{ord}_P(D) \geq 0 \text{ for every } P \in X\}.$$

[b] Some authors use the term *repartition* or *pre-adele* for what is here called an adele, reserving the term adele for when the functions α_P are allowed to lie in the completion of K with respect to the valuation ord_P.

An embedding $K \hookrightarrow A$ is obtained by identifying $f \in K$ with the adele whose every component is equal to f. In particular, let f/Q for $Q \in X$ denote the adele $\alpha \in A$ defined by

$$\alpha_P = \begin{cases} f : P = Q. \\ 0 : P \neq Q. \end{cases}$$

For a divisor D, $A(D) + K$ is an infinite dimensional k-subspace of A, but the quotient space $A/\left(A(D)+K\right)$ is finite dimensional, in fact equal to the index of specialty $i(D)$ of D. This fact is implied by the canonical isomorphism (see [23, Prop. II.3], for example)

$$H^1(C, \mathcal{O}_X(D)) \cong \frac{A}{A(D)+K}$$

and can also be established directly, without cohomological arguments [19]. The next proposition records this fact for ease of reference below.

Proposition 4.1. *With the given notation,* $\dim_k A/\left(A(D)+K\right) = i(D)$.

A *pseudo-differential* (also called a *Weil differential*) is a k-linear map $\omega : A \to k$ vanishing on $A(D) + K$ for some divisor D. Note that if ω_i vanishes on $A(D_i) + K$ $(i = 1, 2)$ then $\omega_1 + \omega_2$ vanishes on $A(D) + K$ for any divisor D with $D \leq D_i$ $(i = 1, 2)$. With scalar multiplication defined in the obvious way, the space of all pseudo-differentials becomes a vector space over k, which we denote by $\Omega^s_{K/k}$ following Moreno [19]. The subspace

$$\Omega^s_{K/k}(D) = \{\omega \in \Omega^s_{K/k} : \omega \text{ vanishes on } A(D) + K\}$$

has dimension $i(D)$ by Prop. 4.1. Stichtenoth works out in full detail the correspondence between differentials and pseudo-differentials [28, Chap. IV]. Here we note only that for a given pseudo-differential ω there is a unique divisor W of smallest possible degree with the following property: if ω vanishes on $A(F) + K$ for some divisor F, then $F \leq W$. As expected, W is also the divisor of the corresponding differential.

4.2. *Goppa codes and adeles*

As in subsection 4.1, let $D = P_1 + \cdots + P_n$, where the P_i are k-rational points. Fix another k-rational point Q $(Q \neq P_i)$ and an integer m with $n > m > 2g - 2$. Let $n' = n + g - 1$ and $D_i = D - P_i$ for $1 \leq i \leq n$. Choose $f_i \in \mathcal{L}(n'Q - D_i)$ so that

$$f_i(P_j) = \begin{cases} 1 : i = j. \\ 0 : i \neq j. \end{cases} \tag{1}$$

Such functions f_i exist since $l(n'Q - D_i) \geq 1$. Also, $l(n'Q) = n$ and the f_i are linearly independent, so they form a basis for $\mathcal{L}(n'Q)$. Now consider the linear code

$$C = \{(c_1, \ldots, c_n)) \in k^n : \mathrm{ord}_Q(c_1 f_1 + \cdots + c_n f_n) \geq -m\}.$$

The distance and dimension of C are easy to compute. Choose a non-zero codeword $(c_1, \ldots, c_n)$ and let $f = \sum_i c_i f_i$. Define $I \subset \{1, \ldots, n\}$ so that $c_i = 0 \leftrightarrow i \in I$, and note that $f(P_i) = 0$ for every $i \in I$. Now since $\mathrm{ord}_Q(f) \geq -m$, we know that f has at most m zeros. This means that $|I| \leq m$, so $(c_1, \ldots, c_n)$ is non-zero in at least $n - m$ positions. As for the dimension, $f \in \mathcal{L}(mQ)$ by definition, so $\dim_k C = l(mQ) = m - g + 1$.

In fact, $C = C_L(D, mQ)$. To see this, note that for $f = \sum_i c_i f_i \in \mathcal{L}(mQ)$ we have $f(P_i) = c_i \cdot f_i(P_i) = c_i$. In other words, a codeword $(c_1, \ldots, c_n) \in C$ is obtained by evaluating some $f \in \mathcal{L}(mQ)$ at the points P_i.

Fix a local parameter t at Q. Expanding each f_i around Q, we can write

$$f_i = \sum_{j=-n'}^{\infty} c_{i,j} \cdot t^j$$

with uniquely determined coefficients $c_{i,j} \in k$. A parity check matrix H for the code can be constructed using these coefficients: the i-th column is the vector of coefficients in the expansion of f_i up to (and including) the $t^{-(m+1)}$ term. The kernel of this matrix consists of linear combinations of the functions f_i with at most m poles at Q, that is to say, codewords.

We now proceed to interpret the parity check matrix H in terms of pseudo-differentials by way of the following two lemmas.

Lemma 4.1. *Letting t denote a local parameter at Q, the set $\mathcal{B} = \{t^{-i}/Q : m < i \leq n'\}$ is a basis for $A/\left(A(mQ - D) + K\right)$ as a vector space over k.*

Proof. Consider first an arbitrary adele α. By the Strong Approximation Theorem [28], there is a function $g \in K$ satisfying $\mathrm{ord}_{P_i}(\alpha - g) > 0$ for each point P_i in the support of D, and $\mathrm{ord}_P(\alpha - g) \geq 0$ for every other point of the curve except Q. It follows that $\alpha \equiv (\alpha_Q - g)/Q$ modulo $A(mQ - D) + K$. In particular, $A/\left(A(mQ - D) + K\right)$ has a basis consisting of adeles everywhere zero except at Q.

If the pole order of $f \in K$ at Q is at most m, then $f/Q \in A(mQ - D)$. On the other hand, if f has more that n' poles at Q, say r poles, there is a non-zero $g \in \mathcal{L}(rQ - D)$ with $\mathrm{ord}_Q(f - g) > -r$, and $f/Q - g \in A(mQ - D)$. This implies that if $f/Q \not\equiv 0$ then $-n \leq \mathrm{ord}_Q(f) < -m$.

We have established that there is a basis for $A/(A(mQ-D)+K)$ consisting of adeles of the form f/Q with $-n \leq \mathrm{ord}_Q(f) < -m$. The basis has size $i(mQ-D) = |\mathcal{B}|$ by Prop. 4.1, and we clearly can obtain $\mathcal{B}$ from it by a linear transformation. □

Lemma 4.2. *With the functions f_i as defined in (1), we have* $1/P_i \equiv f_i/Q \bmod A(mQ-D)+K$ *for* $1 \leq i \leq n$.

Proof. Define $\alpha_i \in A(mQ-D)$ by

$$(\alpha_i)_P = \begin{cases} 0 : P = Q. \\ f_i + 1 : P = P_i. \\ f_i : \text{otherwise.} \end{cases}$$

Then $\alpha_i - f_i = 1/P_i - f_i/Q$, so $1/P_i \equiv f_i/Q$ as claimed. □

A pseudo-differential $\omega \in \Omega^s_{K/k}(mQ-D)$ is determined by a vector

$$\hat{a} = (a_{m+1}, a_{m+2}, \ldots, a_{n'}) \in k^{n'-m}$$

describing the action of ω on elements of $\mathcal{B}$; i.e., $\omega : t^{-i}/Q \mapsto a_i$. In particular, $\omega(1/P_i)$ can be computed as the inner product of $\hat{a}$ and the i-th column of H, the parity check matrix for $C_L(D, mQ)$. And since a parity check matrix of a code is a generator matrix for its dual, we can define the dual code to $C_L(D, mQ)$ purely in terms of adeles by

$$C(D, mQ)^{\perp} = \Big\{ (\omega(1/P_1), \ldots, \omega(1/P_n)) : \omega \in \Omega^s_{K/k}(mQ-D) \Big\}$$

We close the circle by noting that from the correspondence between pseudo-differentials and differentials it can be shown that an arbitrary pseudo-differential maps the adele $1/P$ (for any $P \in X$) to the residue at P of the corresponding differential. Consequently, $C_\Omega(D, mQ)$ as defined in the first section is dual to $C_L(D, mQ)$, which we have established using the theory of adeles and pseudo-differentials and without appeal to the residue theorem. As noted earlier, since our extension space is isomorphic to $H^1(X, \mathcal{O}_X(mQ-D))$, it can be identified with the adelic space $A/(A(mQ-D)+K)$. One angle from which we propose to study rank-2 extensions and their maximal subbundles over finite fields is through this connection to adeles. We showed that every adele is equivalent, modulo $A(mQ-D)+K$, to an adele of the form f/Q; in fact, each such f determines a transition function for a rank-2 bundle in our space of extensions.

Acknowledgements

Both authors are thankful for partial research-travel support under NSA grant MDA904-03-1-0119 [*any opinions, findings, and conclusions or recommendations expressed in this material are those of the authors and do not necessarily reflect the views of the National Security Agency*]. E.P. is currently benefiting from the scholarly atmosphere of the Institut Mittag-Leffler in the *Moduli Spaces* program and is deeply grateful for the hospitality extended to her.

References

[1] A. Ashikhmin, A. Barg, S. Vladut, Linear codes with exponentially many light vectors, *J. Combin. Theory Ser. A* **96** (2001), no. 2, 396-399.
[2] M.F. Atiyah, Vector bundles over an elliptic curve, *Proc. London Math. Soc.* (3) **7** (1957), 414–452.
[3] H.F. Baker, *An introduction to the theory of multiply-periodic functions.* University Press XVI , Cambridge, 1907.
[4] T. Bouganis and D. Coles, A geometric view of decoding AG codes, in *Applied algebra, algebraic algorithms and error-correcting codes (Toulouse, 2003)*, pp. 180–190, Lecture Notes in Comput. Sci., **2643**, Springer, Berlin, 2003.
[5] D. Coles, Vector bundles and codes on the Hermitian curve, *IEEE Trans. Inform. Theory* **51** (2005), no. 6, 2113–2120.
[6] D. Coles, On constructing AG codes without basis functions for Riemann-Roch spaces, in Lecture Notes in Comput. Sci., **3857**, Springer, 2006, pp. 108–117.
[7] A. Cossidente, G. Korchmáros and F. Torres, Curves of large genus covered by the Hermitian curve, *Comm. Algebra* **28** (2000), no. 10, 4707–4728.
[8] H.M. Farkas and I. Kra, Branched two-sheeted covers, *Israel J. Math.* bf 74 (1991), no. 2-3, 169–197.
[9] G. van der Geer, Curves over finite fields and codes, in *European Congress of Mathematics, Vol. II (Barcelona, 2000)*, pp. 225–238, Progr. Math., **202**, Birkhäuser, Basel, 2001.
[10] M.L. Green, On the analytic solution of the equation of fifth degree, *Compositio Math.* **37** (1978), no. 3, 233–241.
[11] R. Hartshorne, *Algebraic geometry*, Graduate Texts in Mathematics, No. **52**. Springer-Verlag, New York-Heidelberg, 1977.
[12] T. Johnsen, Rank two bundles on algebraic curves and decoding of Goppa codes, *Int. J. Pure Appl. Math.* **4** (2003), no. 1, 33–45.
[13] E. Kani, Elliptic curves on abelian surfaces, *Manuscripta Math.* **84** (1994), no. 2, 199–223.
[14] F. Klein, On the order-seven transformation of elliptic functions, in *Math. Sci. Res. Inst. Publ.*, **35**, *The eightfold way*, pp. 287–331, Cambridge Univ. Press, Cambridge, 1999.
[15] H. Lange and S. Narasimhan, Maximal subbundles of rank two vector bundles on curves, *Math. Ann.* **266** (1983), no. 1, 55–72.

[16] K. Magaard, S. Shpectorov and H. Völklein, A GAP package for braid orbit computation and applications, *Experiment. Math.* **12** (2003), no. 4, 385–393.
[17] R. Miranda, Triple covers in algebraic geometry, *Amer. J. Math.* **107** (1985), no. 5, 1123–1158.
[18] K. Miura and H. Yoshihara, Field Theory for Function Fields of Plane Quartic Curves, *J. Algebra* **226** (2000), no. 1, 283–294.
[19] C. Moreno, *Algebraic Curves Over Finite Fields*, Cambridge Univ. Press, 1991.
[20] W.M. Oxbury, Varieties of maximal line subbundles, *Math. Proc. Cambridge Philos. Soc.* **129** (2000), no. 1, 9–18.
[21] D.T. Prapavessi, On the Jacobian of the Klein curve, *Proc. Amer. Math. Soc.* **122** (1994), no. 4, 971–978.
[22] H.E. Rauch and J. Lewittes, The Riemann surface of Klein with 168 automorphisms, in *Problems in Analysis (papers dedicated to Salomon Bochner, 1969)*, Princeton Univ. Press, Princeton, NJ, 1970, pp. 297–308.
[23] J.-P. Serre, *Algebraic Groups and Class Fields*, Springer Graduate Texts in Mathematics, 1988.
[24] T. Shaska, Curves of genus 2 with (N, N) decomposable Jacobians, *J. Symbolic Comput.* **31** (2001), no. 5, 603–617.
[25] T. Shaska, Genus 2 curves with $(3, 3)$-split Jacobian and large automorphism group, in: *Algorithmic number theory* (Sydney, 2002), 205–218, Lecture Notes in Comput. Sci., **2369**, Springer, Berlin, 2002.
[26] T. Shaska, Genus 2 fields with degree 3 elliptic subfields, *Forum Math.* **16** (2004), no. 2, 263–280.
[27] T. Shaska and H. Völklein, Elliptic subfields and automorphisms of genus 2 function fields, in: *Algebra, arithmetic and geometry with applications* (West Lafayette, IN, 2000), 703–723, Springer, Berlin, 2004.
[28] H. Stichtenoth, *Algebraic Function Fields and Codes*, Springer-Verlag 1993.
[29] E.W. von Tschirnhaus, Acta Eruditorium (1683).
[30] S. Tufféry, Automorphismes d'ordre 3 et 7 sur une courbe de genre 3, *Exposition. Math.* **11** (1993), no. 2, 159–162.
[31] S. Tufféry, Les automorphismes des courbes de genre 3 de caractéristique 2, *C. R. Acad. Sci. Paris Sér. I Math.* **321** (1995), no. 2, 205–210.
[32] S. Tufféry, Déformations de courbes avec action de groupe. II, *Forum Math.* **8** (1996), no. 2, 205–218.
[33] P. Tzermias, The group of automorphisms of the Fermat curve, *J. Number Theory* **53** (1995), no. 1, 173–178.

Remarks on s-extremal codes

Jon-Lark Kim

Department of Mathematics,
University of Louisville,
Louisville, KY 40292, USA
E-mail: jl.kim@louisville.edu
www.math.louisville.edu/~jlkim

We study s-extremal codes over $\mathbb{F}_4$ or over $\mathbb{F}_2$. A Type I self-dual code over $\mathbb{F}_4$ or over $\mathbb{F}_2$ of length n and minimum distance d is s-extremal if the minimum weight of its shadow is largest possible. The purpose of this paper is to give some results which are missing in a series of papers by Bachoc and Gaborit [2], by Gaborit [6], and by Bautista, et. al. [1]. In particular, we give an explicit formula for the numbers of the first four nonzero weights of an s-extremal code over $\mathbb{F}_4$. We improve a bound on the length for which there exists an s-extremal code over $\mathbb{F}_4$ (res. $\mathbb{F}_2$) with even minimum distance d (resp. $d \equiv 0 \pmod 4$), and give codes related to s-extremal binary codes.

Keywords: Additive self-dual codes; Ninary self-dual codes; s-extremal codes.

1. Introduction

Binary self-dual codes have been of great interest since the beginning of the coding theory partly because many good linear block codes are either self-orthogonal or self-dual. Furthermore, they have nice properties; in particular, the weight enumerator of a binary self-dual code is invariant under a certain finite group, which often restricts the minimum distance of such a code. We refer to the chapter of self-dual codes [12] for a full discussion of self-dual codes.

It was Conway and Sloane [4] who introduced the notion of the shadow of a binary self-dual code in order to get additional constraints in the weight enumerator of a singly-even binary self-dual code C. The *shadow S of C* is defined as

$$S := C_0^{\perp} \setminus C.$$

Let d be the minimum distance of C and s the minimum weight of S. Bachoc and Gaborit [2] showed that $2d + s \leq \frac{n}{2} + 4$, except in the case

$n \equiv 22 \pmod{24}$ and $d = 4[n/24] + 6$, where $2d + s = n/2 + 8$. Binary codes attaining these bounds are called *s-extremal* [2]. Elkies [5] studied binary s-extremal codes for $d = 2$ and $d = 4$, and Bachoc and Gaborit considered the case when $d = 6$.

Rains [11] gave additional constraints of the weight enumerator of the shadow of an additive self-dual Type I code over $\mathbb{F}_4$ and derived the best known upper bound on the highest possible minimum distance of these codes. Let d_I (d_{II}) be the minimum weight of an additive self-dual Type I (Type II, respectively) code of length $n > 1$. Then

$$d_I \leq \begin{cases} 2\left\lfloor \frac{n}{6} \right\rfloor + 1 \text{ if } n \equiv 0 \pmod{6} \\ 2\left\lfloor \frac{n}{6} \right\rfloor + 3 \text{ if } n \equiv 5 \pmod{6} \\ 2\left\lfloor \frac{n}{6} \right\rfloor + 2 \text{ otherwise} \end{cases}$$

$$d_{II} \leq 2\left\lfloor \frac{n}{6} \right\rfloor + 2.$$

A code meeting the appropriate bound is called *extremal.*

Following the ideas of Bachoc and Gaborit, Bautista, et. al. [1] introduce the notion of an s-extremal additive $\mathbb{F}_4$ code. The authors [1] show that if there is an s-extremal $\mathbb{F}_4$ code of length n with even minimum distance d, then $n < 3d$; they relate s-extremal $\mathbb{F}_4$ codes to other s-extremal codes or extremal $\mathbb{F}_4$ codes.

In this article, we give an explicit formula for $A_d, \cdots, A_{d+3}$, the numbers of the first four nonzero weights of an s-extremal code over $\mathbb{F}_4$. We hope that this formula can be used to verify the nonexistence of certain s-extremal $\mathbb{F}_4$ codes. In particular, we show that for an s-extremal $\mathbb{F}_4$ code of length n with even d, $n \leq 3d - 2$, improving slightly $n \leq 3d - 1$ of [1] and providing the optimality of n. We also briefly consider binary s-extremal codes. We observe that if there is a binary s-extremal code with parameters (s, d) of length n and $d \equiv 0 \pmod{4}$, then $n \leq 6d - 4$, improving $n \leq 6d - 2$ of [6]. Furthermore we relate a binary s-extremal code of length $6d$ to another s-extremal code of that length, and produce extremal Type II codes from certain s-extremal codes. One sees the parallelism between s-extremal codes over $\mathbb{F}_4$ and those over $\mathbb{F}_2$.

2. *s*-Extremal Additive $\mathbb{F}_4$ Codes

We recall basic definitions on additive $\mathbb{F}_4$ codes [3, 7].

An *additive* $\mathbb{F}_4$ *code* C *of length* n is a subset $C \subset \mathbb{F}_4^n$ which is a vector space over $\mathbb{F}_2$. We say that C is an $(n, 2^k)$ *code* if it has 2^k codewords. If

$\mathbf{c} \in C$, the weight of $\mathbf{c}$, denoted by $\mathrm{wt}(\mathbf{c})$, is the Hamming weight of $\mathbf{c}$ and the minimum distance (or minimum weight) d of C is the smallest weight among any non-zero codeword in C. We call C an *$(n, 2^k, d)$ code.*

Let $\mathbf{x} = (x_1, \ldots, x_n)$, $\mathbf{y} = (y_1, \ldots, y_n) \in \mathbb{F}_4^n$. The *trace inner product* of $\mathbf{x}$ and $\mathbf{y}$ is given by

$$\langle \mathbf{x}, \mathbf{y} \rangle := \sum_{i=1}^{n} \mathrm{Tr}(x_i y_i^2)$$

where $\mathrm{Tr} : \mathbb{F}_4 \to \mathbb{F}_2$ is the trace map $\mathrm{Tr}(\alpha) = \alpha + \alpha^2$.

If C is an additive code, its *dual*, denoted $C^\perp$, is the additive code $\{\mathbf{x} \in \mathbb{F}_4^n \mid \langle \mathbf{x}, \mathbf{c} \rangle = 0 \text{ for all } \mathbf{c} \in C\}$. If C is an $(n, 2^k)$ code, then $C^\perp$ is an $(n, 2^{2n-k})$ code. C is *self-orthogonal* if $C \subseteq C^\perp$ and *self-dual* if $C = C^\perp$. If C is self-dual, it is an $(n, 2^n)$ code. For an additive self-dual code over $\mathbb{F}_4$, if all codewords have even weight, the code is *Type II*; otherwise it is *Type I*.

Definition 2.1. Let C be an additive $\mathbb{F}_4$ code of length n which is self-dual with respect to the trace inner product. The *shadow* $S = S(C)$ of C is given by

$$S = \{\mathbf{w} \in \mathbb{F}_4^n \mid \langle \mathbf{v}, \mathbf{w} \rangle \equiv \mathrm{wt}(\mathbf{v}) \pmod 2 \quad \text{for every } \mathbf{v} \in C\}.$$

If C is Type II, $S(C) = C$, while if C is Type I, $S(C)$ is a coset of C (see p. 203 of [12]).

The next theorem, which is the $\mathbb{F}_4$-analog of Theorem 1 of [2], was first given in [1].

Theorem 2.1. *Let C be a Type I additive $\mathbb{F}_4$ code of length n, self-dual with respect to the trace inner product, let $d = d_{\min}(C)$ be the minimum distance of C, let $S = S(C)$ be the shadow of C, and let $s = \mathrm{wt}_{\min}(S)$ be the minimum weight of S. Then $2d + s \leq n + 2$ unless $n = 6m + 5$ and $d = 2m + 3$, in which case $2d + s = n + 4$.*

Theorem 2.1 motivates the next definition [1].

Definition 2.2. Let C be a Type I additive $\mathbb{F}_4$ code of length n, self-dual with respect to the trace inner product, let $d = d_{\min}(C)$ be the minimum distance of C, let $S = S(C)$ be the shadow of C, and let $s = \mathrm{wt}_{\min}(S)$ be the minimum weight of S. We say C is *s-extremal* if the bound of Theorem 2.1 is met, i.e., if $2d + s = n + 2$ except when $n = 6m + 5$ and $d = 2m + 3$, in which case $2d + s = n + 4$.

Remark 2.1. It is interesting to note that the weight enumerator of any s-extremal code is uniquely determined and can be explicitly computed from the values of n and d (or n and s) [1].

Gleason's Theorem for additive $\mathbb{F}_4$-codes holds as follows; see [1, 9, 11] for details.

Theorem 2.2. *Let C be an additive $\mathbb{F}_4$ code of length n which is self-dual with respect to the trace inner product. Let $S = S(C)$ be the shadow of C, and let $C(x,y)$ and $S(x,y)$ be the homogeneous weight enumerators of C and S, respectively. Then*

$$S(x,y) = \frac{1}{|C|} C(x+3y, y-x),$$

and there are polynomials

$$P(X,Y) = \sum_{i=0}^{\lfloor \frac{n}{2} \rfloor} u_i X^{n-2i} Y^i \text{ and } Q(X,Y) = \sum_{i=0}^{\lfloor \frac{n}{2} \rfloor} v_i X^{n-2i} Y^i$$

over $\mathbb{R}$ such that

$$C(x,y) = P(x+y, x^2+3y^2) = Q(x+y, y(x-y))$$

and

$$S(x,y) = P(2y, x^2+3y^2) = Q(2y, \frac{y^2-x^2}{2}).$$

Certain coefficients of $P(x,y)$ and $Q(x,y)$ are 0 as follows; see [1] for details.

Lemma 2.1. *Let C be an additive $\mathbb{F}_4$ code of length n which is self-dual with respect to the trace inner product. Let $S = S(C)$ be the shadow of C. Every vector in S has weight congruent to n modulo 2. Moreover, if we let $s = \mathrm{wt}_{\min}(S)$ be the minimum weight of S and write $s = n - 2r$, then the coefficients u_i and v_i in the polynomials $P(X,Y)$ and $Q(X,Y)$ of Theorem 2.2 are 0 for $r+1 \leq i \leq \lfloor \frac{n}{2} \rfloor$.*

Let C be an s-extremal $\mathbb{F}_4$ code of length n with minimum distance d and the minimum weight s of the shadow of C. In what follows, we derive an explicit formula for $A_d, \cdots A_{d+3}$, where A_i is the number of codewords in C of weight i.

Using the second equation of $C(x,y)$ in Theorem 2.2 and Lemma 2.1, we have

$$C(1,y) = 1 + A_d y^d + A_{d+1} y^{d+1} + A_{d+2} y^{d+2} + A_{d+3} y^{d+3} + \cdots$$
$$= \sum_{i=0}^{d-1} v_i (1+y)^{n-2i} (y(1-y))^i.$$

Dividing both sides by $(1+y)^n$, we get

$$\frac{1}{(1+y)^n}(1 + A_d y^d + A_{d+1} y^{d+1} + A_{d+2} y^{d+2} + A_{d+3} y^{d+3} + \dots)$$
$$= \sum_{i=0}^{d-1} v_i \left(\frac{y(1-y)}{(1+y)^2} \right)^i.$$

Write $f(y) = \frac{1}{(1+y)^n}$ and $g(y) = \frac{y(1-y)}{(1+y)^2}$. Then we have

$$\sum_{i=0}^{d-1} v_i g^i(y) = f(y) + f(y) A_d y^d + \cdots + f(y) A_{d+3} y^{d+3} + O(y^{d+4}). \qquad (1)$$

We expand $f(y)$ as a power series of y as follows.

$$f(y) = \frac{1}{(1+y)^n} = \sum_{i=0}^{\infty} (-1)^i \binom{n+i-1}{i} y^i$$
$$= 1 - ny + \binom{n+1}{2} y^2 - \binom{n+2}{3} y^3 + O(y^4).$$

We plug this in Equation (1) to get the following lemma.

Lemma 2.2. *Under the above notations,*

$$\sum_{i=0}^{d-1} v_i g^i(y) = f(y) + A_d \left\{ 1 - ny + \binom{n+1}{2} y^2 - \binom{n+2}{3} y^3 + O(y^4) \right\} y^d$$
$$+ A_{d+1} \left\{ 1 - ny + \binom{n+1}{2} y^2 - \binom{n+2}{3} y^3 + O(y^4) \right\} y^{d+1}$$
$$+ \cdots$$
$$= f(y) + A_d y^d + \{-nA_d + A_{d+1}\} y^{d+1}$$
$$+ \left\{ A_d \binom{n+1}{2} - nA_{d+1} + A_{d+2} \right\} y^{d+2}$$
$$+ \left\{ -A_d \binom{n+2}{3} + A_{d+1} \binom{n+1}{2} - nA_{d+2} + A_{d+3} \right\} y^{d+3}$$
$$+ O(y^{d+4}).$$

Our next step is to rewrite $g^i(y)$ for $i \geq d$ as a power series in y. Since $g(y) = \frac{y(1-y)}{(1+y)^2}$, we have

$$\begin{aligned}
g^d(y) &= \frac{y^d(1-y)^d}{(1+y)^{2d}} \\
&= y^d \left(\sum_{j=0}^{d} (-1)^j \binom{d}{j} y^j \right) \left(\sum_{i=0}^{\infty} (-1)^i \binom{2d+i-1}{i} y^i \right) \\
&= y^d - 3dy^{d+1} + \left\{ \binom{d}{2} + d\binom{2d}{1} + \binom{2d+1}{2} \right\} y^{d+2} \\
&\quad + \left\{ -\binom{d}{3} - \binom{d}{2}\binom{2d}{1} - d\binom{2d+1}{2} - \binom{2d+2}{3} \right\} y^{d+3} \\
&\quad + O(y^{d+4}).
\end{aligned}$$

Similarly we obtain the following.

$$\begin{aligned}
g^{d+1}(y) &= y^{d+1} + \left\{ -\binom{d+1}{1} - \binom{2d+2}{1} \right\} y^{d+2} \\
&\quad + \left\{ \binom{2d+3}{2} + \binom{d+1}{1}\binom{2d+2}{1} + \binom{d+1}{2} \right\} y^{d+3} \\
&\quad + O(y^{d+4}), \\
g^{d+2}(y) &= y^{d+2} + \left\{ -\binom{2d+4}{1} - \binom{d+2}{1} \right\} y^{d+3} + O(y^{d+4}), \\
g^{d+3}(y) &= y^{d+3} + O(y^{d+4}).
\end{aligned}$$

Now we can rewrite y^i for $i = d, d+1, d+2, d+3$ in terms of $g^i(y)$ and $O(y^{d+4})$ as follows.

$$\begin{aligned}
y^{d+3} &= g^{d+3}(y) + O(y^{d+4}), \\
y^{d+2} &= g^{d+2}(y) + \left\{ \binom{d+2}{1} + \binom{2d+4}{1} \right\} g^{d+3}(y) + O(y^{d+4}), \\
y^{d+1} &= g^{d+1}(y) + \left\{ \binom{d+1}{1} + \binom{2d+2}{1} \right\} g^{d+2}(y) \\
&\quad + (3d+3)(3d+6)g^{d+3}(y) \\
&\quad - \left\{ \binom{2d+3}{2} + \binom{d+1}{1}\binom{2d+2}{1} + \binom{d+1}{2} \right\} g^{d+3}(y) + O(y^{d+4}),
\end{aligned}$$

$$
\begin{aligned}
y^d &= g^d(y) + 3dg^{d+1}(y)\\
&+ \left\{3d(3d+3) - \binom{d}{2} - d\binom{2d}{1} - \binom{2d+1}{2}\right\} g^{d+2}(y)\\
&+ 3d(3d+3)(3d+6)g^{d+3}(y)\\
&- 3d\left\{\binom{2d+3}{2} + (d+1)(2d+2) + \binom{d+1}{2}\right\} g^{d+3}(y)\\
&- (3d+6)\left\{\binom{d}{2} + d\binom{2d}{1} + \binom{2d+1}{2}\right\} g^{d+3}(y)\\
&+ \left\{\binom{d}{3} + \binom{d}{2}\binom{2d}{1} + d\binom{2d+1}{2} + \binom{2d+2}{3}\right\} g^{d+3}(y)\\
&+ O(y^{d+4}).
\end{aligned}
$$

Now let κ_i be the coefficient of $g^i(y)$ in

$$f(y) = \sum_{i\geq 0} \kappa_i g(y)^i$$

Then we plug the above calculations in Lemma 2.2 to obtain the following relation between κ_i and A_i's.

Lemma 2.3.

Under the notations as above,

$$
\begin{aligned}
&(i)\quad \kappa_d = -A_d,\\
&(ii)\ \kappa_{d+1} = -((3d-n)A_d + A_{d+1}),\\
&(iii)\ \kappa_{d+2} = -\frac{1}{2}A_d(9d^2 + 17d - 6dn - 5n + n^2)\\
&\qquad - A_{d+1}(3d+3-n) - A_{d+2},\\
&(iv)\ \kappa_{d+3} = -A_d\left\{\frac{9}{2}d^3 + \frac{51}{2}d^2 + 37d + n(-\frac{9}{2}d^2 - \frac{35}{2}d - 13)\right\}\\
&\qquad - A_d\left\{(3d+6)\binom{n+1}{2} - \binom{n+2}{3}\right\}\\
&\qquad - A_{d+1}\left\{\frac{9}{2}d^2 + \frac{35}{2}d + 13 - n(3d+6) + \binom{n+1}{2}\right\}\\
&\qquad - A_{d+2}(3d+6-n) - A_{d+3}.
\end{aligned}
$$

On the other hand, we can evaluate κ_i using the Bürman-Lagrange Theorem. We recall the Bürman-Lagrange Theorem (as stated in [11]): If

$f(x)$ and $g(x)$ are formal power series with $g(0) = 0$ and $g'(0) \neq 0$ and the coefficients κ_i are defined by

$$f(x) = \sum_{i \geq 0} \kappa_i g(x)^i,$$

then

$$\kappa_i = \frac{1}{i}\left(\text{coefficient of } x^{i-1} \text{ in } f'(x)\left(\frac{x}{g(x)}\right)^i\right).$$

Our functions $f(y)$ and $g(y)$ satisfy these hypotheses, and we have

$$\begin{aligned}
\kappa_d &= \frac{1}{d}\left(\text{coefficient of } y^{d-1} \text{ in } f'(y)\left(\frac{y}{g(y)}\right)^d\right) \\
&= \frac{1}{d}\left(\text{coefficient of } y^{d-1} \text{ in } \left(\frac{-n}{(1+y)^{n+1}}\right)\left(\frac{(1+y)^2}{1-y}\right)^d\right) \\
&= \frac{-n}{d}\left(\text{coefficient of } y^{d-1} \text{ in } \frac{1}{(1+y)^{n-3d+1}(1-y^2)^d}\right).
\end{aligned}$$

Similarly we obtain the following.

$$\begin{aligned}
\kappa_{d+1} &= \frac{-n}{d+1}\left(\text{coefficient of } y^{d} \text{ in } \frac{1}{(1+y)^{n-3d-2}(1-y^2)^{d+1}}\right) \\
\kappa_{d+2} &= \frac{-n}{d+2}\left(\text{coefficient of } y^{d+1} \text{ in } \frac{1}{(1+y)^{n-3d-5}(1-y^2)^{d+2}}\right) \\
\kappa_{d+3} &= \frac{-n}{d+3}\left(\text{coefficient of } y^{d+2} \text{ in} \frac{1}{(1+y)^{n-3d-8}(1-y^2)^{d+3}}\right)
\end{aligned}$$

Corollary 2.1. *If d is even and $n = 3d - 1$ then there is no s-extremal $\mathbb{F}_4$ code.*

Proof. If $n = 3d-1$, then $\kappa_d = \frac{-n}{d}\left(\text{coefficient of } y^{d-1} \text{ in } \frac{1}{(1-y^2)^d}\right)$. Hence $\kappa_d = 0$ as d is even. Then by (i) of Lemma 2.3, $A_d = 0$, which is a contradiction. □

Note that an s-extremal code of length n with minimum distance d must satisfy $n \geq 3d - 5$, and that if d is even, $3d - 4 \leq n < 3d$ [1]. Hence we have the following.

Corollary 2.2. *An s-extremal code of length n with even minimum distance d must satisfy $n = 3d - 4, 3d - 3,$ or $3d - 2$.*

When d is even, Corollary 2.2 fully explains Table 2 of [1] where a possible range of n for which an s-extremal code exists is displayed. After simplifications, we further have an explicit formula for the κ_i's as follows. Therefore combining this with Lemma 2.3 we get an explicit formula for $A_d, A_{d+1}, A_{d+2}, A_{d+3}$.

Proposition 2.1. *Suppose that C is an s-extremal $\mathbb{F}_4$ code of length $n \geq 3d-5$ with minimum distance d. If $m := 3d-n-1 > 0$, then for any nonnegative integer $i = 0, 1, 2, 3, \cdots$,*

$$k_{d+i} = \frac{-n}{d+i}\left(\sum_{j=0, j\equiv 1+i(2)}^{m+3i} \binom{m+3i}{j}\binom{d+i+\frac{d-1+i-j}{2}-1}{d-1+i}\right) \quad \textit{if } d \textit{ even,}$$

and

$$k_{d+i} = \frac{-n}{d+i}\left(\sum_{j=0, j\equiv i(2)}^{m+3i} \binom{m+3i}{j}\binom{d+i+\frac{d-1+i-j}{2}-1}{d-1+i}\right) \quad \textit{if } d \textit{ odd.}$$

Example 2.1. Let C be an s-extremal additive $\mathbb{F}_4$ code of length $n=6$ with minimum distance $d=3$. For example, take as generator matrix of C the 6 by 6 circulant matrix whose first row is $(1, \omega, 1, 0, 0, 0)$ (see [8]). This code is equivalent to the odd Hexacode $\mathcal{O}_6$ [9]. We show how to compute the weight distribution A_3, A_4, A_5, and A_6 by finding corresponding κ_i's. As $\kappa_3 = -2\left(\text{ coefficient of } y^2 \text{ in } (1+y)^2\frac{1}{(1-y^2)^3}\right) = -2\left(\binom{2}{0}\binom{3}{2}+\binom{2}{2}\binom{2}{2}\right) = -8$, we have $A_3 = 8$ by Lemma 2.3. Similarly, we compute $\kappa_4 = -45$, $\kappa_5 = -270$, and $\kappa_6 = -1683$. This implies $A_4 = 21$, $A_5 = 24$, and $A_6 = 10$. This weight distribution also appears in Table 1 of [1].

3. s-Extremal Binary Codes

In this section we consider binary self-dual codes and produce related codes from s-extremal binary codes, as an $\mathbb{F}_2$ analogue of [1].

Let C be a binary Type I self-dual code of length n and let S be its shadow. Let C_0 be the doubly-even subcode of C and let $C_0^{\perp} = C_0 \cup C_2 \cup C_1 \cup C_3$, where $C = C_0 \cup C_2$ and $S = C_1 \cup C_3$. It is well known that $C^{(1)} = C_0 \cup C_1$ and $C^{(3)} = C_0 \cup C_3$ are self-dual if and only if $\mathbf{1} \in C_0$, i.e., $4|n$ (see Lemma 9.4.6, Theorems 9.4.7 and 9.4.10 in [10]).

Conway and Sloane [4] showed that there exist $c_0, \cdots, c_{[n/8]} \in \mathbb{R}$ such that:

$$C(x,y) = \sum_{i=0}^{[n/8]} c_i(x^2+y^2)^{\frac{n}{2}-4i}\{x^2y^2(x^2-y^2)^2\}^i, \tag{2}$$

$$S(x,y) = \sum_{i=0}^{[n/8]} c_i(-1)^i 2^{\frac{n}{2}-6i}(xy)^{\frac{n}{2}-4i}(x^4-y^4)^{2i}. \tag{3}$$

Applying the Bürman-Lagrange Theorem to certain transformed equations of the above equations, Bachoc and Gaborit [2] introduced a notion of an s-extremal binary code; for details see [2].

Theorem 3.1. *Let C be a binary Type I self-dual code of length n and minimum distance d, and let S be its shadow of minimum weight s. Then $2d+s \leq 4+\frac{n}{2}$, unless $n \equiv 22 \pmod{24}$ and $d = 4[n/24]+6$, in which case $2d+s = 8+\frac{n}{2}$.*

Definition 3.1. *A binary Type I code meeting the above bound is called* s-extremal.

When $d \equiv 0 \pmod 4$, the length of a s-extremal code is bounded by $6d$ as follows.

Theorem 3.2. *Let C be an s-extremal code with parameters (s,d) of length n. If $d \equiv 0 \pmod 4$, then $n < 6d$ (i.e., $n \leq 6d-2$).*

Following the proof of the above theorem in [6], we have that for any binary s-extremal code,

$$A_d = \frac{n}{d}\left(\text{coefficient of } y^{d-2} \text{ in } \frac{1}{(1+y^2)^{\frac{n}{2}-3d+1}(1-y^4)^d}\right).$$

In particular, if $n = 6d-2$, then $\frac{n}{2}-3d+1 = 0$. Further if $d \equiv 0 \pmod 4$ then $d-2 \equiv 2 \pmod 4$. Therefore $A_d = 0$, which is impossible. Therefore we obtain the following.

Proposition 3.1. *Let C be an s-extremal code with parameters (s,d) of length n. If $d \equiv 0 \pmod 4$, then $n \leq 6d-4$.*

Codes related to s-extremal additive $\mathbb{F}_4$ codes are described in Propositions 7.1 and 7.3 in [1], whose proofs however contain minor errors. We show that this can be done analogously for binary s-extremal codes in the following propositions. We include the proofs for completeness.

Proposition 3.2. *Let C be a binary s-extremal code of length n and minimum distance d satisfying $2d+s = 4+\frac{n}{2}$. If $d \equiv 2 \pmod 4$ and $s = d+4$,* *then $C^{(1)}$ and $C^{(3)}$ are also s-extremal with minimum distance $d' = d + 2$ and the minimum shadow weight $s' = d$.*

Proof. Since $d \equiv 2 \pmod 4$, $\frac{n}{2} = 3d$ is congruent to 2 (mod 4). Thus all weights in S are congruent to 2 (mod 4) (by Theorem 9.4.7 [10] or by looking at the powers of y in Equation (3)). Thus both $C^{(1)}$ and $C^{(3)}$ are Type I self-dual codes; self-duality follows as $4|n$. We may assume that C_1 contains a vector $\mathbf{x}$ of minimum weight s. There are two possibilities for $d(C_0)$: Either $d(C_0) = d + 2$ or $d(C_0) > d + 2$.

If $d(C_0) > d + 2$, then $d(C_0) \geq d + 6$ as C_0 is doubly-even. So in this case, the minimum distance of $C^{(1)}$ is $d' = \min\{d(C_0), d(C_1)\} = \min\{\geq d + 6, s\} = d + 4$. The shadow of $C^{(1)}$ is $C_2 \cup C_3$, and its minimum weight is $s' = \min\{d(C_2), d(C_3)\} = \min\{d, \geq s\} = d$. As $2d' + s' = 2(d + 4) + d = 3d + 8 = \frac{n}{2} + 8$, n must be congruent to 22 (mod 24). This contradicts the condition that $\frac{n}{2} \equiv 2 \pmod 4$.

Thus we have $d(C_0) = d + 2$. Following the above arguments, we obtain $d' = d + 2$ and $s' = d$. As $2d' + s' = 2(d + 2) + d = 3d + 4 = \frac{n}{2} + 4$, $C^{(1)}$ is s-extremal.

As $d(C_0) = d + 2$, one can show that $C^{(3)}$ is s-extremal in a similar manner. □

Proposition 3.3. *Let C be a binary s-extremal code of length $n = 24\mu + 8$ and minimum distance d. If $d \equiv 2 \pmod 4$ and $s = d + 2$, then both $C^{(1)}$ and $C^{(3)}$ are extremal Type II codes with minimum distance $d + 2$. Moreover the weight enumerators of C_1 and C_3 are the same and are explicitly determined.*

Proof. As $\frac{n}{2} \equiv 0 \pmod 4$, $4|n$ and all weights in S are doubly-even. So $C^{(1)}$ and $C^{(3)}$ are Type II. Since $d(S) = s$, the minimum distance of $C^{(i)}$ is $\min\{d(C_0), d(C_i)\} \geq s$ for $i = 1, 3$, and either $d(C_1) = s$ or $d(C_3) = s$, but not necessarily both. We know that $d(C_0) \geq d + 2 = s$ as $d \equiv 2 \pmod 4$ and C_0 is doubly-even. So $d(C^{(i)}) = \min\{d(C_0), d(C_i)\} \geq s$ for $i = 1, 3$. Since $2d + s = 4 + \frac{n}{2} = 12\mu + 8$ and $d = s - 2$, we have $2(s - 2) + s = 12\mu + 8$ implying $s = 4\mu + 4$. As extremal Type II codes of length $n = 24\mu + 8$ have minimum weight at most $4\mu + 4$, $d(C^{(i)})$ must be $s = d + 2$ and each $d(C^{(i)})$ is extremal.

Finally note that the weight enumerator of a binary extremal Type II code is uniquely determined by its length and that the weight enumerator of C_0 is explicitly determined by that of C. Thus the weight enumerator of C_1 and C_3 are the same and are explicitly determined. □

4. Conclusion

We have obtained results concerning both s-extremal $\mathbb{F}_4$ codes and s-extremal binary codes. In particular, we find an explicit formula for $A_d, \cdots, A_{d+3}$ of an s-extremal $\mathbb{F}_4$ code. This formula is of use for knowing a possible weight distribution of an s-extremal $\mathbb{F}_4$ code. Furthermore, this formula might be used for the upper bound on the length of an s-extremal $\mathbb{F}_4$ code when its minimum distance is odd. For example, suppose that there exists an s-extremal $\mathbb{F}_4$ code with odd d and $n = 3d + 2$. Then $\kappa_{d+1} = 0$ from Proposition 2.1. Hence $A_{d+1} = 2A_d$ by Lemma 2.3. All the weight enumerators of s-extremal $\mathbb{F}_4$ codes [1] do not have this relation. On the other hand we remark that one can obtain a formula for $A_d, \cdots, A_{d+3}$ of a binary s-extremal code in an analogous manner.

We also have improved a bound on the length for which there exists an s-extremal code over $\mathbb{F}_4$ or over $\mathbb{F}_2$ with even minimum distance, and gave binary s-extremal or extremal codes constructed from binary s-extremal codes.

Acknowledgments

The author would like to thank Professor W. Cary Huffman and Dr. Sunghyu Han for invaluable comments on the submitted manuscript, in particular for suggesting improved proofs of Propositions 3.2 and 3.3. The author acknowledges support by a Project Completion Grant from the University of Louisville.

References

[1] E.P. Bautista, P. Gaborit, J.-L. Kim, and J.L. Walker, s-extremal additive $\mathbb{F}_4$ codes, *Advances in Mathematics of Communication,* **1** 111-130 (2007).

[2] C. Bachoc and P. Gaborit, Designs and self-dual codes with long shadows, *J. Combin. Theory Ser. A*, **105** 15–34 (2004).

[3] A. R. Calderbank, E. M. Rains, P. W. Shor, and N. J. A. Sloane. Quantum error correction via codes over $GF(4)$, *IEEE Trans. Inform. Theory,* **44** 1369–1387 (1998).

[4] J. H. Conway and N. J. A. Sloane, A new upper bound on the minimal distance of self-dual codes, *IEEE Trans. Inform. Theory,* **36** 1319–1333 (1990).

[5] N. D. Elkies, Lattices and codes with long shadows, *Math. Res. Lett.*, **2** 643–651 (1995).

[6] P. Gaborit, A bound for certain s-extremal lattices and codes, *preprint.* available at **http://www.unilim.fr/pages_perso/philippe.gaborit/**

[7] P. Gaborit, W. C. Huffman, J.-L. Kim, and V. Pless, On additive GF(4) codes, in *Codes and association schemes (Piscataway, NJ, 1999)*, volume 56

of *DIMACS Ser. Discrete Math. Theoret. Comput. Sci.*, pages 135–149. Amer. Math. Soc., Providence, RI, 2001.

[8] T. A. Gulliver and J.-L. Kim, Circulant based extremal additive self-dual codes over GF(4), *IEEE Trans. Inform. Theory*, **50** 359–366 (2004).

[9] G. Höhn, Self-dual codes over the Kleinian four group, *Math. Ann.*, **327** 227–255 (2003).

[10] W. C. Huffman and V. S. Pless, *Fundamentals of Error-correcting Codes*. Cambridge: Cambridge University Press, 2003.

[11] E. M. Rains, Shadow bounds for self-dual codes, *IEEE Trans. Inform. Theory*, **44** 134–139 (1998).

[12] E. M. Rains and N. J. A. Sloane, Self-dual codes, in *Handbook of Coding Theory*, ed. V. S. Pless and W. C. Huffman. Amsterdam: Elsevier, pp. 177–294, 1998.

Automorphism groups of generalized Reed-Solomon codes

David Joyner, Amy Ksir, Will Traves

Mathematics Dept.
U.S. Naval Academy
Annapolis, MD 21402
E-mails: wdj@usna.edu, ksir@usna.edu, traves@usna.edu

We look at AG codes associated to $\mathbb{P}^1$, re-examining the problem of determining their automorphism groups (originally investigated by Dür in 1987 using combinatorial techniques) using recent methods from algebraic geometry. We classify those finite groups that can arise as the automorphism group of an AG code and give an explicit description of how these groups appear. We give examples of generalized Reed-Solomon codes with large automorphism groups G, such as $G = PSL(2, q)$, and explicitly describe their G-module structure.

Keywords: generalized Reed-Solomon codes, permutation automorphism groups, algebraic-geometric codes, G-modules

1. Introduction

Reed-Solomon codes are popular in applications because fast encoding and decoding algorithms are known for them. For example, they are used in compact discs (more details can be found in §5.6 in Huffman and Pless [4]).

In this paper we study which groups can arise as automorphism groups of a related collection of codes, the algebraic geometry (AG) codes on $\mathbb{P}^1$. These codes are monomially equivalent to generalized Reed-Solomon (GRS) codes. Their automorphism groups were first studied by Dür [2] in 1987 using combinatorial techniques. Huffman [3] gives an excellent exposition of Dür's original work. In this paper, using recent methods from algebraic geometry (due to Brandt and Stichenoth [12], Valentini and Madan [14], Kontogeorgis [9]), we present a method for computing GRS codes with "large" permutation automorphism groups. In contrast to Dür's results, we indicate exactly how these automorphism groups can be obtained.

The paper is organized as follows. In section 2 we review some background on AG codes and GRS codes. In section 3 we review some known results on automorphisms of AG codes, and then prove our main result, characterizing the automorphism groups of AG codes. In section 4 we use these results to give examples of codes with large automorphism groups. In section 5, we discuss the structure of these group representations as G-modules, in some cases determining it explicitly.

2. AG codes and GRS codes

We recall some well-known background on AG codes and GRS codes.

Let X be a smooth projective curve over a field F and let $\overline{F}$ denote a separable algebraic closure of F. We will generally take F to be finite of order q. Let $F(X)$ denote the function field of X (the field of rational functions on X). Recall that a **divisor** on X is a formal sum, with integer coefficients, of places of $F(X)$. We will denote the group of divisors on X by $\mathrm{Div}(X)$. The rational points of X are the places of degree 1, and the set of rational points is denoted $X(F)$.

AG codes associated to a divisor D are constructed from the Riemann-Roch space

$$L(D) = L_X(D) = \{f \in F(X)^\times \mid \mathrm{div}(f) + D \geq 0\} \cup \{0\},$$

where $\mathrm{div}(f)$ denotes the (principal) divisor of the function $f \in F(X)$. The Riemann-Roch space is a finite dimensional vector space over F, whose dimension is given by the Riemann-Roch theorem.

Let $P_1, ..., P_n \in X(F)$ be distinct points and $E = P_1 + ... + P_n \in \mathrm{Div}(X)$. Assume these divisors have disjoint support, $\mathrm{supp}(D) \cap \mathrm{supp}(E) = \emptyset$. Let $C(D, E)$ denote the AG code

$$C(D, E) = \{(f(P_1), ..., f(P_n)) \mid f \in L(D)\}. \qquad (1)$$

This is the image of $L(D)$ under the evaluation map

$$\begin{aligned} \mathrm{eval}_E : L(D) &\to F^n, \\ f &\longmapsto (f(P_1), ..., f(P_n)). \end{aligned} \qquad (2)$$

The following is well-known (a proof can be found in Joyner and Ksir [7]).

Lemma 2.1. *If* $\deg(D) > \deg(E)$, *then* eval_E *is injective.*

In this paper, we restrict to the case where X is the projective line $\mathbb{P}^1$ over F. In this case, if $\deg D \geq 0$ then $\dim L(D) = \deg D + 1$, and otherwise $\dim L(D) = 0$. Thus we will be interested in the case where $\deg D \geq 0$.

In the special case when D is a positive integer multiple of the point at infinity, then this construction gives a Reed-Solomon code. More generally,

$$C = \{(\alpha_1 f(P_1), ..., \alpha_n f(P_n)) \mid f \in L(\ell \cdot \infty)\},$$

is called a **generalized Reed-Solomon code** (or GRS code), where $\alpha_1, ..., \alpha_n$ is a fixed set of non-zero elements in F (called "multipliers").

In fact, for a more general D, this construction gives a code which is monomially equivalent to a GRS code, and which furthermore is MDS (that is, $n + 1 = k + d$, where n is the length, k is the dimension, and d is the minimum distance of the code). We say that two codes C, C' of length n are **monomially equivalent** if there is an element of the group of monomial matrices $Mon_n(F)$ – those matrices with precisely one non-zero entry in each row and column – (acting in the natural way on F^n) sending C to C' (as F-vector spaces).

Lemma 2.2. *Let $X = \mathbb{P}^1/F$, D be any divisor of positive degree on X, and let $E = P_1 + \ldots + P_n$, where $P_1, \ldots, P_n$ are points in $X(F)$ and $n > \deg D$. Let $C(D, E)$ be the AG code constructed as above. Then $C(D, E)$ is an MDS code which is monomially equivalent to a GRS code (with all scalars $\alpha_i = 1$).*

Proof. This is well-known (see for example Stichtenoth [11], §II.2), but we give the details for convenience. $C(D, E)$ has length n and dimension $k = \deg(D) + 1$. By Theorem 13.4.3 of Huffman and Pless [4], its minimum distance d satisfies

$$n - \deg(D) \leq d,$$

and the Singleton bound says that

$$d \leq n + 1 - k = n - \deg(D).$$

Therefore, $d = n + 1 - k$, and this shows that $C(D, E)$ is MDS.

The monomial equivalence follows from the fact that on $\mathbb{P}^1$, all divisors of a given positive degree are (rationally) equivalent, so D is rationally equivalent to $\deg(D) \cdot \infty$. Thus there is a rational function h on X such that

$$D = \deg(D) \cdot \infty + \operatorname{div}(h).$$

Then for any $f \in L(D)$, fh is in $L(\deg(D) \cdot \infty)$. Thus there is a map

$$M : C(D, E) \to C(\deg(D) \cdot \infty, E)$$
$$(f(P_1), \ldots, f(P_n)) \mapsto (fh(P_1), \ldots, fh(P_n))$$

which is linear and whose matrix is diagonal with diagonal entries $h(P_1), \ldots, h(P_n)$. In particular, M is a monomial matrix, so $C(D, E)$ and the GRS code $C(\deg(D) \cdot \infty, E)$ are monomially equivalent. □

Remark 2.1. The **spectrum** of a code of length n is the list $[A_0, A_1, ..., A_n]$, where A_i denotes the number of codewords of weight i. The **dual code** of a linear code $C \subset F^n$ is the dual of C as a vector space with respect to the Hamming inner product on F^n, denoted $C^{\perp}$. We say C is **formally self-dual** if the spectrum of $C^{\perp}$ is the same as that of C. The spectrum of any MDS code is known (see §7.4 in Huffman and Pless [4]), and as a consequence of this we have the following

$$A_j = \binom{n}{j}(q-1)\sum_{i=0}^{j-d}(-1)^i\binom{j-1}{i}q^{j-d-i}, \quad d \leq j \leq n,$$

where q is the order of the finite field F. The following is an easy consequence of this and the fact that the dual code of an MDS code is MDS: if C is a GRS code with parameters $[n, k, d]$ satisfying $n = 2k$ then C is formally self-dual. We will see later some examples of formally self-dual codes with large automorphism groups.

3. Automorphisms

The action of a finite group $G \subset \mathrm{Aut}(X)$ on $F(X)$ is defined by restriction to G of the map

$$\begin{array}{rcl} \rho : \mathrm{Aut}(X) & \longrightarrow & \mathrm{Aut}(F(X)), \\ g & \longmapsto & (f \longmapsto f^g) \end{array}$$

where $f^g(x) = (\rho(g)(f))(x) = f(g^{-1}(x))$.

Note that $Y = X/G$ is also smooth and the quotient map

$$\psi : X \to Y \tag{3}$$

yields an identification $F(Y) = F(X)^G := \{f \in F(X) \mid f^g = f, \ \forall g \in G\}$.

Of course, G also acts on the group $\mathrm{Div}(X)$ of divisors of X. If $g \in \mathrm{Aut}(X)$ and $d_P \in \mathbb{Z}$, for places P of $F(X)$, then $g(\sum_P d_P P) = \sum_P d_P g(P)$. It is easy to show that $\mathrm{div}(f^g) = g(\mathrm{div}(f))$. Because of this, if $\mathrm{div}(f) + D \geq 0$ then $\mathrm{div}(f^g) + g(D) \geq 0$, for all $g \in \mathrm{Aut}(X)$. In particular, if the action of G on X leaves $D \in \mathrm{Div}(X)$ stable then G also acts on $L(D)$. We denote this action by

$$\rho : G \to \mathrm{Aut}(L(D)).$$

Now suppose that $E = P_1 + \ldots + P_n$ is also stabilized by G. In other words, G acts on the set $\mathrm{supp}(E) = \{P_1, \ldots, P_n\}$ by permutation. Then the group G acts on $C(D,E)$ by $g \in G$ sending $c = (f(P_1), ..., f(P_n)) \in C$ to $c' = (f(g^{-1}(P_1)), ..., f(g^{-1}(P_n)))$, where $f \in L(D)$.

Remark 3.1. Observe that this map sending $c \longmapsto c'$, denoted $\phi(g)$, is well-defined. This is clearly true if eval_E is injective. In case eval_E is not injective, suppose c is also represented by $f' \in L(D)$, so $c = (f'(P_1), ..., f'(P_n)) \in C$. Since G acts on the set $\mathrm{supp}(E)$ by permutation, for each P_i, $g^{-1}(P_i) = P_j$ for some j. Then $f(g^{-1}(P_i)) = f(P_j) = f'(P_j) = f'(g^{-1}(P_i))$, so $(f(g^{-1}(P_1)), ..., f(g^{-1}(P_n))) = (f'(g^{-1}(P_1)), ..., f'(g^{-1}(P_n)))$. Therefore, $\phi(g)$ is well-defined.

The **permutation automorphism group** of the code C, denoted $\mathrm{Perm}(C)$, is the subgroup of the symmetric group S_n (acting in the natural way on F^n) which preserves the set of codewords. More generally, we say two codes C and C' of length n are **permutation equivalent** if there is an element of S_n sending C to C' (as F-vector spaces). The **automorphism group** of the code C, denoted $\mathrm{Aut}(C)$, is the subgroup of the group of monomial matrices $Mon_n(F)$ (acting in the natural way on F^n) which preserves the set of codewords. Thus the permutation automorphism group of C is a subgroup of the full automorphism group.

The map ϕ induces a homomorphism of G into the automorphism group of the code. The image of the map

$$\begin{aligned} \phi : G &\to \mathrm{Aut}(C) \\ g &\longmapsto \phi(g) \end{aligned} \tag{4}$$

is contained in $\mathrm{Perm}(C)$.

Define $\mathrm{Aut}_{D,E}(X)$ to be the subgroup of $\mathrm{Aut}(X)$ which preserves the divisors D and E.

When does a group of permutation automorphisms of the code C induce a group of automorphisms of the curve X? Permutation automorphisms of the code $C(D,E)$ induce curve automorphisms whenever D is very ample and the degree of E is large enough. Under these conditions, the groups $\mathrm{Aut}_{D,E}(X)$ and $\mathrm{Perm}\, C$ are isomorphic.

Theorem 3.1. *(Joyner and Ksir [6]) Let X be an algebraic curve, D be a very ample divisor on X, and $P_1, \ldots, P_n$ be a set of points on X disjoint from the support of D. Let $E = P_1 + \ldots + P_n$ be the associated divisor, and $C = C(D,E)$ the associated AG code. Let G be the group of permutation automorphisms of C. Then there is an integer $r \geq 1$ such that if $n >$*

$r \cdot \deg(D)$, then G can be lifted to a group of automorphisms of the curve X itself. This lifting defines a group homomorphism $\psi : \mathrm{Perm}\, C \to \mathrm{Aut}(X)$. Furthermore, the lifted automorphisms will preserve D and E, so the image of ψ will be contained in $\mathrm{Aut}_{D,E}(X)$.

Remark 3.2. An explicit upper bound on r can be determined (see Joyner-Ksir [6]). In the case where $X = \mathbb{P}^1$, $r = 2$. In addition, any divisor of positive degree on $\mathbb{P}^1$ is very ample. Therefore, as long as $\deg D > 0$ and $n > 2\deg(D)$, the groups $\mathrm{Perm}(C)$ and $\mathrm{Aut}_{D,E}(X)$ will be isomorphic.

Now we would like to describe all possible finite groups of automorphisms of $\mathbb{P}^1$. Valentini and Madan [14] give a very explicit list of possible automorphisms of the associated function field $F(X)$ and their ramifications.

Proposition 3.1. *(Valentini and Madan [14]) Let F be finite field of order $q = p^k$. Let G be a nontrivial finite group of automorphisms of $F(X)$ fixing F elementwise and let $E = F(X)^G$ be the fixed field of G. Let r be the number of ramified places of E in the extension $F(X)/E$ and $e_1, \ldots, e_r$ the corresponding ramification indices. Then G is one of the following groups, with $F(X)/E$ having one of the associated ramification behaviors:*

(1) Cyclic group of order relatively prime to p with $r = 2$, $e_1 = e_2 = |G_0|$.
(2) Dihedral group D_m of order $2m$ with $p = 2$, $(p, m) = 1$, $r = 2$, $e_1 = 2$, $e_2 = m$, or $p \neq 2$, $(p, m) = 1$, $r = 3$, $e_1 = e_2 = 2$, $e_3 = m$.
(3) Alternating group A_4 with $p \neq 2, 3$, $r = 3$, $e_1 = 2$, $e_2 = e_3 = 3$.
(4) Symmetric group S_4 with $p \neq 2, 3$, $r = 3$, $e_1 = 2$, $e_2 = 3$, $e_3 = 4$.
(5) Alternating group A_5 with $p = 3$, $r = 2$, $e_1 = 6$, $e_2 = 5$, or $p \neq 2, 3, 5$, $r = 3$, $e_1 = 2$, $e_2 = 3$, $e_3 = 5$.
(6) Elementary Abelian p-group with $r = 1$, $e = |G_0|$.
(7) Semidirect product of an elementary Abelian p-group of order q with a cyclic group of order m with $m|(q-1)$, $r = 2$, $e_1 = |G_0|$, $e_2 = m$.
(8) $PSL(2, q)$, with $p \neq 2$, $q = p^m$, $r = 2$, $e_1 = \frac{q(q-1)}{2}$, $e_2 = \frac{(q+1)}{2}$.
(9) $PGL(2, q)$, with $q = p^m$, $r = 2$, $e_1 = q(q-1)$, $e_2 = q + 1$.

The following result of Brandt can be found in §4 of Kontogeorgis and Antoniadis [8]. It provides a more detailed explanation of the group action on $\mathbb{P}^1$ than the previous Proposition, giving the orbits explicitly in each case.

Notation: In the result below, let $i = \sqrt{-1}$. Also, if $S \subset T$ then let $T - S$ denote the subset of elements of T not in S.

Proposition 3.2. *(Brandt [1]) If the characteristic p of the algebraically closed field of constants F is zero or $p > 5$ then the possible automorphism groups of the projective line are given by the following list.*

(1) Cyclic group of order δ.

(2) $D_\delta = \langle\sigma, \tau\rangle$, $(\delta, p) = 1$ where $\sigma(x) = \xi x$, $\tau(x) = 1/x$, ξ is a primitive δ-th root of one. The possible orbits of the D_δ action are $B_\infty = \{0, \infty\}$, $B^- = \{\text{roots of } x^\delta - 1\}$, $B_+ = \{\text{roots of } x^\delta + 1\}$, $B_a = \{\text{roots of } x^{2\delta} + x^\delta + 1\}$, where $a \in F - \{\pm 2\}$.

(3) $A_4 = \langle\sigma, \mu\rangle$, $\sigma(x) = -x$, $\mu(x) = i\frac{x+1}{x-1}$, $i^2 = -1$. The possible orbits of the action are the following sets: $B_0 = \{0, \infty, \pm 1, \pm i\}$, $B_1 = \{\text{roots of } x^4 - 2i\sqrt{3}x^2 + 1\}$, $B_2 = \{\text{roots of } x^4 - 2i\sqrt{3}x^2 + 1\}$, $B_a = \{\text{roots of } \prod_{i=1}^{3}(x^4 + a_i x^2 + 1)\}$, where $a_1 \in F - \{\pm 2, \pm 2i\sqrt{3}\}$, $a_2 = \frac{2a_1+12}{2-a_1}$, $a_3 = \frac{2a_1-12}{2+a_1}$.

(4) $S_4 = \langle\sigma, \mu\rangle$, $\sigma(x) = ix$, $\mu(x) = i\frac{x+1}{x-1}$, $i^2 = -1$. The possible orbits of the action are the following sets: $B_0 = \{0, \infty, \pm 1, \pm i\}$, $B_1 = \{\text{roots of } x^8 + 14x^4 + 1\}$, $B_2 = \{\text{roots of } (x^4+1)(x^8 - 34x^4 + 1)\}$, $B_a = \{\text{roots of } (x^8 + 14x^4 + 1)^3 - a(x^5 - x)^4\}$, $a \in F - \{108\}$.

(5) $A_5 = \langle\sigma, \rho\rangle$, $\sigma(x) = \xi x$, $\mu(x) = -\frac{x+b}{bx+1}$, where ξ is a primitive fifth root of one and $b = -i(\xi^4 + \xi)$, $i^2 = -1$. The possible orbits of the action are the following sets: $B_\infty = \{0, \infty\} \cup \{\text{roots of } f_0(x) := x^{10} + 11ix^5 + 1\}$, $B_0 = \{\text{roots of } f_1(x) := x^{20} - 228ix^{15} - 494x^{10} - 228ix^5 + 1\}$, $B_0^* = \{\text{roots of } x^{30} + 522ix^{25} + 10005x^{20} - 10005x^{10} - 522ix^5 - 1\}$, $B_a = \{\text{roots of } f_1(x)^3 - af_0(x)^5\}$, where $a \in F - \{-1728i\}$.

(6) Semidirect products of elementary Abelian groups with cyclic groups: $(\mathbb{Z}/p\mathbb{Z} \times ... \times \mathbb{Z}/p\mathbb{Z}) \times \mathbb{Z}/m\mathbb{Z}$ of order $p^t m$, where $m|(p^t - 1)$. Suppose we have an embedding of a field of order p^t into k. Assume $GF(p^t)$ contains all the m-th roots of unity. The possible orbits of the action are the following sets: $B_\infty = \{\infty\}$, $B_0 = \{\text{roots of } f(x) := x\prod_{j=1}^{(p^t-1)/m}(x^m - b_j)\}$, where b_j are selected so that all the elements of the additive group $\mathbb{Z}/p\mathbb{Z} \times ... \times \mathbb{Z}/p\mathbb{Z}$ (t times), when viewed as elements in F, are roots of $f(x)$, $B_a = \{\text{roots of } f(x)^m - a\}$, where $a \in F - B_0$.

(7) $PSL(2, p^t) = \langle\sigma, \tau, \phi\rangle$, $\sigma(x) = \xi^2 x$, $\tau(x) = -1/x$, $\phi(x) = x + 1$, where ξ is a primitive $m = p^t - 1$ root of one. The orbits of the action are $B_\infty = \{\infty, \text{roots of } x^m - x\}$. $B_0 = \{\text{roots of } (x^m - x)^{m-1} + 1\}$, $B_a = \{\text{roots of } ((x^m - x)^{m-1} + 1)^{(m+1)/2} - a(x^m - x)^{m(m-1)/2}\}$, where $a \in F^\times$.

(8) $PGL(2, p^t) = \langle\sigma, \tau, \phi\rangle$, $\sigma(x) = \xi x$, $\tau(x) = 1/x$, $\phi(x) = x + 1$, where ξ is a primitive $m = p^t - 1$ root of one. The orbits of the action are

$B_\infty = \{\infty, \text{roots of } x^m - x\}$. $B_0 = \{\text{roots of } (x^m - x)^{m-1} + 1\}$, $B_a = \{\text{roots of } ((x^m - x)^{m-1} + 1)^{m+1} - a(x^m - x)^{m(m-1)}\}$, *where* $a \in F^\times$.

Proof. Brandt [1], Stichtenoth [12]. □

Let $Y = X/G$ be the curve associated to the field E in Proposition 3.1, and let $\pi : X \to Y$ be the quotient map.

Corollary 3.1. *Assume that (1) the finite field F has characteristic > 5, (2) π is defined over F, (3) for each $p_1 \in X(F)$, all the points p_0 in the fiber $\pi^{-1}(p_1)$ are rational: $p_0 \in X(F)$, and (4) F is so large that the orbits described in Proposition 3.2 are complete. Then the above Proposition 3.2 holds over F.*

Proof. Under the hypotheses given, the inertia group is always equal to the decomposition group and the action of the group G of automorphisms commutes with the action of the absolute Galois group $\Gamma = \mathrm{Gal}(\overline{F}/F)$. □

The following is our main result.

Theorem 3.2. *Assume C is a GRS code constructed from a divisor D with positive degree and defined over a sufficiently large finite field F (as described in Corollary 3.1). Then the automorphism group of C must be one of the groups in Proposition 3.1.*

In fact, the action can be made explicit using Proposition 3.2.

Corollary 3.2. *Each GRS code over a sufficiently large finite field is monomially equivalent to a code whose automorphism group is one of the groups in Proposition 3.1.*

Proof. (of theorem) We assume the field is as in Corollary 3.1. Use Theorem 3.1 and Lemma 2.2. □

It would be interesting to know if this result can be refined in the case when $n = 2k$, as that might give rise to a class of easily constructable self-dual codes with large automorphism group.

4. Examples

Pick two distinct orbits $\mathcal{O}_1$ and $\mathcal{O}_2$ of G in $X(F)$. Assume that D is the sum of the points in the orbit $\mathcal{O}_1$ and let $\mathcal{O}_2 = \{P_1, ..., P_n\} \subset X(F)$. Define the associated code of length n by

$$C = \{(f(P_1), ..., f(P_n)) \mid f \in L(D)\} \subset F^n.$$

This code has a G-action, by $g \in G$ sending $(f(P_1), ..., f(P_n))$ to $(f(g^{-1}P_1), ..., f(g^{-1}P_n))$, so is a G-module. Indeed, by construction, the action of G is by permuting the coordinates of C.

Example 4.1. *Let F be a finite field of characteristic > 5 which contains (1) all 4th and 5th roots of unity, (2) all the roots of $x^{10} + 11ix^5 + 1$, (3) all the roots B_0 of $x^{20} - 228ix^{15} - 494x^{10} - 228ix^5 + 1$, and (4) all the roots B_0^* of $x^{30} + 522ix^{25} + 10005x^{20} - 10005x^{10} - 522ix^5 - 1$. Furthermore, let $B_\infty = \{0, \infty\} \cup \{\text{roots of } x^{10} + 11ix^5 + 1\}$. Let $E = \sum_{P \in B_0} P$ and let $D = \sum_{P \in B_0^* \cup B_\infty} P$. Then $\deg(E) = 20$ and $\deg(D) = 42$. Then $C = C(D, E)$ is a formally self-dual code with parameters $n = 42$, $k = 21$, $d = 22$, and automorphism group A_5.*

This follows from (5) of Proposition 3.2 and Remark 2.1.

Example 4.2. *Let $F = GF(q)$ be a finite field of characteristic $p > 5$ for which $q \equiv 1 \pmod 8$ and for which F contains (1) all the roots of $x^{q-1} - x$, and (2) all the roots B_1 of $((x^{q-1} - x)^{q-2} + 1)^{q/2} - (x^{q-1} - x)^{(q-1)(q-2)/2}$. If $B_\infty = \{\infty, \text{roots of } x^{q-1} - x\}$, then let $D = \frac{(q-1)(q-2)}{4} \sum_{P \in B_\infty} P$, $E = \sum_{P \in B_1} P$, and $C = C(D, E)$. Then C is a formally self-dual code with parameters $n = \frac{q(q-1)(q-2)}{2}$, $k = n/2$, $d = n + 1 - k$, and permutation automorphism group $G = PSL(2, q)$.*

This follows from (7) of Proposition 3.2.

5. Structure of the representations

We study the possible representations of finite groups G on the codes $C(D, E)$. As noted in Lemma 3.1, when E is large enough, this is the same as the representation of G on $L(D)$. Therefore we study the possible representations of G on $L(D)$. For simplicity we will restrict to the case where the support of D is rational, i.e. $D = \sum_{i=1}^{s} a_i P_i$, where $P_1, \ldots, P_s$ are rational points on $\mathbb{P}^1$.

We can give the representation explicitly by finding a basis for $L(D)$. For a divisor D with rational support on $X = \mathbb{P}^1$, it is easy to find a basis for $L(D)$, as follows. Let $\infty = [1 : 0] \in X$ denote the point corresponding to the localization $\overline{F}[x]_{(1/x)}$, and $[p : 1]$ denote the point corresponding to the localization $\overline{F}[x]_{(x-p)}$, for $p \in \overline{F}$. For notational simplicity, let

$$m_P(x) = \begin{cases} x, & P = [1:0] = \infty, \\ \frac{1}{(x-p)}, & P = [p:1]. \end{cases}$$

Then $m_P(x)$ is a rational function with a simple pole at the point P, and no other poles.

Lemma 5.1. *Let $D = \sum_{i=1}^{s} a_i P_i$ be a divisor with rational support on $X = \mathbb{P}^1$, so $a_i \in \mathbb{Z}$ and $P_i \in X(F)$ for $0 \leq i \leq s$.*

(a) If D is effective, then

$$\{1, m_{P_i}(x)^k \mid 1 \leq k \leq a_i, 1 \leq i \leq s\}$$

is a basis for $L(D)$.

(b) If D is not effective but $\deg(D) \geq 0$, then D can be written as $D = D_1 + D_2$, where D_1 is effective and $\deg(D_2) = 0$. Let $q(x) \in L(D_2)$ (which is a 1-dimensional vector space) be any non-zero element. Let $D_1 = \sum_{i=1}^{s} a_i P_i$. Then

$$\{q(x), m_{P_i}(x)^k q(x) \mid 1 \leq k \leq a_i, 1 \leq i \leq s\}$$

is a basis for $L(D)$.

(c) If $\deg(D) < 0$, then $L(D) = \{0\}$.

Proof. This is an easy application of the Riemann-Roch theorem. Note that the first part appears as Lemma 2.4 in Lorenzini [10].

By the Riemann-Roch theorem, $L(D)$ has dimension $\deg D + 1$ if $\deg(D) \geq 0$ and otherwise $L(D) = \{0\}$, proving part (c) and the existence of $q(x)$ in part (b). For part (a), since $m_{P_i}(x)^k$ has a pole of order k at P_i and no other poles, it will be in $L(D)$ if and only if $k \leq a_i$. Similarly, for part (b), $m_{P_i}(x)^k$ will be in $L(D_1)$ if and only if $k \leq a_i$; therefore $m_{P_i}(x)^k q(x)$ will be in $L(D_1 + D_2) = L(D)$ under the same conditions. In each of parts (a) and (b), the set of functions given is linearly independent, so by a dimension count must form a basis for $L(D)$. □

Now let G be a finite group acting on $X = \mathbb{P}^1$ and let D be a divisor with rational support, stabilized by G. Let $S = \operatorname{supp}(D)$ and let

$$S = S_1 \cup S_2 \cup \ldots \cup S_m$$

be the decomposition of S into primitive G-sets. Then we can write D as

$$D = \sum_{k=1}^{m} a_k S_k = \sum_{k=1}^{m} a_k \sum_{i=1}^{s} P_{ik},$$

where for each k, $P_{1k} \ldots P_{sk}$ are the points in the orbit S_k. Then G will act by a permutation on the points $P_{1k} \ldots P_{sk}$ in each orbit, and therefore on the corresponding functions $m_{P_{sk}}(x)$.

Theorem 5.1. *Let X, F, $G \subset \text{Aut}(X) = PGL(2, \overline{F})$, and D be as above. Let $\rho : G \to \text{Aut}(L(D))$ denote the associated representation.*

(a) If D is effective then

$$\rho \cong \mathbf{1} \oplus_{k=1}^{m} a_k \rho_k,$$

where $\mathbf{1}$ denotes the trivial representation, and ρ_k is the permutation representation on the subspace

$$V_k = \text{ span } \{m_P(x) \mid P \in S_k\}.$$

(b) If $\deg(D) > 0$ but D is not effective then $L(D)$ is a sub-G-module of $L(D^+)$, where D^+ is a G-invariant effective divisor satisfying $D^+ \geq D$.

The groups and orbits which can arise are described in Proposition 3.1 above.

Proof. (a) By part (a) of Lemma 5.1, $\{1, m_{P_{ik}}(x)^\ell \mid 1 \leq \ell \leq a_i, 1 \leq i \leq s 1 \leq k \leq m\}$ form a basis for $L(D)$. G will act trivially on the constants. For each ℓ, G will act by permutations as described on each set $\{m_{P_{ik}}(x)^\ell \mid P_{ik} \in S_k\}$.

(b) Since D is not effective, we may write $D = D^+ - D^-$, where D^+ and D^- are non-zero effective divisors. The action of G must preserve D^+ and D^-. Since $L(D)$ is a G-submodule of $L(D^+)$, the claim follows. □

Acknowledgements: We thank Cary Huffman for very useful suggestions on an earlier version and for the references to Dur [2] and Huffman [3]. We also thank John Little for valuable suggestions that improved the exposition.

References

[1] R. Brandt, *Über die Automorphismengruppen von Algebaischen Funktionenkörpern*, Ph. D. Univ. Essen, 1988.

[2] A. Dür, *The automorphism groups of Reed-Solomon codes*, Journal of Combinatorial Theory, Series A 44(1987)69-82.

[3] W. C. Huffman, *Codes and Groups*, in the **Handbook of Coding Theory**, (W. C. Huffman and V. Pless, eds.) Elsevier Publishing Co., 1998.

[4] W. C. Huffman and V. Pless, **Fundamentals of error-correcting codes**, Cambridge Univ. Press, 2003.

[5] D. Joyner and A. Ksir, *Decomposing representations of finite groups on Riemann-Roch spaces* - to appear in PAMS (a similar version entitled *Representations of finite groups on Riemann-Roch spaces, II* is available at `http://front.math.ucdavis.edu/`).

[6] ——, *Automorphism groups of some AG codes,* IEEE Trans. Info. Theory, vol 52, July 2006, pp 3325-3329.

[7] ——, *Modular representations on some Riemann-Roch spaces of modular curves X(N),* in **Computational Aspects of Algebraic Curves**, (Editor: T. Shaska) Lecture Notes in Computing, WorldScientific, 2005.

[8] A. Kontogeorgis and J. Antoniadis, *On cyclic covers of the projective line,* Manuscripta Mathematica Volume 121, Number 1 / September, 2006.

[9] A. Kontogeorgis, *The group of automorphisms of cyclic extensions of rational function fields,* Journal of Algebra , Volume 216, June 1999, p 665-706.

[10] D. Lorenzini, **An invitation to arithmetic geometry**, Grad. Studies in Math, AMS, 1996.

[11] H. Stichtenoth, **Algebraic function fields and codes**, Springer-Verlag, 1993.

[12] ——, **Algebraische Funktionenkörper einer Variablen**, Vorlesungen aus dem Fachbereich Mathematik der Universitt Essen [Lecture Notes in Mathematics at the University of Essen], vol. 1, Universität Essen Fachbereich Mathematik, Essen, 1978.

[13] M. A. Tsfasman and S. G. Vladut, **Algebraic-geometric codes**, Mathematics and its Applications, Kluwer Academic Publishers, Dordrechet 1991.

[14] C. R. Valentini and L. M. Madan, *A Hauptsatz of L. E. Dickson and Artin Schreier extensions*, J. Reine Angew. Math., Volume 318, 1980., 156-177.

About the code equivalence

Iliya G. Bouyukliev*

Institute of Mathematics and Informatics,
Bulgarian Academy of Sciences,
P.O.Box 323, 5000 Veliko Tarnovo, Bulgaria
E-mail: iliya@moi.math.bas.bg

In this paper we discuss an algorithm for code equivalence. We reduce the equivalence test for linear codes to a test for isomorphism of binary matrices.

Keywords: Code equivalence, automorphism group, algorithm, canonical form

1. Introduction

In this paper, we consider the algorithm for equivalence which is implemented in the current version of the package $Q - Extension$ [3]. Mainly, this package can be used for classification of linear codes over small fields. Actually, we reduce, as many other algorithms do, the equivalence test for linear codes to a test for isomorphism of binary matrices or bipartite graphs. This allows us to use the developed algorithm for many other combinatorial objects - nonlinear codes, combinatorial designs, Hadamard matrices, etc.

The paper is organized in the following way: In section 2 we give some main definitions related to the code equivalence and the isomorphism of binary matrices. We also show how to transform the problem of code equivalence to the problem of isomorphism of binary matrices. In section 3, we present an important part of the mathematical base of the algorithm. Section 4 contains the main algorithm with detailed pseudo code. In the end of the section we give some additional invariants.

*Partially supported by the Bulgarian National Science Fund under Contract No MM 1304/2003

2. Codes and binary matrices

2.1. *Equivalence of linear codes*

Let $\mathbb{F}_q^n$ be the n-dimensional vector space over the finite field $\mathbb{F}_q$. The *Hamming distance* between two vectors of $\mathbb{F}_q^n$ is defined as the number of coordinates in which they differ. A *q-ary linear $[n,k,d]_q$ code* is a k-dimensional linear subspace of $\mathbb{F}_q^n$ with minimum distance d. A *generator matrix* G of a linear $[n,k]$ code C is any matrix of rank k (over $\mathbb{F}_q$) with rows from C.

Definition 2.1. We say that two linear $[n,k]_q$ codes C_1 and C_2 are **equivalent**, if the codewords of C_2 can be obtained from the codewords of C_1 via a finite sequence of transformations of the following types:

(1) permutation of coordinate positions;

(2) multiplication of the elements in a given position by a non-zero element of $\mathbb{F}_q$;

(3) application of a field automorphism to the elements in all coordinate positions.

An *automorphism* of a linear code C is a finite sequence of transformations of type (1)–(3), which maps each codeword of C onto a codeword of C. The set of automorphisms of a code C forms a group which is called the *automorphism group* of the code C and denoted by $Aut(C)$.

This definition is well motivated as the transformations (1)–(3) preserve the Hamming distance and the linearity (for more details see [5, Chapter 7.3]). The problem of equivalence of codes has been considered in many papers. We distinguish the works of Leon [7] and Sendrier [11]. The complexity of the Code Equivalence Problem is studied in [10].

The algorithm proposed by Sendrier [11] directly uses generator matrices of the linear codes. It works only for codes with specific properties and cannot be used in the general case.

Let C be a linear code over a field with $q > 2$ elements. In our algorithm, we use a subset D of C which is stable under the action of $Aut(C)$ and generates C as a vector space. If the vector $d \in D$ then the vectors λd for $\lambda \in \mathbb{F}_q \setminus \{0\}$ are also in D. Let $D' = \{d'_1, d'_2, \ldots, d'_K\}$ be a subset of D such that no two vectors $d'_i, d'_j \in D'$ are proportional for $i \neq j$, and for any vector $d \in D$ there is a constant $\lambda \in \mathbb{F}_q \setminus \{0\}$ for which $\lambda d \in D'$.

Let A'' be the matrix with rows $d'_1, d'_2, \ldots, d'_K$. We associate to any element $d'_{i,j}$ the matrix

$$D'_{i,j} = \begin{pmatrix} d'_{i,j} & \alpha_2 d'_{i,j} & \dots & \alpha_{q-1} d'_{i,j} \\ \alpha_2 d'_{i,j} & \alpha_2^2 \cdot d'_{i,j} & \dots & \alpha_2 \alpha_{q-1} d'_{i,j} \\ \dots & \dots & \dots & \dots \\ \alpha_{q-1} d'_{i,j} & \alpha_2 \alpha_{q-1} d'_{i,j} & \dots & \alpha_{q-1}^2 d'_{i,j} \end{pmatrix},$$

where $\mathbb{F}_q \setminus \{0\} = \{1, \alpha_2, \dots, \alpha_{q-1}\}$. In this way we obtain a q-ary $(K(q-1)+n) \times n(q-1)$ matrix A'

$$A' = \begin{pmatrix} D'_{1,1} & D'_{1,2} & \dots & D'_{1,n} \\ D'_{2,1} & D'_{2,2} & \dots & D'_{2,n} \\ \dots & \dots & \dots & \dots \\ D'_{K,1} & D'_{K,2} & \dots & D'_{K,n} \\ 11\dots1 & 00\dots0 & \dots & 00\dots0 \\ 00\dots0 & 11\dots1 & \dots & 00\dots0 \\ \dots & \dots & \dots & \dots \\ \underbrace{00\dots0}_{q-1} & \underbrace{00\dots0}_{q-1} & \dots & \underbrace{11\dots1}_{q-1} \end{pmatrix}$$

From this matrix we easily obtain the binary $(K(q-1)+n) \times n(q-1)$ matrix A such that

$$a_{i,j} = 1 \iff a'_{i,j} = 1, \quad a_{i,j} = 0 \text{ otherwise.} \tag{1}$$

For large enough values of K, $Aut(A)$ will be isomorphic to $Aut(C)$ (see definitions 2.3 and 2.4). The last n rows guarantee that an automorphism σ will map any block of $q-1$ columns of A (which corresponds to a column of A'') to another block of $q-1$ columns.

So we reduce our code equivalence problem to an isomorphism test of binary matrices. Moreover, by the permutation which gives an isomorphism of the binary matrices, we can find the coefficients in point (2) in the definition for equivalence of q-ary codes and the field automorphism when q is a power of a prime (see section 2.3).

2.2. *Isomorphism of binary matrices*

Let us denote by Ω the set of all binary $m \times n$ matrices. We define an ordering in the set $\mathbb{F}_2^n$ as follows: For $a = (\alpha_1, \alpha_2, \dots, \alpha_n) \in \mathbb{F}_2^n$ and $b = (\beta_1, \beta_2, \dots, \beta_n) \in \mathbb{F}_2^n$ we have $a < b \iff \alpha_1 = \beta_1, \dots, \alpha_{j-1} = \beta_{j-1}$, $\alpha_j < \beta_j$ for some $j \leq n$. We use it to define a sorted matrix.

Definition 2.2. A **sorted matrix** is a matrix with rows $a_1, a_2, \dots, a_m$ such that $a_1 \geq a_2 \geq \dots \geq a_m$.

Obviously, we can correspond to any matrix $A \in \Omega$ a sorted matrix A^{sort} in a unique way.

We consider the action of the group S_n on the columns of a matrix $A \in \Omega$. If $\sigma \in S_n$, we denote by $A\sigma$ the matrix obtained from A after the permutation of the columns. If the columns of A are $b_1, b_2, \ldots, b_n$ then the columns of $A\sigma$ are $\sigma(b_1) = b_{1\sigma}$, $\sigma(b_2) = b_{2\sigma}, \ldots, \sigma(b_n) = b_{n\sigma}$. Similarly, we consider the action of S_m on the rows of A. For $\tau \in S_m$, we denote by τA the matrix obtained from A after the permutation of the rows. If $a_1, a_2, \ldots, a_m$ are the rows of A then the rows of τA are $\tau(a_1) = a_{1\tau}$, $\tau(a_2) = a_{2\tau}, \ldots, \tau(a_m) = a_{m\tau}$. Obviously, for any matrix $A \in \Omega$, there is a permutation $\gamma \in S_m$ such that the sorted matrix $A^{sort} = \gamma A$.

Definition 2.3. Two matrices of the same size are **isomorphic** if the rows of the second one can be obtained from the rows of the first one by a permutation of the columns.

This definition is based on the natural action of the symmetric group S_n on the set of columns for all elements in Ω. Obviously, the matrices A and B from the set Ω are isomorphic, or $A \sim B$, if their corresponding sorted matrices are isomorphic. This fact allows us to consider only the sorted matrices in Ω.

Any permutation of the columns of A which maps the rows of A into rows of the same matrix, is called an automorphism of A. The set of all automorphisms of A is a subgroup of the symmetric group S_n and we denote it by $Aut(A)$.

The following definition (equivalent to definition 2.3) is based on the action of the symmetric group S_n on the set of columns and the action of the symmetric group S_m on the set of rows to all elements in Ω.

Definition 2.4. Two matrices of the same size are **isomorphic** if the second one can be obtained from the first one by a permutation of the columns and the rows.

We prefer the first one because it is similar to the usual code equivalence definition. Considering the sorted matrices, we have $A \sim B$ if there exists a permutation $\sigma \in S_n$ such that $B^{sort} = (A\sigma)^{sort}$.

We consider two main problems.

Problem 2.1. *Is there a permutation $\sigma \in S_n$ such that for given binary matrices A and B, $B^{sort} = (A\sigma)^{sort}$?*

Problem 2.2. *For a given binary matrix A, compute a set of generators for the automorphism group of A.*

The definition for isomorphism of binary matrices allows us to consider the set Ω as a union of equivalence classes. Matrices which are isomorphic belong to the same equivalence class. Every matrix of an equivalence class can serve as a representative for this equivalence class. In many cases, a canonical representative is used, which is selected based on some specific conditions. This canonical representative is intended to easily make the distinction between distinct equivalence classes. Practically, it reduces the isomorphism testing of matrices to comparing matrices. More precisely, we can define the canonical representative map as follows:

Definition 2.5. A **canonical representative map** is a function $\rho\colon \Omega \mapsto \Omega$ which satisfies the following two properties:

1. for all $X \in \Omega$ it holds that $\rho(X) \sim X$,
2. for all $X, Y \in \Omega$ it holds that $X \sim Y$ implies $\rho(X) = \rho(Y)$.

We say that the matrix X is in **canonical form** if $\rho(X) = X$.

We consider ordering in the set of all binary $m \times n$ matrices. For the matrices $A = (a_1, a_2, \ldots, a_m)^t$ and $B = (b_1, b_2, \ldots, b_m)^t$ we have $A < B \iff a_1 = b_1, \ldots, a_{j-1} = b_{j-1}, a_j < b_j$ for some $j \leq m$. For any two matrices A and B we can say $A < B$, $A > B$ or $A = B$.

Now we will present a way to choose a canonical representative. For the canonical representative of the class of equivalence of the matrix A we can take the matrix B such that $B^{sort} \geq (A\sigma)^{sort}$ for any permutation $\sigma \in S_n$. It is easy to define the canonical representative in this way but quite complicated to find it. Of course, we can try all permutations in S_n. Using comparison of matrices, we can define ordering for the elements in S_n with respect to a binary matrix A: $\gamma_1 < \gamma_2$ with respect to A if $(A\gamma_1)^{sort} < (A\gamma_2)^{sort}$. The general idea of a class of algorithms including ours is to find a minimal (or maximal) element in the set Π of permutations, which depends on the matrix A, where Π has a much smaller number of elements than S_n.

Definition 2.6. Let $A_1, A_2, \ldots, A_s$ be all different $m \times n$ binary matrices which are isomorphic to the matrix in canonical form B. Let $\sigma_i \in S_n$ be a permutation of the columns of the matrix A_i such that $(A_i\sigma_i)^{sort} = B^{sort}$, $i = 1, \ldots, s$. We call the permutation σ_i a **canonical labeling map** for the matrix A_i defined by B.

As $A_i\tau = B \ \forall \tau \in \sigma_i Aut(B)$, the map σ_i is not unique except when $Aut(B) = \{id\}$. A canonical labeling of the columns of the matrix A_i is $(\sigma_i(1), \sigma_i(2), \ldots, \sigma_i(n))$.

An important computational problem is the following:

Problem 2.3. *For a given binary matrix A compute the canonical form B and a canonical labeling $\sigma \in S_n$ such that $B^{sort} = (A\sigma)^{sort}$.*

The aim of our work is to present an algorithm which defines a specific canonical representative map and gives a solution of the three defined problems.

Problems 1,2, and 3, are connected with the graph isomorphism problem. First of all, any binary matrix can be considered as a bipartite graph. In the case of a bipartite graph, the set of vertices is decomposed into two disjoint colored sets (columns and rows) such that no two graph vertices within the same set are adjacent. Hence, solving the isomorphism problems for bipartite graphs and binary matrices is the same.

In other hand, any graph can be made bipartite by replacing each edge by two edges connected with a new vertex. And any two graphs are isomorphic if and only if the transformed bipartite graphs are. Theoretical results for the graph isomorphism problem can be found in [1], [4].

Next, we briefly describe some of the basic setup and give pseudo-code for the algorithm. For further details see [8].

2.3. *The connection between equivalence of linear codes and isomorphism of binary matrices*

Let C be a linear code over a field with $q > 2$ elements and A be the corresponding binary $(K(q-1)+n) \times n(q-1)$ matrix as presented in (1). To any automorphism φ of C there corresponds a permutation σ_φ from $Aut(A)$ in the following way:

(1) If φ is a permutation of the coordinate positions, the permutation σ_φ is the same φ which acts on the blocks of $q-1$ columns corresponding to the coordinates of C.
(2) If φ is a multiplication of the elements in a given position, say i, by a nonzero element $\alpha \in \mathbb{F}_q$, σ_φ is a permutation of the columns in the block of $q-1$ columns corresponding to the position i. This permutation, considered as an element of the symmetric group S_{q-1}, depends only on α; that's why we denote it by $\sigma^{(\alpha)}$. So for all nonzero elements of $\mathbb{F}_q$ we can collect corresponding permutations and from the permutation

easily find the element. Moreover, the set $Y_q = \{\sigma^{(\alpha)} \mid \alpha \in \mathbb{F}_q \setminus \{0\}\}$ forms a cyclic subgroup of S_{q-1} of order $q-1$.

(3) The case when φ is a field automorphism is more complicated. Then the corresponding permutation σ_φ is a permutation of the columns in the blocks. As in the previous case, it can be considered as an element of S_{q-1} and depends only on the field automorphism. As we know, the Galois group of a finite field with $q = p^s$ elements is a cyclic group of order s; that's why the set X_q of the corresponding permutations in S_{q-1} forms a cyclic group of order s.

(4) When φ is a finite sequence of the transformations of the three types, then σ_φ is the product of the corresponding permutations of the columns in A.

Example 2.1. Let C be a quaternary code and $\mathbb{F}_4 = \{0, 1, x, x^2\}$, $x^3 = 1$. We associate the elements of the field with binary 3×3 matrices in the following way:

$$0 \longmapsto \begin{pmatrix} 000 \\ 000 \\ 000 \end{pmatrix}, \quad 1 \longmapsto \begin{pmatrix} 100 \\ 001 \\ 010 \end{pmatrix}, \quad x \longmapsto \begin{pmatrix} 001 \\ 010 \\ 100 \end{pmatrix}, \quad x^2 \longmapsto \begin{pmatrix} 010 \\ 100 \\ 001 \end{pmatrix}. \tag{2}$$

It is easy to see that the multiplication by x corresponds to the permutation (132) of the columns in any of these blocks, and the multiplication by x^2 corresponds to the permutation (123). The only nontrivial automorphism of the field maps the element $a \in \mathbb{F}_4$ to its conjugate $\overline{a} = a^2$. We can represent it as the permutation (23) of the columns combined with the same permutation of the rows. So the transposition $(12) = (23)(132)$ corresponds to the field automorphism combined with a multiplication by x.

Example 2.2. Let consider the field $\mathbb{F}_5 = Z_5$. Then

$$1 \to \begin{pmatrix} 1000 \\ 0010 \\ 0100 \\ 0001 \end{pmatrix}, \quad 2 \to \begin{pmatrix} 0010 \\ 0001 \\ 1000 \\ 0100 \end{pmatrix}, \quad 3 \to \begin{pmatrix} 0100 \\ 1000 \\ 0001 \\ 0010 \end{pmatrix}, \quad 4 \to \begin{pmatrix} 0001 \\ 0100 \\ 0010 \\ 1000 \end{pmatrix}. \tag{3}$$

In this case we have $Y_5 = \{id, \sigma^{(2)} = (1342), \sigma^{(4)} = (14)(23), \sigma^{(3)} = (1243)\}$.

Proposition 2.1. *If the codes C and C' are equivalent, then the corresponding matrices A and A' are isomorphic. Moreover, if $C' = \phi(C)$ then $\sigma = \sigma_\phi$ is a composition of a permutation of the n blocks of $q-1$ columns* *corresponding to the coordinates of the codes, and permutations from a coset* τY_q*, where* $\tau \in X_q$*.*

Proposition 2.2. *If there is a permutation* $\sigma \in S_{n(q-1)}$ *such that* $(A\sigma)^{sort} = (A')^{sort}$*, this permutation is a composition of a permutation of the* n *blocks of* $q-1$ *columns corresponding to the coordinates of the codes, and permutations from a coset* τY_q*, where* $\tau \in X_q$*, then the codes* C *and* C' *are equivalent.*

Proof. The permutation of the n blocks corresponds to a permutation of the coordinates of C. A permutation from the coset τY_q corresponds to a field automorphism followed by a multiplication of the elements in a given position by a nonzero element of the field. □

If the matrices A and A' are isomorphic, they have the same canonical form B. As $B^{sort} = (A\tau)^{sort} = (A'\tau')^{sort}$ for some permutations τ and τ', we can take $\sigma = \tau(\tau')^{-1}$ to be the isomorphism. Obtaining the canonical form of the matrices, we find also their automorphism groups. If the automorphism groups $Aut(A)$ and $Aut(A')$ of two isomorphic matrices consist only of permutations as described in the proposition, then the corresponding codes are equivalent. Really, from the structure of the matrices, it follows that a permutation φ, such that $\varphi(A) = A'$, maps any block of A into a block of A'. Moreover, if $\tau \in S_{q-1}, \tau \notin Y_q$, then $\tau\sigma^{(\alpha)}\tau^{-1} \notin Y_q$. Hence, if φ is not of the type as described in the proposition, the group $Aut(A')$ will also contain elements which are not of this type - but this is not the case.

Proposition 2.3. *Let* C *be a linear code over* $\mathbb{F}_q$ *and* A *be the corresponding binary* $(K(q-1)+n) \times n(q-1)$ *matrix as presented in Eq. (1). If all automorphisms of* A *are of the type described in Proposition 2.2, then* $Aut(C) \cong Aut(A)$*.*

3. Orbits, partitions, invariants

3.1. *Orbits*

The group $Aut(A)$ splits the columns of A into disjoint sets $O_1, O_2, \ldots, O_k$ called *orbits*. Two columns a_1 and b_1 are in the same orbit if and only if there is an automorphism $\sigma \in Aut(A)$ such that $\sigma(a_1) = b_1$. All automorphisms $\gamma \in Aut(A)$ for which $\gamma(a_1) = a_1$ form a group $Aut(A_{a_1})$ called the *stabilizer* of a_1. All the automorphisms which map a_1 to b_1

form a coset of the stabilizer $Aut(A_{a_1})$. Moreover, if a_1 and b_1 are in the same orbit, their stabilizers $Aut(A_{a_1})$ and $Aut(A_{b_1})$ are conjugated, i.e. $Aut(A_{a_1}) = \sigma^{-1} Aut(A_{b_1})\sigma$.

The group $Aut(A_{a_1})$ also splits the columns of A into disjoint orbits, which we denote by $O_1^{a_1}, O_2^{a_1}, \ldots, O_r^{a_1}$, as $O_i^{a_1} \subset O_j$ for a suitable j, $i = 1, \ldots, r$. If a_1 and b_1 are in the same orbit, to any orbit $O_i^{a_1}$, it corresponds in a unique way an orbit $O_i^{b_1}$. We call the orbits $O_i^{a_1}$ and $O_i^{b_1}$ *corresponding* with respect to the fixed columns a_1 and b_1, and denote this by $O_i^{a_1} \wr O_i^{b_1}$. Any two corresponding orbits $O_i^{a_1}$ and $O_i^{b_1}$ ($O_i^{a_1} \wr O_i^{b_1}$) belong to the same orbit induced by $Aut(A)$. Moreover, for any $a_2 \in O_i^{a_1}$ and $b_2 \in O_i^{b_1}$, there exists an automorphism $\sigma \in Aut(A)$ such that $b_1 = \sigma(a_1)$ and $b_2 = \sigma(a_2)$. Conversely, if $a_2 \in O_i^{a_1}$ and $b_2 \in O_i^{b_1}$, but the orbits $O_i^{a_1}$ and $O_i^{b_1}$ are not corresponding, then for any γ, for which $\gamma(a_1) = b_1$, we have $\gamma(a_2) \neq b_2$.

Similarly, for any $a_2 \in O_i^{a_1}$ and $b_2 \in O_i^{b_1}$, $O_i^{a_1} \wr O_i^{b_1}$, we denote the corresponding orbits with respect to the fixed pairs of columns a_1, a_2 and b_1, b_2 by $O_{i_2}^{a_1,a_2} \wr O_{i_2}^{b_1,b_2}$. If $|Aut(A_{a_1,a_2})| > 1$, we can continue to fix columns. In the general case, we denote the stabilizer of the points $a_1, a_2, \ldots, a_k$ by $Aut(A_{a_1,a_2,\ldots,a_k})$. The corresponding orbits are denoted by $O_i^{a_1,a_2,\ldots,a_k} \wr O_i^{b_1,b_2,\ldots,b_k}$.

Let $Aut(A_{a_1,a_2,\ldots,a_k})$ and $Aut(A_{b_1,b_2,\ldots,b_k})$ be conjugated, i.e. there exists an automorphism σ such that $\sigma(a_i) = b_i$, $i = 1, 2, \ldots, k$. If $|Aut(A_{a_1,a_2,\ldots,a_k})| = 1$ then any of the corresponding orbits has only one element and therefore these orbits define the automorphism σ.

If $\gamma \in S_n$ and $O_i^{a_1,a_2,\ldots,a_k} = \{o_1, o_2, \ldots, o_j\}$ is an orbit induced by $Aut(A_{a_1,a_2,\ldots,a_k})$ then $O_i^{\gamma(a_1),\gamma(a_2),\ldots,\gamma(a_k)} = \{\gamma(o_1), \gamma(o_2), \ldots, \gamma(o_j)\}$ is an orbit induced by $Aut(A\gamma_{\gamma(a_1),\gamma(a_2),\ldots,\gamma(a_k)})$.

3.2. *Partitions, ordered partitions*

A *partition* $\pi = (V_1, V_2, \ldots, V_r)$ of a set L is a family of disjoint nonempty subsets $V_1, V_2, \ldots, V_r, V_i \subset L$, called *cells*, such that $V_1 \cup V_2 \cup \cdots \cup V_r = L$. A cardinality of a cell is the number of its elements. A cell is called *discrete* if it consists of only one element, and the partition is discrete if all its cells are discrete.

Any group G of automorphisms, $G \subset Aut(A)$, splits the columns into orbits. But in this case we have no criteria to order the cells. The trivial group $\{id\}$ splits the columns into a discrete partition.

Any automorphism induces a partition of columns with respect to the cyclic group generated by this automorphism. Let G_i be the cyclic group generated by γ_i, and let π_i be the partition which corresponds to the orbits of G_i, $i = 1, 2$. The orbits of the group G, generated by γ_1 and γ_2, form a new partition π, and we can find it in the following rule using π_1 and π_2. If there are two columns which are in different cells V_i and V_j in π_{l_1} and in the same cell in π_{l_2}, $\{l_1, l_2\} = \{1, 2\}$, then the columns of V_i and V_j have to be in one cell in π.

An ordered partition is a partition, for which $V_i < V_j$ or $V_i > V_j$ for any $i \neq j$. We will write the ordered partitions in increasing order, i.e. $V_i < V_j$ for $i < j$.

Let $\pi = (V_1, V_2, \ldots, V_r)$ is a partition and $\gamma \in S_n$. Then $\gamma(\pi) = (\gamma(V_1), \gamma(V_2), \ldots, \gamma(V_r))$ where $\gamma(V_j) = (\gamma(a_1), \gamma(a_2), \ldots, \gamma(a_i))$ for $V_j = [a_1, a_2, \ldots, a_i]$.

3.3. *Definition of invariants*

An invariant of the columns of a matrix A with respect to the group $Aut(A)$ is a function f_1 which maps any column to an element of an ordered set M (for example Z), as $f_1(c_i) = m_i$, $m_i \in M$, such that if $\sigma(c_i) = c_j$, $\sigma \in Aut(A)$, then $f_1(c_i) = f_1(c_j)$. Moreover, $f_1(c)$ has the same value as $f_1(\gamma(c))$ with respect to $\gamma Aut(A\gamma)\gamma^{-1}$ for any permutation $\gamma \in S_n$.

The invariant f_1 induces an ordered partition of the set of columns of the matrix , as $f_1(c_i) = f_1(c_j) \Leftrightarrow c_i, c_j \in V_p$, $f_1(c_i) < f_1(c_j) \Leftrightarrow c_i \in V_p, c_j \in V_q$ for $p < q$. This ordered partition can be considered as:

- arranging the columns in groups - any cell consists of the columns in one or more orbits.

- reordering the columns with respect to the cells and their order. If the partition is discrete, it defines a permutation of the columns in A.

- we can choose a cell as *special*. For example this could be the first largest cell.

The group $Aut(A)$ stabilizes the defined partition π_1. We define invariants with respect to the stabilizer $Aut(A_{a_1,a_2,\ldots,a_k})$ of the columns $a_1, a_2, \ldots, a_k$ in the following way.

Definition 3.1. Let $\pi_k = (V_1, V_2, \ldots, V_{r_k})$ be an ordered partition such that $\sigma(\pi_k) = \pi_k$ for any $\sigma \in Aut(A_{a_1,a_2,\ldots,a_{k-1}})$ and a_k be a column in the special cell V_{j_s}. An invariant of the columns of a matrix A with respect to the group $Aut(A_{a_1,a_2,\ldots,a_{k-1},a_k})$ and the ordered partition π_k is a function f_{k+1}, which maps any column to an element of M, such that:

1. $f_{k+1}(a_k) < f_{k+1}(b)$ for all $b \in V_{j_s} \setminus \{a_k\}$.
2. $f_{k+1}(a) < f_{k+1}(b)$ for any $a \in V_i$, $b \in V_j$, where $V_i, V_j \in \pi_k, i < j$.
3. $f_{k+1}(a) = f_{k+1}(b)$ when a and b are in the same orbit with respect

to the stabilizer $Aut(A_{a_1,a_2,\dots,a_k})$. Moreover, $f_{k+1}(a) = f_{k+1}(\gamma(a))$ with respect to $\gamma Aut((A\gamma)_{\gamma(a_1),\gamma(a_2),\dots,\gamma(a_k)})\gamma^{-1}$ and $\gamma(\pi_k)$ for any $\gamma \in S_n$.

Definition 3.2. We call the set of invariants F **strong** if the columns in different orbits have different values in M.

Let $F = \{f_1, f_2, f_3, \dots\}$ be a set of invariants. To obtain a discrete partition, induced by F and the matrix A, we can use the following algorithm:

```
00]disc_part(inp A:binary matrix; F:set of invariants; π0:partition;
00]     out k:number of fixed columns;
00]         π1, π2, ..., πk:partitions;
00]         w:array of cells; v:vector of fixed columns );
00]   var
00]   i: integer; v: vector;
01]    begin
02]      i := 1;
03]      v[i] := 0;
04]      inv_act; { using f1 and π0 find π1 = (V1, V2 ..., Vr1) and Vs1 ; }
                               {Vs1 is a "special" cell }
05]      w[1]:=Vs1;
06]      while πi is not discrete do
07]      begin
08]         choose ai from Vsi;
09]         fix(ai); { (V1, ..., Vsi, ..., Vri) go to (V1, ..., ai, Vsi \ {ai}, ..., Vri) }
10]         v[i] := ai;
11]         i := i + 1;
12]         inv_act; { using fi find πi = (V1, V2, ..., Vri) and Vsi; }
13]         w[i]:=Vsi;
14]      end;
15]      k := i;
16]    end;
```

We use the following notations in the algorithms: **inp** - input variables, **out** - output variables, **inp_out** - variables, used as input and output (they change in the corresponding algorithm).

After acting with f_1 on the columns of the input partition π_0, the algorithm obtains (step 4) a partition induced by f_1 and a special cell V_{s_1}. If the partition obtained is not discrete, the algorithm chooses a column from the special cell, and collects this column in $v[i]$. In row 9, the algorithm fixes the chosen column, i.e. it splits the special cell V_{s_i} in the partition π_1 into two cells. The first one is discrete and contains only the fixed column. In row 12, using the invariant f_i, the algorithm obtains the next partition. In the end, the variable k keeps the number of fixed columns and the number of levels, and w keeps the special cells in the different levels.

We call a *position* of a cell V_l in the partition $\pi_j = (V_1, \ldots, V_{l-1}, V_l, \ldots, V_r)$ the number $|V_1| + |V_2| + \cdots + |V_{l-1}| + 1$. From the definition 3.1, it follows that any cell V_i in the partition π_j consists of ordered cells in the partition π_{j+1} and the position of the first one is the same as the position of V_i.

The set v of fixed columns in steps one, two, etc., and the algorithm *disc_part* define in a unique way an ordered partition. We call the set v *the vector of the fixed columns.*

If we choose different columns a_i from the special cell V_{s_i}, the algorithm disc_part determines different discrete partitions. Let us denote by Π the set of all different discrete partitions which can be generated using the algorithm disc_part. Let $\pi_k = (V_1, V_2 \ldots, V_{r_i}) \in \Pi$ be a discrete partition. This means that any cell is discrete, $r_i = n$, and $\pi_k = ([c_{i_1}], [c_{i_2}], \ldots, [c_{i_n}])$, where c_j are columns in A.

3.4. *Properties of partitions induced by invariants*

Let A be a binary matrix and F be a set of invariants.

Proposition 3.1. *Let the stabilizers $Aut(A_{a_1,\ldots,a_k})$ and $Aut(A_{b_1,\ldots,b_k})$ be conjugate and the orbits $O_{i_1}^{a_1,\ldots,a_k}$ and $O_{i_1}^{b_1,\ldots,b_k}$ be corresponding ($O_{i_1}^{a_1,\ldots,a_k} \wr O_{i_1}^{b_1,\ldots,b_k}$). If $O_{i_1}^{a_1,\ldots,a_k}$ belongs to the special cell, then $O_{i_1}^{b_1,\ldots,b_k}$ also belongs to the special cell but after fixing $b_1, b_2, \ldots, b_k$.*

Proof. Let $f_2(d) = m_{i_1} \in M$ where d is a column in $O_{i_1}^{a_1}$ or in another orbit in the special cell. By point 2 in definition 3.1, the value of f_2 will be also m_{i_1} for the columns in the corresponding orbits. This means that these corresponding orbits form a special cell after fixing b_1. For $k > 1$, the proposition can be proved trivially by induction with respect to the number of fixed columns. □

Corollary 3.1. *Let π' and π'' be two partitions obtained in the row 12 of disc_part and their corresponding vectors of fixed columns v' and v'' have k elements. If there exists an automorphism σ such that $v'' = \sigma(v')$, then $\pi'' = \sigma(\pi')$. If $\pi \in \Pi$ and $\sigma \in Aut(A)$ then $\sigma(\pi) \in \Pi$.*

The discrete partition $\pi_k = ([c_{i_1}], [c_{i_2}], \ldots, [c_{i_n}])$ determines the permutation of the columns $\widehat{\pi}_k = (1 \to c_{i_1}, 2 \to c_{i_2}, \ldots, n \to c_{i_n})$. Conversely, for any permutation we have a unique discrete partition.

We compare discrete partitions π_A and π_B of the matrices A and B, respectively, in the following way: $\pi_A < \pi_B \Leftrightarrow (A\widehat{\pi}_A)^{sort} < (B\widehat{\pi}_B)^{sort}$,

$\pi_A \asymp \pi_B \Leftrightarrow (A\widehat{\pi}_A)^{sort} = (B\widehat{\pi}_B)^{sort}$.

Lemma 3.1. *Consider the matrices A and $B = A\tau$ for $\tau \in S_n$ and the sets Π_A and Π_B of all discrete partitions obtained for A and B using the algorithm disc_part. Then there is an one-to-one correspondence between Π_A and Π_B. Moreover, for any discrete partition $\pi_A \in \Pi_A$ there is a discrete partition $\pi_B \in \Pi_B$ such that $\pi_B \asymp \pi_A$.*

Proof. Let $\pi_A = ([c_{i_1}], [c_{i_2}], \ldots, [c_{i_n}])$ be a discrete partition in Π_A with vector of the fixed columns v. From the properties of the orbits and definition 3.1, it follows that $\tau(v) = (\tau(v_1), \tau(v_2), \ldots, \tau(v_k))$ is the vector of the fixed columns of the partition $\tau(\pi_A) = ([\tau(c_{i_1})], [\tau(c_{i_2})], \ldots, [\tau(c_{i_n})])$. Actually, the columns c_{i_l} and $\tau(c_{i_l})$ are the same, $l = 1, 2, \ldots, n$. From proposition 3.1, it follows that $\tau(\pi_A) \in \Pi_B$. Hence $\pi_A \asymp \tau(\pi_A)$ with respect to the definition given above. □

For a fixed column a, we call the position of the corresponding discrete cell in the partition a *position of this column*. So if we fix a column a, its position is not changed until the end of the procedure, where we obtain a discrete partition. In the algorithm *disc_part*, we can obtain not only the vector of fixed columns v, but also the vector of their positions vp.

The comparing of the discrete partitions helps us to define a canonical discrete partition.

Lemma 3.2. *The maximal discrete partition c in Π such that $c \asymp \max\{\pi_j; \pi_j \in \Pi\}$, which we call canonical, determines a permutation $\widehat{c}$ which is a canonical labeling map for A.*

Proof. It follows from the definition for canonical labeling map and lemma 3.1. □

Proposition 3.2. *Two discrete partitions $\gamma_1 = ([c_{i_1}], [c_{i_2}], \ldots, [c_{i_n}])$ and $\gamma_2 = ([c_{j_1}], [c_{j_2}], \ldots, [c_{j_n}])$, for which $\gamma_1 \asymp \gamma_2$, define an automorphism $\widehat{\sigma} = (c_{i_1} \to c_{j_1}, c_{i_2} \to c_{j_2}, \ldots, c_{i_n} \to c_{j_n})$, which is $\widehat{\sigma} = \widehat{\gamma}_1 \cdot \widehat{\gamma}_2^{-1}$.*

Proof. $\gamma_1 \asymp \gamma_2 \Rightarrow (A\widehat{\gamma}_1)^{sort} = (A\widehat{\gamma}_2)^{sort}$. □

Lemma 3.3. *Two discrete partitions π' and π'' in Π with vectors of fixed columns v' and v'' of length k are equal ($\pi' \asymp \pi''$) if and only if v'_j and v''_j belong to corresponding orbits for any $j \leq k$ (if $v'_j \in O_j^{v'_1,\ldots,v'_{j-1}}$ and $v''_j \in O_j^{v''_1,\ldots,v''_{j-1}}$ for $j \leq k$, then $O_j^{v'_1,\ldots,v'_{j-1}} \wr O_j^{v''_1,\ldots,v''_{j-1}}$).*

Proof. If $\pi' \asymp \pi''$ then there is an automorphism $\widehat{\sigma} = \widehat{\pi'} \cdot \widehat{\pi''}^{-1}$ which maps the first partition to the second one. Hence $\widehat{\sigma}(v') = v''$ and so the corresponding vectors of fixed columns are in corresponding orbits.

Conversely, if v'_j and v''_j belong to corresponding orbits for any $j \leq k$, then there exists an automorphism σ such that $\sigma(v'_j) = v''_j$ for $j = 1, 2, \ldots, k$ and therefore $\pi'' = \sigma(\pi')$. □

Theorem 3.1. *Let $T_1 \subset \Pi$ and $T_2 \subset \Pi$ be the sets of all discrete partitions with vectors of fixed columns $(v_1, \ldots, v_j, a, \ldots)$ and $(v_1, \ldots, v_j, b, \ldots)$, and a and b be in the same orbit with respect to $Aut(A_{v_1,\ldots,v_j})$. Then any element $\pi_{T_2} \in T_2$ can be presented as $\pi_{T_2} = \sigma(\pi_{T_1})$ for some $\sigma \in Aut(A_{v_1,\ldots,v_j})$. The permutation σ is an automorphism which means that π_{T_1} and π_{T_2} are equal.*

Proof. Let $\pi_{T_1} \in T_1$ is a partition with a vector of fixed columns $v = (v_1, \ldots, v_j, a, v_{j+2}, \ldots, v_k)$. There is a permutation $\sigma \in Aut(A_{v_1,\ldots,v_j})$ such that $\sigma(a) = b$. Then $\sigma(v) = (v_1, \ldots, v_j, b, \sigma(v_{j+2}), \ldots, \sigma(v_k))$ is the vector of fixed columns for the partition $\sigma(\pi_{T_1})$. This partition is in T_2 (see proposition 3.1). Now it is trivially to see that any element $\pi_{T_2} \in T_2$ can be presented as $\pi_{T_2} = \sigma(\pi_{T_1})$ for some $\sigma \in A_{v_1,\ldots,v_j}$. □

Corollary 3.2. *If two discrete partitions π' and π'' of a matrix A are equal then their vectors of positions of fixed columns vp' and vp'' are the same.*

Corollary 3.3. *If all the invariants in F are strong (i.e. every special cell consists of one orbit) then all discrete partitions in Π are equal.*

3.5. *Invariants of columns and rows*

Let us consider the second definition for isomorphism of matrices. In analogy to the definition for the columns invariants, we can define row invariants which induce ordered partitions π_k^{row} with respect to the stabilizer of the columns $Aut(A_{a_1,a_2,\ldots,a_{k-1}})$ and the previous row partition π_{k-1}^{row}.

Now on, we denote by π an ordered partition which consists of π^{column} and π^{row}, or $\pi = (\pi^{column}, \pi^{row})$. We denote the cells of π^{row} by $V'_i = [a'_1, \ldots, a'_j]$.

There are invariants of columns which are very effective and recursively depend on rows invariants.

Definition 3.3. We call **distance** between b and V the number of ones in a row b and the columns in a set V and denote $d(b, V)$. Similarly, we define distance between a column b and a set of rows V.

Now we consider an invariant which is based on the following trivial fact.

Lemma 3.4. *Let us consider the set V_{column} of columns of a binary matrix which consists of one or a few orbits with respect to a group of automorphisms G. Then a necessary condition two rows a and b from the set of rows V_{rows} to be in the same orbit with respect to G is $d(a, V_{column}) = d(b, V_{column})$. Similarly, this works for two columns and a set of rows.*

This claim is also true in the case when G is a stabilizer of columns $Aut(A_{a_1,a_2,\ldots,a_k}) \subset Aut(A)$.

We give an example to show how to use lemma 3.4 to obtain an invariant and the induced by it partition. We denote by R an ordered partition of rows or columns, which we use for comparing. Actually, R can be an ordered partition of some of the rows and columns (not of all rows and columns) or even the empty set.

Example 3.1. Let us consider the matrix

$$A = \langle 1111000, 0101100, 0010110, 0001011, 1000111, 1100010, 0110001\rangle$$

In the beginning, we have the trivial partition of the columns

$$\pi_{column} = R_{column} = (V_1), \quad V_1 = [1, 2, 3, 4, 5, 6, 7],$$

and the trivial partition of the rows $\pi_{row} = R_{row} = (V_1')$, for $V_1' = [1', 2', 3', 4', 5', 6', 7']$.

The number of ones in the rows is $1' - 4, 2' - 3, 3' - 3, 4' - 3, 5' - 4, 6' - 3, 7' - 3$, or $d(1', V_1) = 4$, $d(1', V_1) = 4$, $d(2', V_1) = 3$, $d(3', V_1) = 3$, $d(4', V_1) = 3$, $d(5', V_1) = 4$, $d(6', V_1) = 3$, $d(7', V_1) = 3$. This means that the set of the rows has at least 2 orbits with respect to $Aut(A)$. The number of ones in the rows (or the distance to the set of all columns V_1) induces the following ordered partition: $\pi_{row} = (V_1', V_2') = ([2', 3', 4', 6', 7'], [1', 5'])$.

In the second step we use the obtained partition π_{row} as R_{row} and compare the distances from the columns to the cells of R_{row}. So we obtain the following distances from the columns to the cells of R_{row}:

	1	2	3	4	5	6	7
$d(*, V_1')$	1	3	2	2	2	3	2
$d(*, V_2')$	2	1	1	1	1	1	1

These distances induce the next ordered partition of the columns. $\pi_{column} = (V_1, V_2, V_3) = ([1], [3,4,5,7], [2,6])$. In the third step, we compare the distances between the rows and the obtained π_{column}:

	$1'$	$2'$	$3'$	$4'$	$5'$	$6'$	$7'$
$d(*, V_1)$	1	0	0	1	1	1	0
$d(*, V_2)$	2	2	2	2	2	1	2
$d(*, V_3)$	1	1	1	1	1	2	1

Hence, after this step we have

$$\pi_{row} = (V_1', V_2', V_3') = ([2', 3', 4', 7'], [6'], [1', 5']).$$

In step 4, for the columns we obtain following distances

	1	2	3	4	5	6	7
$d(*, V_1')$	0	2	2	2	2	2	2
$d(*, V_2')$	1	1	0	0	0	1	0
$d(*, V_3')$	2	1	1	1	1	2	1

There is no new splitting of cells and therefore the process stops. We can generalize all the calculations for the columns in the following way: to any column we correspond in a unique way a polynomial with integer coefficients:

$$f(1) = 1 + Y(1 + 2x) + Y^2(x + 2x^2)$$
$$f(2) = f(6) = 1 + Y(3 + x) + Y^2(2 + x + x^2)$$
$$f(3) = f(4) = f(5) = f(7) = 1 + Y(2 + x) + Y^2(2 + x^2)$$

The coefficients for Y^0 is one because all the columns are in the same cell in the beginning. The coefficients for Y and Y^2 depend on the distances to the corresponding cells of R_{column} in the steps two and four.

Actually, we repeat some of the calculations in this procedure. In step two, we look for distances between all columns and the rows in the set $[2', 3', 4', 6', 7']$. In step 4, we look for the distances to $[2', 3', 4', 7']$ and $[6']$. It is clear that in step 4 we can obtain the same splitting of columns if we compute only the distances to the cell $[6']$ or to cell of rows $[2', 3', 4', 7']$. Generally, it is necessary to calculate the distances to all cells except one.

We skip the first largest cell (with maximum cardinality) for efficiency. If there is only one cell, there is no reason to compare again with it.

To obtain the final partition of columns and rows, we use the following algorithm:

stable(**inp** A:bin_mat; **inp_out** π:partition; **inp** πh:partition; **inp** copy:string);
00] **var** i: integer;
00] $\pi_{columns}$, π_{rows}: partition; $\pi h_{columns}$, πh_{rows}: partition;
01] **begin**
02] init $\pi_{columns}$ and π_{rows} using π;
03] init $R_{columns}$ and R_{rows} using πh;
04] split(**inp_out:** π_{rows}, **inp:** $R_{columns}$, **out:** R_{rows}, **inp:** copy);
05] copy:='some';
06] split(**inp_out:** $\pi_{columns}$, **inp:** R_{rows}, **out:** $R_{columns}$, **inp:** copy);
07] **while not** $((|R_{columns}| = 0)$ **and** $(|R_{rows}| = 0))$ **do**
08] **begin**
10] split(**inp_out:** π_{rows}, **inp:** $R_{columns}$, **out:** R_{rows}, **inp:** copy);
11] split(**inp_out:** $\pi_{columns}$, **inp:** R_{rows}, **out:** $R_{columns}$, **inp:** copy);
12] **end**;
13] $\pi_{columns}$ and π_{rows} form π;
14] **end**;

Split partitions π_{rows} with respect to $R_{columns}$ and copy the result in R_{rows}, which will be used in the next step to partition the columns.

split(**inp_out** π:partition of rows (or columns);
inp R^{now}:partition of columns (or rows);
out R^{next}:partition of rows (or columns);
inp copy:string);
begin
R^{next}:=(); { empty }
for every cell V in π **do**
begin
partition V in $V_1, \ldots, V_g$ such that $a \in V_i$ and $b \in V_j$ for $i < j \Leftrightarrow$
$d(a, R_r^{now}) = d(b, R_r^{now})$ for $r = 1, \ldots, l-1$, and $d(a, R_l^{now}) < d(b, R_l^{now})$ for some l
replace V in π with $V_1, \ldots, V_g$ in that order;
if copy = 'every' **then**
add all $V_1, \ldots, V_g$ in R^{next} in that order **else** {copy='some' }
add all $V_1, \ldots, V_g$ without V_t (V_t is the first largest cell) in R^{next} in that order;
end;
end;

The algorithm *stable* has four parameters. The first one is the binary matrix which we consider. The second one is the input partition whose cells the algorithm will split depending on the distances to the cells of the ordered partition πh. The final result (output of the algorithm) is also written in π. The parameter *some* takes two values: 'some' and 'every'.

The algorithm *stable* skips the mentioned above additional calculations when the parameter 'copy' has the value 'some'.

The partition $\pi_1 = (V_1^{column}, \ldots, V_{s_1}^{column}, \ldots, V_{r_1}^{column}; V_1^{row}, \ldots, V_{r_1'}^{row})$, which we have obtained as a result of the algorithm $stable(A, \pi, \pi h = \pi, copy =' every')$, can be considered as induced by the invariant f_1. We can find the special cell $V_{s_1}^{column}$ as it is said in the definition.

This algorithm can be used to obtain the partition, induced by f_2, in the following way: Let fix a_1 in π_1, i.e. $\pi_2 = (V_1^{column}, \ldots, [a_1], V_{s_1}^{column} \setminus \{a_1\}, \ldots, V_{r_1}^{column}; V_1^{row}, \ldots, V_{r_1'}^{row})$ and $\pi h = ([a_1])$. Then we run the algorithm *stable* with parameters $(A, \pi := \pi_2, \pi h, copy =' some')$. The process continues until the step when we obtain a discrete partition.

The suggested algorithm is proper to be used in rows 4 and 12 in the algorithm *disc_part* in the following form:

inv_act(**inp_out** π:partition; **inp** πh:partition; **out** *sp_cell*:cell; **inp** copy:string);
begin
 stable(**inp:**A, **inp_out:**π, **inp:**πh, **inp:**copy);
 find a special cell *sp_cell*;
end;

In the first step of the algorithm *disc_part*, row 4, the parameters of *inv_act* have to be $\pi h = \pi$ and $copy = every$, and in the other steps of *disc_part*, in row 12, the parameter πh is a partition with one cell and it has only one column - this is the last fixed column. In all these steps $copy = some$.

Let A be a matrix without repeated rows. It is easy to see that any discrete partition for the columns leads to a discrete partition of the rows. If the matrix A contains repeated rows, a discrete partition of the columns leads to a partition of the rows with discrete cells or cells with repeated rows. Without lost of generality, we can split a cell with repeated rows into discrete cells. Hence, as an output of the algorithm *disc_part* we obtain a discrete partition of the columns and of the rows. Thus, we have the following lemma:

Lemma 3.5. *Any discrete partition of the columns obtained by disc_part defines in a unique way a discrete partition of the rows.*

Remark 3.1. The ordered discrete partitions obtained with the algorithm *disc_part*, which uses in rows 4 and 12 the algorithm *inv_act*, allows us to compare binary matrices instead of sorted binary matrices.

Remark 3.2. This type of invariant is related to 'equitable partition' or 'stable partition' of graphs. Algorithms for one-stable partition can be found in [8] and [6]. A good survey and additional results for one-stable, two-stable and k-stable partitions can be found in [2].

4. Main algorithm

The strategy of the algorithm is similar to the McKay's algorithm [8]. Let Υ be the set of all vectors of fixed columns which can be obtained in row 10 of the algorithm *disc_part*. We can define a tree with these vectors. The root of the tree is the empty set. In the first level, the nodes are different columns from the special cell in the partition induced by f_1. We fix these columns. The fixed column a_1 determines the columns from the special cell induced by f_2. These columns form nodes in the second level, which are successors of a_1, and so on. The leaves of the tree correspond to the discrete partitions from Π.

Our algorithm visits all nonequal discrete partitions in Π with backtrack search (step by step, try out all the possibilities) to the tree. It also finds the maximal (canonical) discrete partition among them. When the algorithm has discovered automorphisms it collects and uses them to prune the search tree. All these automorphisms generate the automorphism group of the matrix.

We call a discrete partition $first$ in Π if the corresponding vector of fixed columns v_fdisc is lexicographically smallest (the left leaf in the search tree). The first discrete partition is very important for the algorithm. We compare any new obtained discrete partition with the first one and with the maximal found so far. The number of columns which are in the same orbit with the columns in the vector of fixed columns v_fdisc is counted. In this way the algorithm calculates the order of automorphism group (using that $|Aut(A)| = |O(a)||Aut(A_a)|$).

The main variables, used by the algorithm, are:

- $fdisc$: partition – the first discrete partition with vector of fixed columns v_fdisc.
- $orbits$: partition – The orbits of G, $G \leq A_{v\_fdisc_1,\ldots,v\_fdisc_{h-1}}$. If the algorithm has discovered the automorphisms $\gamma_1, \ldots, \gamma_l$, this partition consists of cells which correspond to the orbits of the group $G \leq Aut(A)$ with generators $\gamma_1, \ldots, \gamma_l$. In the beginning $G = \{id\}$. Then in some steps G coincides with $A_{v\_fdisc_1,\ldots,v\_fdisc_h}$ for $h = |v\_fdisc|, |v\_fdisc| - 1, \ldots, 1$. In the end of the algorithm $G = Aut(A)$.
- k: integer – the current depth of the backtrack search.

- h: integer – shows the smallest depth reached by the backtrack search. The algorithm looks for the columns which are in the same orbit with $v\_fdisc[h]$ with respect to $Aut(A_{v\_fdisc_1,\ldots,v\_fdisc_{h-1}})$ in the special cell $w[h]$. In the beginning $h = |v\_fdisc| - 1$. After visiting all columns in the nondiscrete special cell obtained in the process of generating of the first discrete partition, h takes values $h-1$ and so on.
- sp_cell: cell – the special cell obtained after the action of the corresponding invariant.
- w: array of cells – If $k > h$, $w[k]$ consists of the first columns (with smallest index) from the orbits of $Aut(A_{v_1,v_2,\ldots,v_{k-1}})$ which are in the special cell. If $k \leq h$, then $w[k]$ consists of all columns from the special cell.
- *tree*: array of integers – $tree[k]$ shows the number of columns in $w[k]$ which are visited so far.
- π: array of partitions – $\pi[k]$ is the current partition with vector of fixed columns v.
- *cdisc*: partition – keeps the maximal discrete partition to the current point of the execution (candidate for canonical) with vector of fixed points v_cdisc.
- *ind*: integer – the number of columns in the orbit of $Aut(A_{v\_fdisc_1,\ldots,v\_fdisc_{h-1}})$, which contains $v\_fdisc[h]$.
- *size*: integer – The order of the group $Aut(A_{v\_fdisc_1,\ldots,v\_fdisc_h})$. In the end $size = |Aut(A)|$.
- list_of_aut contains discovered generators of the automorphism group $Aut(A)$.
- πh: partition with one cell with one element $v[k-1]$.

```
00]canon(inp A:bin_mat; inp: π0:partition; out: cdisc:partition;);
00]var  orbits, fdisc, πh: partition;
00]         π: array of partitions;
00]         w: array of cells;
00]         sp_cell: cell;
00]         k, size, ind, h: integer;
00]         tree, v, v_fdisc, v_cdisc: array of integer;
00]         γ: automorphism;
00]         list_of_aut;
00]begin
01]  gen_f(inp: π0; out: π, k, w, v);
02]  fdisc := π[k];
03]  v_fdisc := v;
04]  v_cdisc := v;
05]  cdisc := π[k];
06]  k := k − 1; orbits = ([1], [2], . . . , [n]);
```

```
07] h := k;
08] size := 1;
09] ind := 1;
10] for i := 1 to k do tree[i] := 1;
11] while k <> 0 do
12] begin
13]   if |w[k]| − tree[k] > 0 then
14]     begin
15]       tree[k] := tree[k] + 1;
16]       find_next_v[k]_in_w[k](inp:w[k]; out:v[k]);
17]       if k = h then if(v_fdisc[h] and v[k] are in the same orbit) then ind := ind + 1;
18]       if (k > h) or (k = h and (v[k] is first element of an orbit)) then
19]       begin {if depth}
20]        k := k + 1; tree[k] := 0; v[k] := 0; w[k] := []; {empty}
21]        int_to_part(inp: v[k − 1]; out: πh); {set πh }
22]        π[k] := π[k − 1]; fix(inp: v[k − 1]; inp_out: π[k]);
23]        inv_act(inp_out: π[k]; inp: πh,k; out: sp_cell);
24]        if ifdiscrete(π[k]) then
25]          begin
26]                if π[k] = cdisc then
27]                begin
28]                   π[k] and cdisc define an automorphism γ;
29]                   if gama_ext_orbits(inp: γ; inp_out: orbits) then
30]                        begin
31]                           add γ into list_of_aut;
32]                           if (v[h] is not first element in an orbit) then k := h;
33]                           if (v_fdisc[h] and v[h] in the same orbit) then ind := ind + 1;
34]                           gcd is the position of first difference between v and v_cdisc
35]                        end;
36]                        if (k <> h) then  k := gcd;
37]                end;
38]                if π[k] > cdisc then begin cdisc := π[k]; v_cdisc := v; end;
39]                if π[k] = fdisc then
40]                begin
41]                   π[k] and fdisc define an automorphism γ;
42]                   if gama_ext_orbits(inp: γ; inp_out: orbits) then
43]                         begin
44]                               add γ into list_of_aut;
45]                               ind := ind + 1;
46]                         end;
47]                     k := h;
48]                end;
49]          end {end discrete}
50]            else if k > h then restrict(inp: list_of_aut,sp_cell; out: w[k]);
51]       end {if depth}
52]   end {if}
53]   else
54]    begin k := k − 1;
55]       if h > k then begin h := k; size := size · ind; ind := 1; end;
56]    end;
57] end; {end while}
```

58]**end**;

In row 1 the algorithm finds the first discrete partition. The procedure *disc_part* can be used as *gen_f* after changing row 08 (*choose* a_i *from* V_{s_i}) with 08| *choose* a_i – *the column with smallest index in* V_{s_i}.

The discrete partition, obtained in row 1, is the first discrete and maximal discrete partition *cdisc* in this step with a vector of fixed columns $v\_fdisc = v\_cdisc$ (rows 2-5).

While the current level is not 0 (row 11) the backtrack search continues. If the number of columns in $w[k]$ is bigger than the number of the visited columns in $w[k]$ ($tree[k]$), we continue with the next element in $w[k]$ (rows 15, 16, with the procedure find_next_v[k]_in_w[k]).

If $k > h$, or $k = h$ and $v[k]$ is the first element of an orbit, the algorithm continues in depth. In the case when $k = h$ but $v[k]$ is not the first element of an orbit, the next partition is defined by the vector of fixed columns $(v[1], \ldots, v[k-1], v[k])$. But the algorithm already has passed all discrete partitions which are defined by the vector of fixed columns $(v[1], \ldots, v[k-1], v[k]_f, \ldots)$, where $v[k]_f$ is the first in the same orbit. By theorem 3.1, we can skip the current $v[k]$.

Using the last fixed column as the only column in the partition πh, the previous known partition $\pi[k-1]$ after fixing the same column (row 22, procedure fix), and *inv_act*, the algorithm obtains the next partition and the next special cell. If the obtained partition is discrete, the algorithm compares it with the current maximal *cdisc* (row 28) and $fdisc$ (row 39).

In the first comparing we have two cases. If the algorithm has discovered an automorphism, and this automorphism gives new (extended) orbits, then it is collected in *list_of_aut*. We check whether the element $v[h]$ is first in any of the new orbits. If not, the backtracking jumps to the level h, because the first element is already passed ($v\_fdisc[1] = v[1], \ldots, v\_fdisc[h-1] = v[h-1]$). If yes, it jumps to the level of the first difference between v and *v_cdisc*. If the current discrete partition $\pi[k]$ is bigger than *cdisc*, the algorithm takes $\pi[k]$ as maximal (or canonical). In the second comparing, if the algorithm discovers an automorphism, the backtracking jumps to the level h. The discovered automorphisms, which fix the columns $v[1], \ldots, v[k-1]$, form a group G. If the obtained partition is not discrete, the algorithm puts all columns from the special cell *sp_cell* (row 23), which are first elements in orbits with respect to G, in $w[k]$ (with procedure *restrict*(**inp:** list_of_aut,sp_cell; **out:** $w[k]$)).

4.1. *Additional invariants*

There are two general strategies to improve the efficiency of the main algorithm. The first one is to cut the part of the search tree which corresponds to the set of vectors Υ. The next example is based on the fact that the sets of vectors of positions Ψ' and Ψ'' corresponding to vectors of fixed columns Υ' and Υ'' for the matrices A and $\gamma(A)$, $\gamma \in S_n$, are equal: $\Psi' = \Psi''$.

We can redefine the canonical partition to be $c \asymp \max\{\pi_j, \pi_j \in \Pi'\}$, where Π' is the set of discrete partitions which have lexicographically largest vector of positions of fixed columns.

In the main algorithm, we have to compare the vector of positions vp, corresponding to v, with the vector vpf of the positions of the first discrete partition $fdisc$ and the vector of positions vpc of the discrete partition which is a candidate for maximal *cdisc*. If the current vp coincides with vpf or vpc, the backtrack search continues - the algorithm expects an automorphism. If $vp[k] <> vpf[k]$ and $vp[k] < vpc[k]$, the backtracking jumps in the previous level. Another similar approach can be found in [8].

The other strategy is to use proper invariants, which will help us to decrease the number of the discrete partitions in Π. This happens when the number of the orbits in the special cells are smaller than before. In fact, if every spacial cell consists of only one orbit, the algorithm visits only $j+1$ discrete partitions to obtain j generators of the automorphism group. The number of possible generators is bounded by $n-1$ (n is the number of columns). To do this, we can use stronger invariants. Unfortunately, such invariants usually are computationally expensive. There are two options:

If we consider structures with a small group, we use an additional invariant in lower level. We call this level *pointed.* If we expect structures with a large group, we use an additional invariant in levels which depend on given parameters - for example the size of the largest cell in the current partition (we use as pointed levels the levels in which this size is smaller than a given constant). To use additional invariants, we redefine *inv_act*:

```
inv_act(inp_out π: partition; inp πh: partition; k: integer; out sp_cell:cell);
begin
  if k is in a pointed level then
  begin
    partition the special cell sp_cell in πh using additional invariants
    stable(inp A: bin_mat; inp_out π: partition; inp πh: partition; inp copy: string );
  end else
    stable(inp A: bin_mat; inp_out π: partition; inp πh; partition; inp copy: string );
 find a special cell sp_cell;
end;
```

The pointed levels are input for the main algorithm and depend on the user. There are no special cell in the beginning. To use an additional invariant in the first level, we consider the set of all columns of the matrix as a special cell. Of course, we have to use the redefined *inv_act* in the procedure *disc_part*.

The difference here is that we use as an input partition for comparing πh in *stable* not only the fixed column but the partition obtained after splitting the special cell using the additional invariants. Now we describe the type of the additional invariants which we use. Let's consider the following 8×8 matrix

$$\begin{pmatrix} 10001110 \\ 01101001 \\ 01001110 \\ 01010101 \\ 10010101 \\ 10110010 \\ 01110010 \\ 10101001 \end{pmatrix}$$

This matrix has the same number of ones in any row and column, so we can expect that all columns are in the same orbit. But this is not true. If we consider the sum of the first, third and forth columns, we obtain $inv = (1, 1, 0, 1, 2, 3, 2, 2)$. To this vector, we correspond the polynomial

$$Yx^3 + Y^2x^3 + Y^3x,$$

such that Y^ax^b shows that inv has b elements equal to a. Then we calculate the sums of the first column with all pairs of two other columns. So we obtain

$$inv_1(Y, x) = Y(12x^3 + 9x^4) + Y^2(12x^3 + 9x^4) + Y^3(12x)$$

which is the sum of the corresponding polynomials. With $\binom{n}{3} \times m$ operations, we can have similar polynomials for all columns. In this way we obtain

$$inv_1 = inv_2 = Y(12x^3 + 9x^4) + Y^2(12x^3 + 9x^4) + Y^3(12x)$$

and

$$inv_3 = inv_4 = \cdots = inv_8 = Y(8x^3 + 9x^4 + 2x^6) + Y^2(8x^3 + 9x^4 + 2x^6) + Y^3(8x + 12x^2)$$

These polynomials split the set of all columns (with respect to the lexicographic ordering of the corresponding vectors) in two cells and define

$\pi = ([3,4,5,6,7,8],[1,2])$. This means that we have at least two orbits. We call this type of invariants additional 'sum' invariants with complexity 3 in level 1. For the graph invariants you can see [9].

Remark 4.1. Additional invariants are necessary only in cases when the matrix A has a very specific structure. For example, when A is an incidence matrix of a combinatorial design. This algorithm can be used also in the case when we have coloring of the columns.Then the initial partition will depend on the coloring.

5. Efficiency and storage requirements

About the efficiency of the algorithm *stable* for graphs, which is an important part of the main algorithm, we refer to [8]. The efficiency of the main algorithm depends on the size and the structure of the automorphism group and the cardinality of the set of discrete partitions Π. The author does not know a reasonable theoretical bound for this cardinality.

As we mentioned, this implementation needs $m \times n$ units of memory (for the matrix A), which is less than $(m+n) \times (m+n)$ units - the memory used for the corresponding graph. This fact helps us to use easily variables which need a lot of memory. These variables are: 1) partitions of the rows and columns $\pi_1, \pi_2, \ldots, \pi_k$. Of course, $k \leq n$, but if we consider matrix without repeated columns k will be much smaller; 2) the set of special cells w - only for columns; 3) the obtained automorphisms. Actually, we keep the orbits of the columns with respect to the cyclic group generated by the corresponding automorphism. This can be realized with two arrays with length n (see [6]). 4) the first discrete partition $fdisc$ and the current maximal partition *cdisc*. For any of them we need two arrays with length $n+m$.

References

[1] L. Babai, Automorphism groups, isomorphism, reconstruction, Handbook of Combinatorics (R. L. Graham, M. Grötschel, and L. Lovász, Eds.), Vol. II, North-Holland, Amsterdam, pp. 1447–1540, (1995).

[2] O. Bastert, Stabilization Procedures and Applications, Zentrum Mathematik, Technische Universität München, (2000).

[3] I. Bouyukliev and J. Simonis, Some new results for optimal ternary linear codes, *IEEE Trans. Inform. Theory*, vol. 48, No. 4, pp. 981-985, (2002).

[4] M. Goldberg, The graph isomorphism problem, Handbook of Graph Theory (J. L. Gross and J. Yellen, Eds.), CRC Press, pp. 68-78, (2004).

[5] P. Kaski and P. R. Ostergard, Classification Algorithms for Codes and Designs, Springer, (2006).

[6] W. Kocay, On writing isomorphism programs, Computational and Constructive Design Theory (ed. W. D. Wallis), Kluwer, pp. 135-175, (1996).

[7] J. Leon, Computing automorphism groups of error-correcting codes, *IEEE Trans. Inform. Theory*, vol. 28, pp. 496-511, (1982).

[8] B. McKay, Practical graph isomorphism, *Congressus Numerantium*, 30, pp. 45–87, (1981).

[9] B. McKay, *nauty* user's guide (version 1.5). Technical Report TR-CS-90-02, Computer Science Department, Australian National University, (1990).

[10] E. Petrank and R. Roth; Is code equivalence easy to decide? *IEEE Trans. Inform. Theory*, vol. 43, pp. 1602–1604, (1997).

[11] N. Sendrier, The Support Splitting Algorithm, *IEEE Trans, Info. Theory*, vol. 46, pp. 1193-1203, (2000).

Permutation decoding for binary self-dual codes from the graph Q_n where n is even.

J. D. Key

Department of Mathematical Sciences
Clemson University
Clemson SC 29634, U.S.A.
E-mail: keyj@clemson.edu
http://www.ces.clemson.edu/~keyj

P. Seneviratne

Department of Mathematical Sciences
Clemson University
Clemson SC 29634, U.S.A.

The binary self-dual $[2^n, 2^{n-1}, n]_2$ codes from the adjacency matrices of the n-cubes Q_n, where $n \geq 6$ and is even, are examined and 2- and 3-PD-sets of size $n2^n$ are found.

Keywords: graphs, codes, permutation decoding

1. Introduction

For $n \geq 2$, the graph with vertices the 2^n vectors of $\mathbb{F}_2^n$ and two vertices adjacent if their coordinates differ in precisely one place, is called the n-cube, denoted by Q_n. We examine the binary code obtained from the row span of an adjacency matrix for Q_n over the field $\mathbb{F}_2$, and show that when n is even it is self-dual and can be used for permutation decoding. Our main result obtaining 3-PD-sets is as follows:

Theorem 1.1. *For n even and $n \geq 8$, let*

$$T_n = \{T(w)t_i \mid w \in \mathbb{F}_2^n, 1 \leq i \leq n\},$$

where $T(w)$ is the translation by $w \in \mathbb{F}_2^n$, $t_i = (i, n)$ for $i < n$ is a transposition in the symmetric group S_n, and t_n is the identity map. Then T_n is a 3-PD-set of size $n2^n$ for the self-dual $[2^n, 2^{n-1}, n]_2$ code C_n from an *adjacency matrix for the n-cube Q_n, with the information set*

$$\mathcal{I} = [\mathbf{0}, \mathbf{1}, \ldots, \mathbf{2^{n-1}} - \mathbf{3}, \mathbf{2^n} - \mathbf{2}, \mathbf{2^n} - \mathbf{1}].$$

This is proved in Section 4, with the notation for T_n and $\mathcal{I}$ given in Section 3. Background definitions and notions are in Section 2 and general properties of the graph Q_n, the symmetric design obtained from it, and its binary codes, are in Section 3.

2. Background and terminology

The notation for designs and codes is as in [1]. An incidence structure $\mathcal{D} = (\mathcal{P}, \mathcal{B}, \mathcal{J})$, with point set $\mathcal{P}$, block set $\mathcal{B}$ and incidence $\mathcal{J}$ is a t-(v, k, λ) design, if $|\mathcal{P}| = v$, every block $B \in \mathcal{B}$ is incident with precisely k points, and every t distinct points are together incident with precisely λ blocks. The design is **symmetric** if it has the same number of points and blocks. The **code C_F of the design $\mathcal{D}$** over the finite field F is the space spanned by the incidence vectors of the blocks over F. If $\mathcal{Q}$ is any subset of $\mathcal{P}$, then we will denote the incidence vector of $\mathcal{Q}$ by $v^{\mathcal{Q}}$. If $\mathcal{Q} = \{P\}$ where $P \in \mathcal{P}$, then we will write v^P instead of the more cumbersome $v^{\{P\}}$. Thus $C_F = \langle v^B \mid B \in \mathcal{B} \rangle$, and is a subspace of $F^{\mathcal{P}}$, the full vector space of functions from $\mathcal{P}$ to F.

All the codes here are **linear codes**, and the notation $[n, k, d]_q$ will be used for a q-ary code C of length n, dimension k, and minimum weight d, where the weight $\mathrm{wt}(v)$ of a vector v is the number of non-zero coordinate entries. The distance $\mathrm{d}(u, v)$ between two vectors u, v is the number of places in which they differ, i.e. $\mathrm{wt}(u - v)$. A **generator matrix** for C is a $k \times n$ matrix made up of a basis for C, and the **dual** code $C^\perp$ is the orthogonal under the standard inner product $(,)$, i.e. $C^\perp = \{v \in F^n | (v, c) = 0 \text{ for all } c \in C\}$. A **check matrix** for C is a generator matrix for $C^\perp$. The all-one vector will be denoted by $\jmath$, and is the vector with all entries equal to 1. Two linear codes of the same length and over the same field are **isomorphic** if they can be obtained from one another by permuting the coordinate positions. An **automorphism** of a code C is an isomorphism from C to C. The automorphism group will be denoted by $\mathrm{Aut}(C)$. Any code is isomorphic to a code with generator matrix in so-called **standard form**, i.e. the form $[I_k \mid A]$; a check matrix then is given by $[-A^T \mid I_{n-k}]$. The first k coordinates are the **information symbols** and the last $n - k$ coordinates are the **check symbols**.

The **graphs**, $\Gamma = (V, E)$ with vertex set V and edge set E, discussed here are undirected with no loops. A graph is **regular** if all the vertices have the same valency. The **adjacency matrix** A of a graph of order n is an $n \times n$ matrix with entries a_{ij} such that $a_{ij} = 1$ if vertices v_i and v_j are adjacent, and $a_{ij} = 0$ otherwise.

Permutation decoding was first developed by MacWilliams [7] and involves finding a set of automorphisms of a code called a PD-set. The method is described fully in MacWilliams and Sloane [8, Chapter 16, p. 513] and Huffman [4, Section 8]. In [5] and [6] the definition of PD-sets was extended to that of s-PD-sets for s-error-correction:

Definition 2.1. If C is a t-error-correcting code with information set $\mathcal{I}$ and check set $\mathcal{C}$, then a **PD-set** for C is a set $\mathcal{S}$ of automorphisms of C which is such that every t-set of coordinate positions is moved by at least one member of $\mathcal{S}$ into the check positions $\mathcal{C}$.

For $s \leq t$ an s-**PD-set** is a set $\mathcal{S}$ of automorphisms of C which is such that every s-set of coordinate positions is moved by at least one member of $\mathcal{S}$ into $\mathcal{C}$.

That a PD-set will fully use the error-correction potential of the code follows easily and is proved in Huffman [4, Theorem 8.1]. That an s-PD-set will correct s errors follows in the same way (see [5, Result 2.3]).

The algorithm for permutation decoding is as follows: we have a t-error-correcting $[n, k, d]_q$ code C with check matrix H in standard form. Thus the generator matrix $G = [I_k|A]$ and $H = [-A^T|I_{n-k}]$, for some A, and the first k coordinate positions correspond to the information symbols. Any vector v of length k is encoded as vG. Suppose x is sent and y is received and at most s errors occur, where $s \leq t$. Let $\mathcal{S} = \{g_1, \ldots, g_m\}$ be an s-PD-set. Compute the syndromes $H(yg_i)^T$ for $i = 1, \ldots, m$ until an i is found such that the weight of this vector is s or less. Compute the codeword c that has the same information symbols as yg_i and decode y as cg_i^{-1}.

3. Binary codes of cubic graphs

For $n \geq 2$ let Q_n denote the n-cube (see [9]) and $\mathcal{D}_n$ the symmetric 1-design obtained by defining the 2^n vertices (i.e. vectors in $\mathbb{F}_2^n$) to be the points $\mathcal{P}$, and a block $\bar{v}$ for every point (vector) v by

$$\bar{v} = \{w \mid w \in \mathcal{P} \text{ and } w \text{ adjacent to } v \text{ in } Q_n\}.$$

Then $\mathcal{D}_n$ is a 1-$(2^n, n, n)$ symmetric design with the property that two distinct blocks meet in zero or two points and similarly any two distinct points are together on zero or two blocks.

We will use the following notation: for $r \in \mathbb{Z}$ and $0 \leq r \leq 2^n - 1$, if $r = \sum_{i=1}^{n} r_i 2^{i-1}$ is the binary representation of r, let $\mathbf{r} = (r_1, \ldots, r_n)$ be the corresponding vector in $\mathbb{F}_2^n$, i.e. point in $\mathcal{P}$.

The complement of $v \in \mathcal{P}$ will be denoted by v_c. Thus $v_c(i) = 1 + v(i)$ for $1 \leq i \leq n$, where $v(i)$ denotes the ith coordinate entry of v. Similarly, for $\alpha \in \mathbb{F}_2$, $\alpha_c = \alpha + 1$. Clearly $v_c = v + \mathbf{2^n - 1}$.

The binary code C_n of the design $\mathcal{D}_n$ is the same as the row span over $\mathbb{F}_2$ of an adjacency matrix for Q_n, and for n even and $n \geq 4$, it is a $[2^n, 2^{n-1}, n]_2$ self-dual code. Before showing this, we show why the case for n odd is not of interest.

Proposition 3.1. *For n odd, the binary code C_n of $\mathcal{D}_n$ is the full space $\mathbb{F}^{2^n}$.*

Proof: For n odd, it can be verified directly that

$$v^{(x_1,\ldots,x_n)} = \overline{v^{(x_1,\ldots,(x_n)_c)}} + \sum_{i=1}^{n-1} \overline{v^{(x_1,\ldots,(x_i)_c,\ldots,x_{n-1},x_n)}}$$

for all choices of $x = (x_1, \ldots, x_n)$. Thus C_n contains all the vectors of weight 1 and is the full space. ■

The automorphism group of the design and of the code contains (properly, for $n \geq 4$) the automorphism group $TS_n = T \rtimes S_n$ of the graph (see [9]), where T is the translation group of order 2^n and S_n is the symmetric group acting on the n coordinate positions of the points $v \in \mathcal{P}$. We will write, for each $w \in \mathcal{P}$, $T(w)$ for the automorphism of C_n defined by the translation on $\mathbb{F}_2^n$ given by $T(w) : v \mapsto v + w$ for each $v \in \mathbb{F}_2^n$. The identity map will be denoted by $\iota = T(0)$. Then $T = \{T(w) \mid w \in \mathcal{P}\}$.

Lemma 3.1. *The group TS_n acts imprimitively on the points of the design $\mathcal{D}_n$ for $n \geq 4$ with $\{v, v_c\}$, for each $v \in \mathbb{F}_2^n$, a block of imprimitivity.*

Proof: We need only show that for $g \in TS_n$, and any $v \in \mathbb{F}_2^n$, $v_c g = (vg)_c$, which will make the set $\{v, v_c\}$ a block of imprimitivity. Clearly TS_n is transitive on points. For $g \in S_n$ the assertion is clear. If g is the translation $T(u)$, where $T(u) : v \mapsto v+u$, then $v_c g = v_c T(u) = v + \mathbf{2^n - 1} + u = vT(u) + \mathbf{2^n - 1} = (vg)_c$. Thus for any $g \in TS_n$ and any $v \in \mathbb{F}_2^n$, $v_c g = (vg)_c$. ■

For each i such that $1 \leq i < n$ let $t_i = (i, n) \in S_n$, i.e. the automorphism of C_n defined by the transposition of the coordinate positions. For $n \geq 4$

let

$$P_n = \{t_i \mid 1 \leq i \leq n-1\} \cup \{\iota\} \tag{1}$$

$$T_n = TP_n. \tag{2}$$

Since the translation group T is normalized by S_n, elements of the form $T(w)t_iT(u)$ are all in T_n, i.e. $\sigma^{-1}T(u)\sigma = T(u\sigma^{-1})$, so that for transpositions t, $tT(u) = T(ut)t$.

Proposition 3.2. *For n even, $n \geq 4$, C_n is a $[2^n, 2^{n-1}, n]_2$ self-dual code with*

$$\mathcal{I} = [\mathbf{0}, \mathbf{1}, \ldots, \mathbf{2^{n-1}} - \mathbf{3}, \mathbf{2^n} - \mathbf{2}, \mathbf{2^n} - \mathbf{1}]$$

as an information set.

Proof: Using the natural ordering for the points and blocks, the incidence matrix for Q_n has the form

$$B_n = \begin{pmatrix} B_{n-2} & I_{2^{n-2}} & I_{2^{n-2}} & 0 \\ I_{2^{n-2}} & B_{n-2} & 0 & I_{2^{n-2}} \\ I_{2^{n-2}} & 0 & B_{n-2} & I_{2^{n-2}} \\ 0 & I_{2^{n-2}} & I_{2^{n-2}} & B_{n-2} \end{pmatrix} \tag{3}$$

where B_{n-2} is the incidence matrix of the graph Q_{n-2}. It is easy to prove that the matrix has rank 2^{n-1} and it can be shown by induction that the minimum weight is n. That the code is self-dual follows from the earlier observation that blocks meet in 0, 2 or n points.

To show that $\mathcal{I}$ is an information set, let B_n^* be the first 2^{n-1} rows of B_n. Clearly B_n^* has rank 2^{n-1} and generates the same code as B_n. We want to switch the column indexed by $\mathbf{2^{n-1}} - \mathbf{2}$ with that indexed by $\mathbf{2^n} - \mathbf{2}$, and the column indexed by $\mathbf{2^{n-1}} - \mathbf{1}$ with that indexed by $\mathbf{2^n} - \mathbf{1}$. Notice that $\mathbf{2^{n-1}} - \mathbf{2} \in \overline{\mathbf{2^{n-1}} - \mathbf{1}}$, so the 2×2 submatrix of B_n^* from the $(2^{n-1}-2)^{th}$ and $(2^{n-1}-1)^{th}$ rows and columns has the form $\begin{bmatrix} 0 & 1 \\ 1 & 0 \end{bmatrix}$, while the corresponding 2×2 submatrix from the same rows but the last two columns is just I_2. Thus the column interchanges described will give the information set $\mathcal{I}$. ■

If $\mathcal{I}$ is as in the proposition, the corresponding check set is $\mathcal{C}$. We will write

$$\mathcal{I}_1 = [\mathbf{0}, \mathbf{1}, \ldots, \mathbf{2^{n-1}} - \mathbf{3}] \tag{4}$$

$$\mathcal{C}_1 = [\mathbf{2^{n-1}}, \mathbf{2^{n-1}} + \mathbf{1}, \ldots, \mathbf{2^n} - \mathbf{3}] \tag{5}$$

$$\mathcal{I}_2 = [\mathbf{2^n} - \mathbf{2}, \mathbf{2^n} - \mathbf{1}] \tag{6}$$

$$\mathcal{C}_2 = [\mathbf{2^{n-1}} - \mathbf{2}, \mathbf{2^{n-1}} - \mathbf{1}] \tag{7}$$

and

$$a = \mathbf{2^n} - \mathbf{2} = (0, 1, \ldots, 1, 1) \ , \ b = \mathbf{2^n} - \mathbf{1} = (1, 1, \ldots, 1, 1) \quad (8)$$

$$A = \mathbf{2^{n-1}} - \mathbf{2} = (0, 1, \ldots, 1, 0) \ , \ B = \mathbf{2^{n-1}} - \mathbf{1} = (1, 1, \ldots, 1, 0) \quad (9)$$

Notice that the points a and b are placed in $\mathcal{I}$ in order to have points and their complements in $\mathcal{I}$ since under any automorphism $g \in TS_n$ of the design, if $vg = w$ then $v_c g = w_c$, by Lemma 3.1. Thus we have $a_c = \mathbf{1}$ and $b_c = \mathbf{0}$, $A_c = \mathbf{1} + \mathbf{2^{n-1}}$, $B_c = \mathbf{2^{n-1}}$, and $v + v_c = b$ for any vector $v \in \mathcal{P}$.

4. 3-PD-sets

In this section we prove the main result, Theorem 1.1, obtaining 3-PD-sets. Since the minimum weight is n, the code cannot correct three errors if $n < 8$. However the proof of the theorem holds for $n = 4, 6$ as well.

Proof of Theorem 1.1: Let $\mathcal{T} = \{x, y, z\}$ be a set of three points in $\mathcal{P}$. We need to show that there is an element in T_n that maps $\mathcal{T}$ into $\mathcal{C}$. We consider the various possibilities for the points in $\mathcal{T}$. If $\mathcal{T} \subseteq \mathcal{C}$ then use ι. Thus suppose at least one of the points is in $\mathcal{I}$ and, by using a translation, suppose that one of the points, say z, is $\mathbf{0}$. If $\mathcal{T} \subseteq \mathcal{I}$, then $T(\mathbf{2^{n-1}})$ will work. Now we consider the other cases.

(1) $x \in \mathcal{I}_1$, $y \in \mathcal{C}_1$

Then there are i_x, i_y such that $2 \leq i_x, i_y \leq n-1$ such that $x(i_x) = y(i_y) = 0$. If $i_x = i_y = i$, then $\mathcal{T}t_i \subseteq \mathcal{I}$, unless $yt_i \in \{A, B\}$, so $t_i T(\mathbf{2^{n-1}})$ will work unless $yt_i \in \{A, B\}$. If $yt_i = A$, then $y(1) = y(i) = 0$, $y(j) = 1$ otherwise. If $x(1) = 0$, then $t_1 T(\mathbf{2^{n-1}})$ will work. If $x(1) = 1$, then take any $j \neq 1, i, n$, and use $T(\mathbf{2^{j-1}}) t_i T(\mathbf{2^{n-1}})$. If $yt_i = B$, then $y(i) = 0$ and $y(j) = 1$ otherwise. Here we can take any $j \neq 1, i, n$, and use $T(\mathbf{2^{j-1}}) t_i T(\mathbf{2^{n-1}})$.

If x and y have no common zero, then if $y = x_c$, so that $x + y = b$, we can use $T(x)T(\mathbf{2^{n-1}})$. If $x(i) = y(i) = 1$, where $1 \leq i \leq n-1$, then $t_i T(\mathbf{2^{n-1}} - \mathbf{1})$ can be used.

(2) $x \in \mathcal{I}_1$, $y \in \mathcal{C}_2$

Since $x \in \mathcal{I}_1$, $x(i) = 0$ for some i such that $2 \leq i \leq n-1$. If there is a j such that $j \neq i$ and $2 \leq j \leq n-1$ with $x(j) = 0$, then $T(\mathbf{2^{i-1}} + \mathbf{2^{n-1}})$ can be used.

If there is no such j, then either $x(1) = x(i) = x(n) = 0$ and $x(j) = 1$ for $j \notin \{1, i, n\}$, or $x(i) = x(n) = 0$ and $x(j) = 1$ for $j \notin \{i, n\}$. In either case, take $j \neq i$, $2 \leq j \leq n-1$. Then the map $T(\mathbf{2^{j-1}} + \mathbf{2^{n-1}})$ can be used.

(3) $x \in \mathcal{I}_2$, $y \in \mathcal{C}_1$

(a) $x = a$: since $y \in \mathcal{C}_1$, there is a j such that $2 \leq j \leq n-1$ with $y(j) = 0$. If $y(i) = 1$ for $i \neq j$ and $1 \leq i \leq n$, or if $y(1) = 0$ and $y(i) = 1$ for $i \neq j$ and $2 \leq i \leq n$, then $T(A)$ will work. If there is an $i \neq j$ such that $y(i) = y(j) = 0$ where $2 \leq i, j \leq n-1$, then $t_j T(\mathbf{2^{n-1}})$ can be used.

(b) $x = b$: this follows exactly as in the $x = a$ case except that in the first two cases for y use $T(B)$ instead of $T(A)$.

(4) $x \in \mathcal{I}_2$, $y \in \mathcal{C}_2$

(a) $x = a$, $y = A$: use $T(a)t_2 T(\mathbf{2^{n-1}})$.
(b) $x = a$, $y = B$: use $t_{n-1}T(B)$.
(c) $x = b$, $y = A$: use $t_{n-1}T(B)$.
(d) $x = b$, $y = B$: use $t_1 T(\mathbf{1} + \mathbf{2^{n-1}})$.

(5) $x, y \in \mathcal{C}$

(a) $x, y \in \mathcal{C}_1$: if $x + y = B$ then $T(B)$ will work. Otherwise $x(i) = y(i)$ for some i such that $1 \leq i \leq n-1$. Again $T(B)$ will work unless x or y are $(0, \ldots, 0, 1)$ or $(1, 0, \ldots, 0, 1)$. If $x = (0, \ldots, 0, 1)$ then $y(i) = 0$ for some i such that $2 \leq i \leq n-1$. Then $t_i T(\mathbf{2^{n-1}})$ can be used unless $y(j) = 1$ for all $j \neq i$, or $y(1) = y(i) = 0$ and $y(j) = 1$ for $j \neq 1, i$; in these cases $t_i T(\mathbf{2^{i-1}} + \mathbf{2^{n-1}})$ can be used. The same arguments hold if $x = (1, 0, \ldots, 0, 1)$.

(b) $x \in \mathcal{C}_1$, $y \in \mathcal{C}_2$: since $x \in \mathcal{C}_1$, there is a j such that $2 \leq j \leq n-1$ with $x(j) = 0$. Then $t_j T(\mathbf{2^{j-1}} + \mathbf{2^{n-1}})$ can be used.

(c) $x, y \in \mathcal{C}_2$: $T(\mathbf{2^{n-2}} + \mathbf{2^{n-1}})$ will work.

This completes all the cases and proves the theorem. ■

Note that this result also shows that the set T_n is a 2-PD-set for C_n for $n = 6$. However, this set T_n with this information set $\mathcal{I}$ will not give a 4-PD-set, since it is quite easy to verify that the set of four points $\{\mathbf{0}, \mathbf{2}, \mathbf{2^n} - \mathbf{2}, \mathbf{2^{n-1}} - \mathbf{1}\}$ cannot be moved by any element of T_n into the check positions.

5. Discussion

The automorphism group of the symmetric 1-design is much larger than that of the graph. In particular, it will contain any invertible $n \times n$ matrix over $\mathbb{F}_2$ with the property that the sum of any two of its rows has weight 2. In fact, if $v \in \mathcal{P}$ has an even number of entries equal to 1, then the matrix A having for rows the points in $\bar{v}$, will be be an automorphism of $\mathcal{D}_n$ that also preserves the blocks of imprimitivity. If v has an odd number of entries equal to 1, it will not be invertible. There are also other, non-linear, automorphisms, of the design, and that also preserve these blocks of imprimitivity, as is indicated by computations with Magma [2, 3].

It is possible to arrange more interchanges so that more instances of a point and its complement in the information set occur. Thus s-PD-sets for $s > 3$ seem possible in general.

References

[1] E. F. Assmus, Jr and J. D. Key. *Designs and their Codes.* Cambridge: Cambridge University Press, 1992. Cambridge Tracts in Mathematics, Vol. 103 (Second printing with corrections, 1993).

[2] W. Bosma, J. Cannon, and C. Playoust. The Magma algebra system I: The user language. *J. Symb. Comp.*, 24, 3/4:235–265, 1997.

[3] J. Cannon, A. Steel, and G. White. Linear codes over finite fields. In J. Cannon and W. Bosma, editors, *Handbook of Magma Functions*, pages 3951–4023. Computational Algebra Group, Department of Mathematics, University of Sydney, 2006. V2.13, http://magma.maths.usyd.edu.au/magma.

[4] W. Cary Huffman. Codes and groups. In V. S. Pless and W. C. Huffman, editors, *Handbook of Coding Theory*, pages 1345–1440. Amsterdam: Elsevier, 1998. Volume 2, Part 2, Chapter 17.

[5] J. D. Key, T. P. McDonough, and V. C. Mavron. Partial permutation decoding of codes from finite planes. *European J. Combin.*, 26:665–682, 2005.

[6] Hans-Joachim Kroll and Rita Vincenti. PD-sets related to the codes of some classical varieties. *Discrete Math.*, 301:89–105, 2005.

[7] F. J. MacWilliams. Permutation decoding of systematic codes. *Bell System Tech. J.*, 43:485–505, 1964.

[8] F. J. MacWilliams and N. J. A. Sloane. *The Theory of Error-Correcting Codes.* Amsterdam: North-Holland, 1983.

[9] Gordon Royle. Colouring the cube. Preprint.

The Sum-Product Algorithm on Small Graphs

M. E. O'Sullivan

Dept. of Mathematics and Statistics
San Diego State University
San Diego, CA, 92182-7720
E-mail: mosulliv@math.sdsu.edu

J. Brevik

Dept. of Mathematics and Statistics
California State University, Long Beach
Long Beach, CA, 90840
E-mail: jbrevik@csulb.edu

R. Wolski

Dept. of Computer Science University of California
Santa Barbara, CA 75275-0338
E-mail: rich@cs.ucsb.edu

Keywords: Sum-product algorithm, low-density parity-check codes, finite length bipartite graphs.

1. Introduction

One of the great achievements in coding theory in the last decade or so has been the discovery that iterative decoding methods, such as the sum-product algorithm, can be used to achieve Shannon capacity; see [7, 11]. Although there are provable asymptotic results for the performance of the sum-product algorithm, there is little that can be said for finite length codes. In this article we focus on very simple cases for which we can derive exact formulas for convergence of the sum-product algorithm. By establishing some simple, but provable, results we hope to build a foundation for further algebraic analysis. These examples may also enhance the intuitive understanding of the algorithm and thereby yield improved heuristic methods for code construction.

Given a binary matrix H, the sum-product algorithm is defined by using the bipartite graph of H. It is to be expected that the sum-product algorithm will yield better decoding performance on some bipartite graphs than on others. What makes one graph (or, equivalently, matrix) better than another? Several properties have been proposed, some of them based on other decoding algorithms. *Short cycles* in the bipartite graph of the parity-check matrix are considered problematic; see [16]. The reasoning is that inaccurate received estimates of bit values passed to the decoding algorithm are self-reinforcing in the presence of short cycles. Recent work suggests that short cycles are particularly problematic when the degrees of the nodes involved are low; see [14]. An erasure correction algorithm that is similar to belief propagation fails exactly when it arrives at a *stopping set*; see [1]. These sets also seem to foil the belief-propagation algorithm. The experiments by MacKay and Postol in [8] with the Margulis group-theoretic construction led them to attribute decoding failure in the error floor region to *near-codewords*. These are vectors v such that Hv has low weight. Richardson in [10] calls near-codewords *trapping sets*; he also sees them as a cause of error floors. *Pseudo-codewords* arise from codewords in a code for a covering graph of the bipartite graph of the check matrix; see [6]. The closure of the set of pseudo-codewords is a polytope in $\mathbb{R}^n$ where n is the dimension of the code. The articles [4, 15] investigate their relevance for sum-product decoding. Pseudo-codewords are directly relevant in another approach to decoding due to Feldman [2, 3] that uses linear programming. This algorithm attempts to maximize a linear functional over the polytope of pseudo-codewords, and the vertices of the polytope are the possible solutions to the problem. Bipartite graphs with good *expansion* properties were shown to be asymptotically good for a low-complexity decoding algorithm presented in [12].

In the simple cases that we examine, pseudo-codewords, near-codewords and stopping sets do not play a role and expansion is not meaningful because the graphs are very small. We do see a difference between graphs that are very similar, but differ in one aspect, the existence of short cycles. The difference in performance of the sum-product algorithm yields some surprises.

Section 2 introduces the bipartite graphs under investigation and some experiments with the performance of the sum-product algorithm. In Section 3 we give our algebraic analysis of the sum-product algorithm for a restricted set of bipartite graphs, those in which all check nodes have degree 2. Section 4 applies our algebraic results to the bipartite graphs of Section 2 and explains the differences in performance therein.

2. Experimental Results

Figures 1 and 3 show several bipartite graphs. Following common practice, the circular nodes (shaded) are called bit nodes and the square nodes are called check nodes. A code is defined by allowing bit nodes to take values in $\mathbb{Z}/2$, such that each check node is connected to an even number of 1s. It is readily seen that all the codes defined by these graphs are *repetition codes*, that is, codes whose only two codewords are the vector of repeated 0s and that of repeated 1s.

Consider the two graphs in Figure 1, each of which determines the repetition code of length four. The em 4-Choose-2 graph is constructed by creating one check for each two element subset of the four bit nodes. As we will show in the next section, the *Two-to-One* graph is a two-to-one cover of the complete bipartite graph on 2 bit-nodes and 3 check-nodes (in fact, it is the unique connected cover).

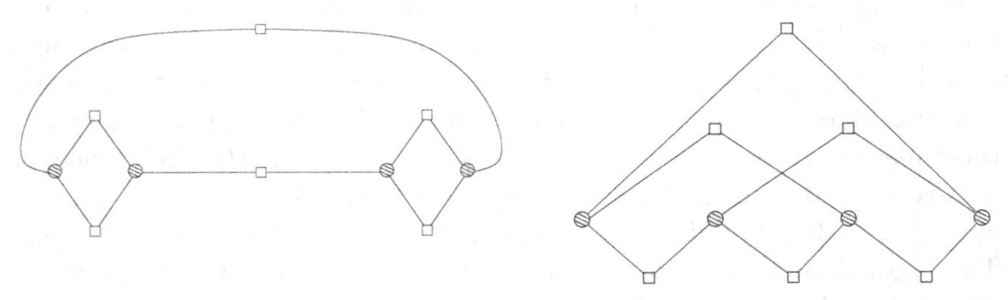

Fig. 1. Two graphs defining the repetition code of length four. *Two-to-One* on the left, and *4-Choose-2* on the right.

Figure 2 shows the performance of the sum-product algorithm on each graph, for each of two different termination criteria. The "fine" case used a threshold of 10^{-20}, while the "coarse" case used a threshold of 10^{-3}. It is evident that the 4-Choose-2 graph has superior performance, and that it is less affected by the degradation of performance under a coarser threshold. It would be tempting to attribute the superior performance to the larger girth of the 4-Choose-2 graph.

Consider the three graphs in Figure 3, which are all 3-to-1 covers of the complete bipartite graph on 2 bit-nodes and 3 check-nodes. One can show that any connected 3-to-1 cover is one of these three. One of the graphs has girth 8, one has three 4-cycles, and one has two 4-cycles.

Figure 4 shows the performance of the sum-product algorithm on these graphs, for the same two termination criteria as used above. Perhaps sur-

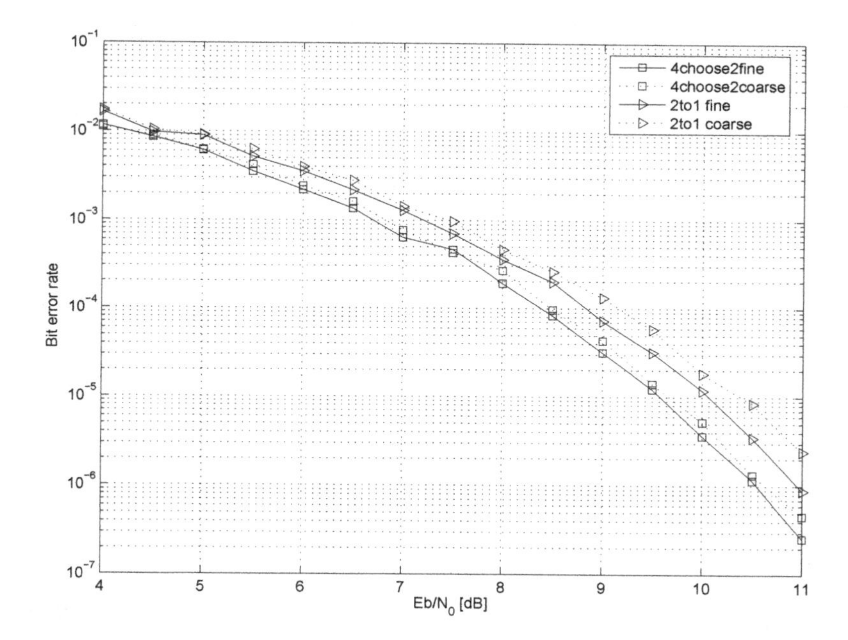

Fig. 2. The performance of the sum-product algorithm on the graphs in Figure 1, using two different termination criteria.

prisingly, the performance of the sum-product algorithm is the same when the threshold is fine. On the other hand, with a coarse threshold we see greater degradation of performance corresponding to a greater number of 4-cycles.

In the following sections we will derive formulas for convergence which

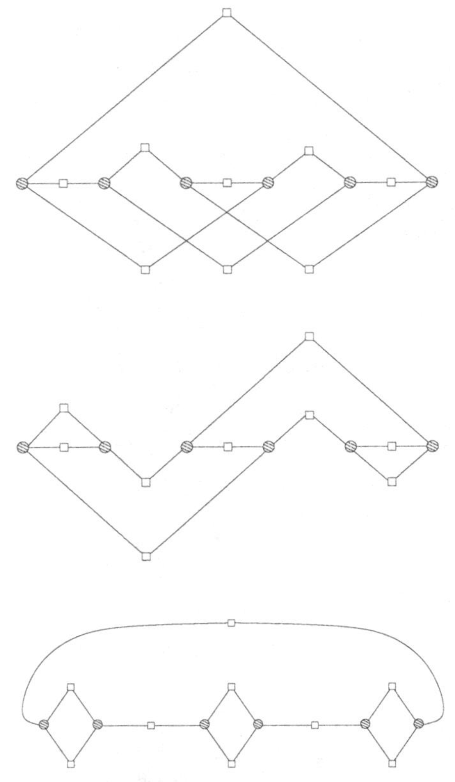

Fig. 3. Three graphs defining the repetition code of length 6. From the top *No 4-cycles*, *Two 4-cycles*, and *Three 4-cycles*.

will explain the performance in these examples.

3. Analysis of the Sum-Product Algorithm

The principal goal of this section is to develop our algebraic analysis of the sum-product algorithm for graphs on which all check nodes have degree 2. One may readily check that the code defined by such a graph is a repetition code. We start with a discussion of maps of bipartite graphs, including covering maps and automorphisms. We then present the version of the sum-product algorithm that we use and show how it is affected by an automorphism. Finally we show that the algorithm simplifies dramatically when all check nodes have degree 2.

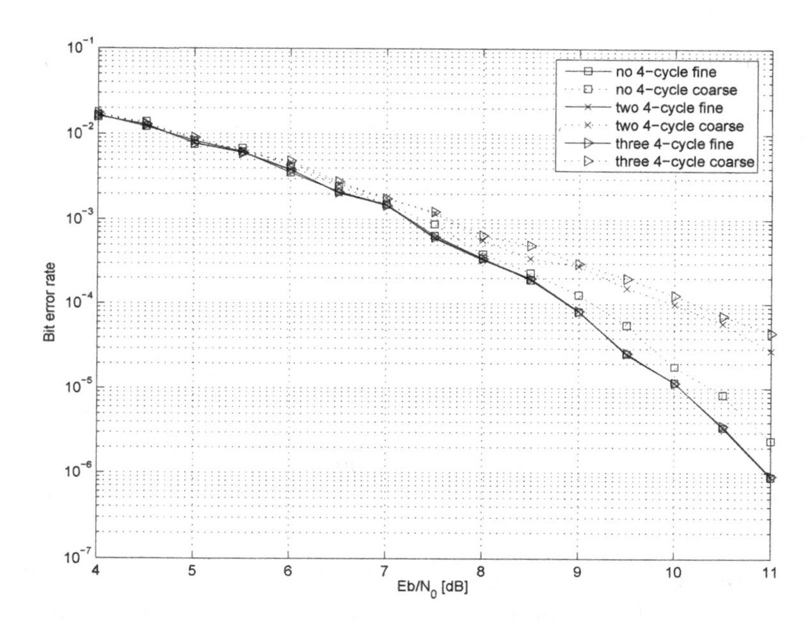

Fig. 4. The performance of the sum-product algorithm on the graphs in Figure 3, using two different termination criteria.

Bipartite Graphs

Definition 3.1. A *bipartite graph* consists of an edge set E and two sets of nodes L and R with two structural maps $\lambda : E \to L$ and $\rho : E \to R$ giving the ends of each edge E. A *codeword* is an association of 0 or 1 to each $\ell \in L$ such that each $r \in R$ is connected to an even number of nonzero bits. The elements of L are typically called *bit nodes* and the elements of R

check nodes.

A binary matrix H yields a bipartite graph by taking R to be the set of rows of H, L the set of columns of H and E enumerating the nonzero entries of H, so that for e the edge associated to the nonzero entry $H_{r\ell}$, $\lambda(e) = \ell$ and $\rho(e) = r$.

Definition 3.2. A *map* of bipartite graphs $\sigma : (\overline{E}, \overline{L}, \overline{R}, \overline{\lambda}, \overline{\rho}) \longrightarrow (E, L, R, \lambda, \rho)$ is a triple of functions $\sigma_E : \overline{E} \to E$, $\sigma_L : \overline{L} \to L$, and $\sigma_R : \overline{R} \to R$ such that $\lambda(\sigma_E(e)) = \sigma_L(\overline{\lambda}(e))$ and similarly $\rho(\sigma_E(e)) = \sigma_R(\overline{\rho}(e))$.

We say σ is a *covering map* if for each $e \in E$, $\ell \in L$, and $r \in \mathbb{R}$,

$$\begin{aligned} |\sigma_E^{-1}(e)| &= |\sigma_L^{-1}(\ell)| \\ &= |\sigma_R^{-1}(r)| \end{aligned}$$

and for each $\overline{\ell} \in \overline{L}$ and $\overline{r} \in \overline{R}$ we have

$$\begin{aligned} |\overline{\lambda}^{-1}(\overline{\ell})| &= |\lambda^{-1}(\sigma_L(\overline{\ell}))| \qquad \text{and} \\ |\overline{\rho}^{-1}(\overline{r})| &= |\rho^{-1}(\sigma_R(\overline{r}))|. \end{aligned}$$

If $n = |\sigma_E^{-1}(e)|$, we say the map is an n-fold cover.

An *automorphism* of a bipartite graph is a map σ from a bipartite graph to itself such that σ_E, σ_L and σ_R is a bijection.

Example 3.1. Let $L = \{0, 1\}$, $R = \{A, B, C\}$ and let $E = L \times R$. The projections of E onto each factor define a bipartite graph which we will call *2-bits-3-checks.* Figure 5 shows the graph. The automorphism group is $S_3 \times S_2$ where S_n is the symmetric group on n objects. The action of S_3 permutes the check nodes while fixing the bit nodes, whereas the action of S_2 is reflection through the central axis of the diagram.

The Two-to-One graph in Figure 1 maps to 2-bits-3-checks by taking the leftmost bits of each diamond to 0, the rightmost bits to 1, the top checks of each diamond to A, the bottom checks of each diamond to C and the other two checks to B. The map is a two-to-one cover. The reader may verify that the 4-Choose-2 graph in Figure 5 does not map to 2-bits-3-checks.

Each of the graphs in Figure 3 also maps to 2-bits-3-checks yielding 3-to-1 covers.

The Sum-Product Algorithm

The following algorithm is the sum-product algorithm, expressed using the notation for a bipartite graph introduced above. We also use positive real

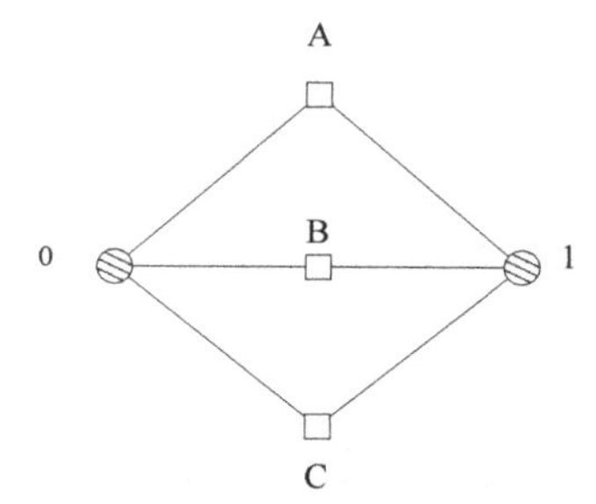

Fig. 5. The bipartite graph *2-bits-3-checks*.

numbers to represent the probability distributions in the algorithm. The input data for bit ℓ is the "odds" that the actual intended or transmitted value for that bit was 1, expressed as the likelihood ratio $u_\ell = p_\ell(1)/p_\ell(0)$. Likewise, the messages along the edges of the graph produced by the algorithm are expressed as the odds of 1. The algorithm uses the transform from the "odds of 1" domain to the difference domain in which a probability distribution p is represented as $p(0) - p(1)$, which is in the interval $[-1, +1]$. The function $s : \mathbb{R} \cup \{\infty\} \longrightarrow \mathbb{R} \cup \{\infty\}$ defined by $s(x) = \frac{1-x}{1+x}$ transforms from one domain to the other. Notice that $s(s(x)) = x$. In the literature, the ratio $p(0)/p(1)$ is sometimes used rather than $p(1)/p(0)$. We prefer the latter, since with this notation the same function s is used to translate in each direction.

Algorithm 3.1 (Sum-Product Algorithm).

INPUT: *For each* $\ell \in L$, $u_\ell \in (0, \infty)$. *Termination criteria* $\epsilon > 0$.
DATA STRUCTURES: *For each* $e \in E$, $x_e, y_e \in (0, \infty)$.
INITIALIZATION: *Set* $y_e \leftarrow 1$ *for all* $e \in E$.

ALGORITHM:

BIT-TO-CHECK STEP: *For each* $e \in E$, *set*

$$x_e \leftarrow u_{\lambda(e)} \prod_{\substack{f:\lambda(f)=\lambda(e) \\ f \neq e}} y_f$$

CHECK-TO-BIT STEP: *For each* $e \in E$, *set*

$$y_e \leftarrow s\left(\prod_{\substack{f:\rho(f)=\rho(e) \\ f \neq e}} s(x_f) \right)$$

NEW ESTIMATE STEP: *Set*

$$\hat{u}_\ell \leftarrow u_\ell \prod_{e \in \lambda^{-1}(l)} y_e$$

TERMINATION AND OUTPUT: *If either $\hat{u}_\ell < \epsilon$ or $\hat{u}_\ell > 1/\epsilon$ for all $\ell \in L$ then output the hard decision based on $\hat{u}_\ell$: Vector $w \in \mathbb{F}^L$ such that*

$$w_\ell = \begin{cases} 1 & \text{if } \hat{u}_\ell > 1 \\ 0 & \text{else} \end{cases}$$

When analyzing the algorithm it will sometimes prove useful to indicate the iteration using a superscript. Thus for example, we will sometimes write

$$x_e^{(t+1)} \leftarrow u_{\lambda(e)} \prod_{\substack{f:\lambda(f)=\lambda(e) \\ f \neq e}} y_f^{(t)}$$

Let σ be an automorphism of the bipartite graph. Then the update step of the algorithm says that

$$x_{\sigma_E(e)} \leftarrow u_{\lambda(\sigma_E(e))} \prod_{\substack{f:\lambda(f)=\lambda(\sigma_E(e)) \\ f \neq \sigma_E(e)}} y_f$$

Since $\{f \in E : \lambda(f) = \lambda(\sigma_E(e))\} = \{\sigma_E(f) : f \in E, \lambda(f) = \lambda(e)\}$, and since $\lambda\sigma_E = \sigma_L\lambda$ the update may be rewritten

$$x_{\sigma_E(e)} \leftarrow u_{\sigma_L(\lambda(e))} \prod_{\substack{f:\lambda(f)=\lambda(e) \\ f \neq e}} y_{\sigma_E(f)} \tag{1}$$

Proposition 3.1. *Let σ be an automorphism of the graph and suppose that $u_{\sigma_L(\ell)} = u_\ell$ for all $\ell \in L$. Then $x_e = x_{\sigma(e)}$ and $y_e = y_{\sigma(e)}$ at each iteration of the algorithm.*

Proof. At initialization $y_e^{(0)} = y_{\sigma_E(e)}^{(0)}$ for all $e \in E$ since $y_e(0) = 1$. We proceed by induction assuming that the statement holds at iteration t.

$$x_{\sigma_E(e)}^{(t+1)} = u_{\sigma_L(\lambda(e))} \prod_{\substack{f:\lambda(f)=\lambda(e) \\ f \neq e}} y_{\sigma_E(f)}^{(t)} \tag{2}$$

$$= u_{\lambda(e)} \prod_{\substack{f:\lambda(f)=\lambda(e) \\ f \neq e}} y_f^{(t)} \tag{3}$$

$$= x_e^{(t+1)} \tag{4}$$

The analogous argument is used to show that $y_{\sigma(e)}^{(t+1)} = y_e^{(t+1)}$. □

Reduction of the SPA to a local sum algorithm when check nodes have degree 2

We now restrict attention to a fixed bipartite graph in which each check node has degree 2. We will also assume that the graph is connected. One may readily check that the code defined by such a graph is a repetition code. The sum-product algorithm simplifies dramatically because at the check to bit step there is only one term in the product.

Proposition 3.2. *If all right nodes have degree* 2*, then then all edge messages are monomials in the* u_ℓ.

Proof. Clearly, at initialization $y_e^{(0)} = 1$ is a monomial as claimed. If all $y_e^{(t)}$ are monomial then all $x_e^{(t+1)}$ are as well, since the bit-to-check step just involves multiplication. Each right node has degree 2, so the product in the check-to-bit step has only one term. Since s is an involution, $y_e^{(t+1)} = x_{e'}^{(t+1)}$ where e' is the unique edge distinct from e sharing the same right node. Thus we may establish the proposition by induction. $\square$

Notation 3.1. For an edge e let e' be the unique edge, distinct from e, with $\rho(e) = \rho(e')$. Let us use $\boldsymbol{a}_e \in \mathbb{N}^L$ to denote the (vector of) exponents appearing in x_e so $x_e = \prod_{\ell \in L} u_\ell^{\boldsymbol{a}_{e,\ell}}$. We will abbreviate this product as as $\boldsymbol{u}^{\boldsymbol{a}_e}$. When we want to specify the tth iteration we will write $\boldsymbol{a}_e^{(t)}$.

Let $\boldsymbol{0} \in \mathbb{N}^L$ be the zero vector and let $\boldsymbol{\delta}_\ell \in \mathbb{N}^L$ be the vector which is 1 in the ℓth component and 0 otherwise.

We may reduce the sum-product algorithm to an algorithm that computes the exponents of the input data u_ℓ for each edge e. Note that no notation is needed for y_e since it is equal to $x_{e'}$.

Algorithm 3.2 (Local Sum Algorithm).

DATA STRUCTURES: *For each* $e \in E$, $\boldsymbol{a}_e, \in \mathbb{N}^L$.
INITIALIZATION: *Set* $\boldsymbol{a}_e \leftarrow \boldsymbol{0}$ *for all* $e \in E$.

ALGORITHM: *Set*

$$\boldsymbol{a}_e \leftarrow \boldsymbol{\delta}_{\lambda(e)} + \sum_{\substack{f:\lambda(f)=\lambda(e) \\ f \neq e}} \boldsymbol{a}_{f'} \tag{5}$$

Since $\boldsymbol{a}$ is a vector of integers doubly indexed by $e \in E$ and $\ell \in L$, we can consider $\boldsymbol{a}$ as an element of the vector space $\mathbb{C}^{|E||L|}$ of dimension $|E||L|$ over the complex numbers $\mathbb{C}$.

Our update function is a linear inhomogeneous map on this space, the inhomogenous part coming from the term $\boldsymbol{\delta}_{\lambda(e)}$. The homogeneous part of the map is represented by the matrix $\boldsymbol{M}$, defined as follows.

$$\boldsymbol{M}_{(e,\ell),(f,m)} = \begin{cases} 1 & \text{if } \lambda(e) = \lambda(f'), f' \neq e, \text{ and } \ell = m \\ 0 & \text{else} \end{cases} \tag{6}$$

We can write the local sum algorithm in a homogeneous way by using a "dummy" variable to supply the necessary $\boldsymbol{\delta}_{\lambda(e)}$ terms. To this end, define $\boldsymbol{T} \in \mathbb{C}^{|E||L|}$ such that

$$\boldsymbol{T}_{e,l} = \begin{cases} 1 & \text{if } \lambda(e) = l \\ 0 & \text{else} \end{cases} \tag{7}$$

Let

$$\overline{\boldsymbol{M}} = \begin{bmatrix} \boldsymbol{M} & \boldsymbol{T} \\ \boldsymbol{0}^T & 1 \end{bmatrix} \tag{8}$$

Then

$$\begin{bmatrix} \boldsymbol{a}^{(t)} \\ 1 \end{bmatrix} = \overline{\boldsymbol{M}} \begin{bmatrix} \boldsymbol{a}^{(t-1)} \\ 1 \end{bmatrix} \tag{9}$$

We have thus reduced analysis of the sum-product algorithm in our restricted case to the problem of understanding the dynamics of the matrix $\boldsymbol{M}$.

Proposition 3.3. *Suppose that all check nodes have degree 2. Let $x_e = \boldsymbol{u}^{\boldsymbol{a}_e}$ and let $\boldsymbol{a}$ be the concatenation of the exponent vectors $\boldsymbol{a}_e$. Then at iteration t,*

$$\begin{bmatrix} \boldsymbol{a}^{(t)} \\ 1 \end{bmatrix} = \overline{\boldsymbol{M}}^t \begin{bmatrix} \boldsymbol{0} \\ 1 \end{bmatrix} \tag{10}$$

We can use the automorphism group of the bipartite graph to reduce the dimensionality of the problem. Consider an automorphism σ of the graph. From (1) we have

$$\begin{aligned} x_{\sigma_E(e)} &\leftarrow u_{\sigma_L(\lambda(e))} \prod_{\substack{f:\lambda(f)=\lambda(e) \\ f\neq e}} y_{\sigma_E(f)} \\ &= u_{\sigma_L(\lambda(e))} \prod_{\substack{f:\lambda(f)=\lambda(e) \\ f\neq e}} x_{\sigma_E(f')} \end{aligned}$$

where $\rho(f') = \rho(f)$, so

$$\boldsymbol{a}_{\sigma_E(e)} \leftarrow \boldsymbol{\delta}_{\sigma_L(\lambda(e))} + \sum_{\substack{f:\lambda(f)=\lambda(e)\\ f\neq e}} \boldsymbol{a}_{\sigma_E(f')} \tag{11}$$

Proposition 3.4. *Let σ be an automorphism of a bipartite graph in which all checks have degree 2. At any iteration of the local sum algorithm,*

$$\boldsymbol{a}_{\sigma_E(e),\sigma_L(\ell)} = \boldsymbol{a}_{e,l}$$

Proof. At initialization, the claim is immediate. We proceed by induction. From the algorithm,

$$\boldsymbol{a}^{(t+1)}_{e,\ell} = \sum_{\substack{f:\lambda(f)=\lambda(e)\\ f\neq e}} \boldsymbol{a}^{(t)}_{f',\ell} + \begin{cases} 1 & \text{if } \lambda(e) = \ell \\ 0 & \text{else} \end{cases}$$

From the action of σ in (11) we have

$$\boldsymbol{a}^{(t+1)}_{\sigma_E(e),\sigma_L(\ell)} = \left(\boldsymbol{\delta}_{\sigma_L(\lambda(e))}\right)_{\sigma_L(\ell)} + \sum_{\substack{f:\lambda(f)=\lambda(e)\\ f\neq e}} \boldsymbol{a}^{(t)}_{\sigma_E(f'),\sigma_L(\ell)}$$

Using the induction hypothesis and observing that $\sigma_L(\lambda(e)) = \sigma_L(\ell)$ if and only if $\lambda(e) = \ell$ we have

$$\begin{aligned} \boldsymbol{a}^{(t+1)}_{\sigma_E(e),\sigma_L(\ell)} &= \sum_{\substack{f:\lambda(f)=\lambda(e)\\ f\neq e}} \boldsymbol{a}^{(t)}_{f',\ell} + \begin{cases} 1 & \text{if } \lambda(e) = \ell \\ 0 & \text{else} \end{cases} \\ &= \boldsymbol{a}^{(t+1)}_{e,\ell} \end{aligned}$$

□

As a consequence of this proposition we may compute just the exponents $\boldsymbol{a}_e$ for one edge from each orbit under the automorphism group of the bipartite graph. The exponent for any edge f may be obtained by applying an appropriate automorphism to the representative from the orbit of f. Instead of using the update matrix (6) we may simplify to a matrix $\boldsymbol{N}$ with one representative edge for each orbit. The entries in $\boldsymbol{N}$ must be derived from (5) replacing $\boldsymbol{a}_{f'}$ with some $\boldsymbol{a}_{\sigma(g)}$ for g the representative for the orbit of f' and σ an automorphism taking g to f'.

4. Examples

In this section we solve the following question for a number of bipartite graphs in which all check nodes have degree 2: Under what conditions on the input values u_ℓ will all values x_e in the sum-product algorithm converge to 0 (or to ∞)? In these examples, we ignore termination criteria and examine the convergence behavior of the infinite sequence $x_e^{(t)}, t = 1, 2, ...$

Our method is to identify a set of representatives for the edge set E under the action of the automorphism of the graph. Let $\boldsymbol{a}$ be the vector of exponents indexed by these representatives and by $\ell \in L$. We derive the update matrix $\boldsymbol{N}$ using one representative from each equivalence class. We then have a dynamical system: for

$$\overline{\boldsymbol{N}} = \begin{bmatrix} \boldsymbol{N} & \boldsymbol{T} \\ \boldsymbol{0}^T & 1 \end{bmatrix} \tag{12}$$

we have

$$\begin{bmatrix} \boldsymbol{a}^{(t)} \\ 1 \end{bmatrix} = \overline{\boldsymbol{N}}^t \begin{bmatrix} \boldsymbol{0} \\ 1 \end{bmatrix} \tag{13}$$

Except for one case, that of a simple cycle, we will write $\begin{bmatrix} \boldsymbol{0} \\ 1 \end{bmatrix} = \sum_{i=1}^t \boldsymbol{w}_i$ as a sum of eigenvectors of $\overline{\boldsymbol{N}}$. Let μ_i be the eigenvalue associated to $\boldsymbol{w}_i$. Then

$$\begin{bmatrix} \boldsymbol{a}^{(t)} \\ 1 \end{bmatrix} = \sum_{i=1}^{t} \mu_i^t \boldsymbol{w}_i \tag{14}$$

We can determine convergence from the fact that for large t the eigenvectors with largest eigenvalues will dominate.

A cycle of length 2m

Consider a cycle of length $2m$ with both bit nodes and check nodes enumerated from 0 to $m-1$. Let b_+ and b_i be the edges such that $\lambda(b_+) = \lambda(b_i) = b$ and $\rho(b_+) = b$, $\rho(b_-) = b-1 \mod m$. It is clear that the symmetry group is the dihedral group D_m and that it is transitive on edges. For the reflection around bit node 0 we have from Proposition 3.4, $\boldsymbol{a}_{0_+,\ell} = \boldsymbol{a}_{0_-,-\ell}$ (computing $-\ell$ modulo m). For rotation by 1 the proposition says

$$\begin{aligned} \boldsymbol{a}_{(k+1)_+,l+1} &= \boldsymbol{a}_{k_+,l} \\ \boldsymbol{a}_{(k+1)_-,l+1} &= \boldsymbol{a}_{k_-,l} \end{aligned}$$

Choosing 0_- as our representative edge, from the algorithm (5)

$$\begin{aligned}\boldsymbol{a}_{0_-} &= (1,0,\ldots,0) + \boldsymbol{a}_{1_-} \\ &= (1,0,\ldots,0) + (\boldsymbol{a}_{0_-,m-1}, \boldsymbol{a}_{0_-,0}, \ldots, \boldsymbol{a}_{0_-,m-2})\end{aligned}$$

so the update matrix N is circulant implementing a shift by 1. For $m = 5$ we have

$$\overline{\boldsymbol{N}} = \begin{bmatrix} 0&0&0&0&1&1 \\ 1&0&0&0&0&0 \\ 0&1&0&0&0&0 \\ 0&0&1&0&0&0 \\ 0&0&0&1&0&0 \\ 0&0&0&0&0&1 \end{bmatrix}$$

One can check that

$$\overline{\boldsymbol{N}}^5 = \begin{bmatrix} 1&0&0&0&0&1 \\ 0&1&0&0&0&1 \\ 0&0&1&0&0&1 \\ 0&0&0&1&0&1 \\ 0&0&0&0&1&1 \\ 0&0&0&0&0&1 \end{bmatrix}$$

More generally we have the following

Proposition 4.1. *Suppose the bipartite graph is a cycle of length $2m$. Let $\mathbf{1}$ be the vector of length m which is 1 in all components and let $\boldsymbol{\delta}_\ell$ be the vector of length m which is 1 in the ℓth component and 0 elsewhere. Write $t = ms + k$ with $k \in \{0, \ldots, m-1\}$. The local sum algorithm produces*

$$\boldsymbol{a}_{0_-}^{(ms+k)} = s\mathbf{1} + \sum_{\ell=0}^{k-1} \boldsymbol{\delta}_\ell$$

The sum-product algorithm converges to 0 when $\prod_\ell u_\ell < 1$, converges to ∞ when $\prod_\ell u_\ell > 1$ and oscillates with period m when $\prod_\ell u_\ell = 1$.

Proof. Let I_k be the circulant matrix corresponding to a cyclic shift by k. The update matrix is

$$\overline{\boldsymbol{N}} = \begin{bmatrix} I_1 & \boldsymbol{\delta}_0 \\ \mathbf{0}^T & 1 \end{bmatrix}$$

One can prove by induction that for $0 \le k < m$

$$\overline{N}^{ms+k} = \begin{bmatrix} I & s\mathbf{1} \\ \mathbf{0}^T & 1 \end{bmatrix} \begin{bmatrix} I_k & \sum_{\ell=0}^{k-1} \boldsymbol{\delta}_\ell \\ \mathbf{0}^T & 1 \end{bmatrix}$$

$$= \begin{bmatrix} I_k & s\mathbf{1} + \sum_{\ell=0}^{k-1} \boldsymbol{\delta}_\ell \\ \mathbf{0}^T & 1 \end{bmatrix}$$

Thus $\boldsymbol{a}_{0_-}^{ms+k} = s\mathbf{1} + \sum_{\ell=0}^{k-1} \boldsymbol{\delta}_\ell$. Since $x_{0_-} = \boldsymbol{u}^{\boldsymbol{a}_{0_-}} = (u_0 \cdots u_{m-1})^s \prod_{\ell=0}^{k-1} u_\ell$, the converge properties are easily verified. □

2 bits n checks

As noted earlier the automorphism group of this graph is $S_3 \times S_2$. There is only a single orbit for the edges under this group action, so we may reduce analysis of the sum-product algorithm to the consideration of a single edge e. Proposition 3.4 shows that all edges leaving the same bit have the same vector of exponents. Edges connected to different bit nodes differ by transposition of the entries. That is, for e an edge with $\lambda(e) = 0$ and f an edge with $\lambda(e) = 1$ we have $\boldsymbol{a}_f = (\boldsymbol{a}_{e,1}, \boldsymbol{a}_{e,0})$. The update of the algorithm is $\boldsymbol{a}_e^{(t+1)} = 2\boldsymbol{a}_f + (1,0) = (\boldsymbol{a}_{e,1}^{(t)}, \boldsymbol{a}_{e,0}^{(t)}) + (1,0)$. Using e as the representative edge, in the matrix equation (12) we have

$$\overline{N} = \begin{bmatrix} 0 & 2 & 1 \\ 2 & 0 & 0 \\ 0 & 0 & 1 \end{bmatrix}$$

The eigenvalues of $\overline{N}$ are $2, -2$, and 1, with respective eigenvectors $\boldsymbol{w}_1 = \begin{bmatrix} 1 \\ 1 \\ 0 \end{bmatrix}$, $\boldsymbol{w}_2 = \begin{bmatrix} 1 \\ -1 \\ 0 \end{bmatrix}$, and $\boldsymbol{w}_3 = \begin{bmatrix} -\frac{1}{3} \\ -\frac{2}{3} \\ 1 \end{bmatrix}$. The initial vector is

$$\boldsymbol{a}^{(0)} = \begin{bmatrix} 0 \\ 0 \\ 1 \end{bmatrix} = \frac{1}{2}\boldsymbol{w}_1 - \frac{1}{6}\boldsymbol{w}_2 + \boldsymbol{w}_3$$

We can calculate

$$\begin{aligned} \boldsymbol{a}^{(t)} &= \frac{1}{2}\cdot 2^t \boldsymbol{w}_1 - \frac{1}{6}(-2)^t \boldsymbol{w}_2 + \boldsymbol{w}_3 \\ &= \begin{bmatrix} 2^{t-1} - \frac{1}{6}(-2)^t - \frac{1}{3} \\ 2^{t-1} + \frac{1}{6}(-2)^t - \frac{2}{3} \\ 1 \end{bmatrix} = \begin{cases} \begin{bmatrix} \frac{2^t-1}{3} \\ \frac{2^{t+1}-2}{3} \\ 1 \end{bmatrix}, & t \text{ even} \\ \begin{bmatrix} \frac{2^{t+1}-1}{3} \\ \frac{2^t-2}{3} \\ 1 \end{bmatrix}, & t \text{ odd.} \end{cases} \end{aligned}$$

Thus, for t even, the message passed by bit 0 at the tth iteration is

$$u_0^{-\frac{1}{3}} u_1^{-\frac{2}{3}} \cdot (u_0 u_1^2)^{\frac{2^t}{3}},$$

which tends toward 0 if $u_0 u_1^2 < 1$ and toward ∞ if $u_0 u_1^2 > 1$. On the other hand, when t is odd, the message looks like

$$u_0^{-\frac{1}{3}} u_1^{-\frac{2}{3}} \cdot (u_0^2 u_1)^{\frac{2^t}{3}},$$

and this quantity tends toward 0 if $u_0^2 u_1 < 1$ and toward ∞ if $u_0^2 u_1 > 1$.

We see that these conditions are symmetric with respect to u_0 and u_1 and therefore the sum-product algorithm

- converges to the codeword $[0, 0]$ if $u_0^2 u_1$ and $u_0 u_1^2$ are both less than 1;
- converges to the codeword $[1, 1]$ if $u_0^2 u_1$ and $u_0 u_1^2$ are both greater than 1;
- fails to converge if $u_1^2 u_2 < 1 < u_1 u_2^2$ or $u_1^2 u_2 < 1 < u_1 u_2^2$.

These results generalize in straightforward way. We will say that two positive real numbers have the *same parity* when they are either both less than 1 or both greater than 1.

Proposition 4.2. *Suppose the bipartite graph has 2 bit nodes and m check nodes. Then the sum-product algorithm converges when $u_0 u_1^{m-1}$ and $u_0^{m-1} u_1$ have the same parity and it divirges otherwise.*

4-Choose-2

The symmetry group in this case is S_4 and is transitive on edges. Let us enumerate the bits from 0 to 3, going left to right, and identify the edges by an ordered pair of bits. Then edge $(0, 1)$ is the edge from 0 to the check for the pair $\{0, 1\}$. Let $\boldsymbol{a}$ be the vector of exponents for the edge $(0, 1)$. The two edges used to update $\boldsymbol{a}$ are the edges $(2,0)$ and $(3,0)$. Proposition 3.4 allows us to write the messages along these edges by making use of the permutations $(0,2,1)$ and $(0,3,1)$. Thus the update is

$$\boldsymbol{a}^{(t+1)} = (1,0,0,0) + (\boldsymbol{a}_1^{(t)}, \boldsymbol{a}_2^{(t)}, \boldsymbol{a}_0^{(t)}, \boldsymbol{a}_3^{(t)}) + (\boldsymbol{a}_1^{(t)}, \boldsymbol{a}_3^{(t)}, \boldsymbol{a}_2^{(t)}, \boldsymbol{a}_0^{(t)})$$

and the matrix

$$\boldsymbol{N} = \begin{bmatrix} 0 & 2 & 0 & 0 \\ 0 & 0 & 1 & 1 \\ 1 & 0 & 1 & 0 \\ 1 & 0 & 0 & 1 \end{bmatrix}$$

The eigenvalues of $\overline{\boldsymbol{N}}$ are 2, $-\frac{1 \pm \sqrt{7}i}{2}$ and 1. The eigenspace for 2 is spanned by $(1,1,1,1,0)$. Since the other eigenvalues have norm $\sqrt{2}$ or 1, the value at any edge is dominated by $u_0u_1u_2u_3$. Thus the sum product algorithm converges if and only if $u_0u_1u_2u_3 \neq 1$.

We skip the analysis of the 2-to1 cover, since it is similar to the 3-to-1 covers done below. The final result is that the sum-product algorithm converges if and only if $u_0u_2(u_1u_3)^2$ and $(u_0u_2)^2u_1u_3$ have the same parity. This explains the superior performance of the 4-Choose-2 graph.

3-*to*-1 ***covers of the complete*** **2-*bits*-3-*checks graph***

Each of the graphs in Figure 3 are 3-to-1 covers of the complete 2-bits-3-checks graph examined above. Let us enumerate the bits from 0 to 5, going from left to right. We briefly summarize the analysis for two of the graphs and do the third in some detail.

The No 4-Cycles graph given in Figure 3 has a very large automorphism group, generated by any automorphism switching the parity of all bit nodes and two copies of S_3, one acting on the odd numbered bits (and fixing the even numbered bits), and the other acting on the even numbered bits (and fixing the odd numbered bits). In each case there are associated permutations of the checks and edges. There is only one orbit of edges under the automorphism group. The analysis in this case is the simplest of the three; the update matrix $\overline{\boldsymbol{N}}$ has dimension only 7×7.

The Three 4-cycles graph of Figure 3 has automorphism group generated by a cyclic shift by 2 of the bit nodes, vertical reflections of each diamond, and reflection around the central axis, which switches parity of the bit nodes. The edges have two orbits under the automorphism group, namely those involved in 4-cycles and those not. Therefore, in this case we need only keep track of two 6-vectors, which we will take to emanate from bit 0, and the update matrix $\overline{\boldsymbol{N}}$ has dimension 13×13.

The Two 4-cycles graph given in Figure 3, has automorphisms generated by σ, the 180-degree rotation around the center; τ, which interchanges the first and second bits, the third and fourth bits, and the fifth and sixth bits (and performs the necessary permutations of the checks and edges), and π, which interchanges the two checks on the far left of the graph and their attendant edges.

The edges of this graph form 4 orbits under its automorphism group, which we can take to be the upper and lower edges from bit 0 and the middle and lower edges from bit 3. The update matrix $\overline{\boldsymbol{N}}$ is therefore 25×25, the first 24 dimensions coming from the exponents along the respective edges and, as in previous examples, the last one "driving" the dynamical system by adding 1 to the respective bits' exponents at each time step. We can describe the matrix as follows: Label the vectors of exponents for the u_i along the edges as follows: $\boldsymbol{a}$ along the upper edge from bit 0; $\boldsymbol{b}$ along the lower edge from bit 0; $\boldsymbol{c}$ along the lower edge from bit 3; and $\boldsymbol{d}$ along the horizontal edge from bit 3. The updating rule is then as follows, where $'$ denotes updated values:

$$\begin{aligned}
(\boldsymbol{a}_0, \boldsymbol{a}_1, \boldsymbol{a}_2, \boldsymbol{a}_3, \boldsymbol{a}_4, \boldsymbol{a}_5)' &= (\boldsymbol{a}_1 + \boldsymbol{c}_0 + 1, \boldsymbol{a}_0 + \boldsymbol{c}_1, \boldsymbol{a}_3 + \boldsymbol{c}_2, \\
&\qquad \boldsymbol{a}_2 + \boldsymbol{c}_3, \boldsymbol{a}_5 + \boldsymbol{c}_4, \boldsymbol{a}_4 + \boldsymbol{c}_5) \\
(\boldsymbol{b}_0, \boldsymbol{b}_1, \boldsymbol{b}_2, \boldsymbol{b}_3, \boldsymbol{b}_4, \boldsymbol{b}_5)' &= (2\boldsymbol{a}_1 + 1, 2\boldsymbol{a}_0, 2\boldsymbol{a}_3, 2\boldsymbol{a}_2, 2\boldsymbol{a}_5, 2\boldsymbol{a}_4) \\
(\boldsymbol{c}_0, \boldsymbol{c}_1, \boldsymbol{c}_2, \boldsymbol{c}_3, \boldsymbol{c}_4, \boldsymbol{c}_5)' &= (\boldsymbol{b}_4 + \boldsymbol{d}_1, \boldsymbol{b}_5 + \boldsymbol{d}_0, \boldsymbol{b}_2 + \boldsymbol{d}_3, \\
&\qquad d_2, \boldsymbol{b}_0 + \boldsymbol{d}_5, \boldsymbol{b}_1 + \boldsymbol{d}_4) \\
(\boldsymbol{d}_0, \boldsymbol{d}_1, \boldsymbol{d}_2, \boldsymbol{d}_3, \boldsymbol{d}_4, \boldsymbol{d}_5)' &= (\boldsymbol{b}_0 + \boldsymbol{b}_4, \boldsymbol{b}_1 + \boldsymbol{b}_5, 2\boldsymbol{b}_2, 2\boldsymbol{b}_3, \boldsymbol{b}_0 + \boldsymbol{b}_4, \boldsymbol{b}_1 + \boldsymbol{b}_5)
\end{aligned}$$

From these equations one derives the 25×25 transition matrix $\overline{\boldsymbol{N}}$.

For each of the three graphs the eigenvalues have norm 2, $\sqrt{2}$ or 1. The initial vector $(0, 0, 0, ..., 1)$ can be decomposed into a sum of eigenvectors as $\frac{1}{6}\boldsymbol{w}_2 + \frac{1}{18}\boldsymbol{w}_{-2} +$ other terms, where $\boldsymbol{w}_2 = (1, 1, 1, ..., 1, 1, 0)$ has eigenvalue 2, $\boldsymbol{w}_{-2} = (1, -1, 1, -1, ..., 1, -1, 0)$ has eigenvalue -2 and the other terms involve eigenvectors associated to smaller eigenvalues. Therefore, as the system evolves, these two terms dominate. At odd iterations, the significant terms look like $2^t \cdot \frac{1}{9}(1, 2, 1, 2...1, 2, 0)$ and at even iterations like $2^t \cdot \frac{1}{9}(2, 1, 2, 1, ..., 2, 1, 0)$. Thus, if $(u_0 u_2 u_4)(u_1 u_3 u_5)^2$ and $(u_0 u_2 u_4)^2(u_1 u_3 u_5)$ have the same parity the sum-product algorithm will converge; otherwise, it will diverge. This explains why the performance curves under the stringent criterion given in Figure 4 are the same for the three graphs. The algorithm under the stringent criterion reflects the convergence behavior of the infinite sequences, which are identical. This is somewhat surprising given the presence of multiple 4-cycles in two of them and desirable "large girth" in the third.

Degradation of performance with the coarse termination criterion

We now consider the difference in performance for the three graphs under the less stringent termination criterion. Let us write the initial vector $(0, 0, \ldots, 1)$ as $\frac{1}{6}\boldsymbol{w}_2 + \frac{1}{18}\boldsymbol{w}_{-2} + \boldsymbol{w}_{\sqrt{2}} + \boldsymbol{w}_1$, where $\boldsymbol{w}_{\sqrt{2}}$ is the contribution from all eigenvectors associated to eigenvalues of norm $\sqrt{2}$ and $\boldsymbol{w}_1$ is the contribution from the eigenvectors associated to eigenvalues of norm 1. Consider inputs $u_0, \ldots, u_5$ such that $(u_0u_2u_4)(u_1u_3u_5)^2$ and $(u_0u_2u_4)^2(u_1u_3u_5)$ have the same parity, say both < 1, so that the sum-product algorithm converges. The sum-product algorithm will terminate early and be incorrect, when at iteration t, some of the $\hat{u}_\ell^{(t)}$ are larger than 10^3 and others are less than 10^{-3}. For this to happen some of the u_ℓ must be larger than 1, and their contribution at the locations ℓ' where $\hat{u}_{\ell'}^{(t)} > 10^3$ must be unusually high. We have $\hat{u}_{\ell'} = \boldsymbol{u}^{\sum_{\lambda(e)=l'} \boldsymbol{a}_e}$ and, at iteration t, each $\boldsymbol{a}_e^{(t)}$ is some subvector of $\overline{\boldsymbol{N}}^t(\boldsymbol{w}_2 + \boldsymbol{w}_{-2} + \boldsymbol{w}_{\sqrt{2}} + \boldsymbol{w}_1)$ (subject to a permutation of indices). The contribution from $\boldsymbol{u}^{\overline{\boldsymbol{N}}^t(\boldsymbol{w}_2+\boldsymbol{w}_{-2})}$ is either $(u_0u_2u_4)(u_1u_3u_5)^2$ or $(u_0u_2u_4)^2(u_1u_3u_5)$ for each of the graphs. Thus early termination is due to the contribution from $\overline{\boldsymbol{N}}^t\boldsymbol{w}_{\sqrt{2}}$. The difference in performance is due to the different dynamics for this expression in the three graphs.

For the No 4-cycle graph the vector of exponents for $\hat{u}_1^{2t+s}$ is $(-2)^t(0, 2/3, 0, -1/3, 0, -1/3)$. Thus $\hat{u}$ may be greater than 10^3 when u_1 is large and u_3 and u_5 are small, or vice-versa. The situation at other bit nodes is similar. For the Three 4-cycle graph the vector of exponents at bit node 1 behaves more chaotically.

$$\begin{array}{ll} (0, 2/3, 0, -1/3, 0, -1/3) & \text{at iteration } 0 \\ (1, 2/3, 0, -1/3, -1, -1/3) & \text{at iteration } 2 \\ (-1, -10/3, 0, 5/3, 1, 5/3) & \text{at iteration } 4 \\ (-3, 2/3, 0, -1/3, 3, 1/3) & \text{at iteration } 6 \end{array}$$

The exact cause of inferior performance for the Three 4-cycle graph is not obvious, but we think two things play a role. First, the L_1-norm of the vector of exponents for $\hat{u}_\ell$ is larger for the Three 4-cycle graph. Second, the chaotic variation of the vector of exponents for $\hat{u}_\ell$ means that there are more conditions that can lead to early, and incorrect, termination.

5. Concluding Remarks

Although the examples considered in this article are very simple and do not define codes of practical interest, there are several interesting results and some properties that may have relevance for realistic codes.

We remark that the usual explanations for decoding failure seem irrelevant for these graphs. For each of the examples the only stopping sets are trivial, either the empty set or the entire set. One does find near-codewords of different weights for the different graphs in Figures 5 and 3, but they don't appear to be directly related to the convergence criteria reported here. Expansion hardly makes sense with such small graphs. The polytope of pseudo-codewords is simply a line segment generated by the all 0 codeword and the all 1 codeword, so the only extremal pseudo-codewords are in fact codewords. Thus pseudo-codewords do not explain decoding failure. These properties may be associated with decoding failure on large graphs, but these examples suggest they are not a cause, but rather correlated to some deeper causative phenomenon.

One explanation for pseudo-codewords causing decoding failure is that the sum-product algorithm on a given graph may be affected by codewords in a covering graph. Our examples suggest the reverse effect, the sum-product algorithm on the covering graphs of 2-bits-3-checks seem to have inherited the convergence behavior of their base graph. Furthermore, the 4-Choose-2 graph shows that one can do better than a covering graph. It would be interesting to see if this effect can be proven, whether for bipartite graphs with all check nodes of degree 2, or more generally. If so, this suggests an inherent weakness in the low-density matrices constructed in several articles, *e.g.* [5, 9, 13], using a block matrix of circulants. It also presents a challenge to find other algebraic methods for constructing bipartite graphs.

References

[1] C. Di, D. Proietti, I. E. Teletar, T. J. Richardson, and R. L. Urbanke. Finite length analysis of low-density parity-check codes on the binary erasure channel. *IEEE Trans. Inform. Theory*, 48(6):1570–1579, 2002.

[2] Jon Feldman. *Decoding Error-Correcting Codes via Linear Programming.* PhD thesis, Massachusetts Institute of Technology, 2003.

[3] Jon Feldman, M. J. Wainwright, and D. R. Karger. Using linear programming to decode linear codes. *IEEE Trans. Inform. Theory*, 51(3):954–972, Mar. 2005.

[4] G. D. Forney, R. Koetter, F. R.Kschischang, and A. Reznik. On the effective weights of pseudocodewords for codes defined on graphs with cycles. In *Proc. IMA workshop on codes systems and graphical models*, pages 101–112, 2001.

[5] M. P. C. Fossorier. Quasi-cyclic, low-density parity-check codes from circulant permutation matrices. *IEEE Trans. Inform. Theory*, 50(8):1788–1793, 2004.

[6] R. Koetter and P. O. Vontobel. Graph covers and iterative decoding of finite length codes. In *Proc. 3rd Int. Conf. on Turbo Codes and Related Topics*, pages 75–82, Sept. 2003.

[7] D. J. C. MacKay. Good error-correcting codes based on very sparse matrices. *IEEE Trans. Inform. Theory*, 45(2):399–431, 1999.

[8] David J. C. MacKay and M. J. Postol. Weaknesses of Margulis and Ramanujan–Margulis low-density parity-check codes. In *Proceedings of MFCSIT2002, Galway*, volume 74 of *Electronic Notes in Theoretical Computer Science*. Elsevier, 2003.

[9] M. E. O'Sullivan. Algebraic construction of sparse matrices with large girth. *IEEE Trans. Inform. Theory*, 52:718–727, 2006.

[10] T. Richardson. Error floors of LDPC codes. In *Proc. of the 42-th Annual Allerton Conference on Communication, Control, and Computing*, pages 1426–1435, 2004.

[11] T. Richardson, A. Shokrollahi, and R. Urbanke. Design of capacity-approaching irregular low-density parity-check codes. *IEEE Trans. Inform. Theory*, 47(2):619–639, 2001.

[12] M. Sipser and D. A. Spielman. Expander codes. *IEEE Trans. Inform. Theory*, 42(6, part 1):1710–1722, 1996.

[13] R. M. Tanner. On graph constructions for LDPC codes by quasi-cyclic extensions. In M. Blaum, P. Farrell, and H. C. .A. van Tilborg, editors, *Information, Coding and Mathematics*, pages 209–219. Kluwer Academic, 2002.

[14] T. Tian, C. Jones, J. Villasenor, and R. Wesel. Selective avoidance of cycles in irregular ldpc code construction. *IEEE Trans. on Comm.*, 52(8):1242–1247, Aug. 2004.

[15] P. O. Vontobel and R. Koetter. Lower bounds on the minimum pseudoweight of linear codes. In *2004 IEEE Int. Symp. Infor. Theory*, page 70, 2004.

[16] N. Wiberg. *Codes and Decoding on General Graphs*. PhD thesis, Linköping University, Sweden, 1996.

On the extremal graph theory for directed graphs and its cryptographical applications

V. A. Ustimenko

University of Maria Curie-Sklodowska,
Lublin, Poland,
E-mail: vasyl@golem.umcs.lublin.pl

The paper is devoted to the graph based cryptography. The girth of a directed graph (girth indicator) is defined via its smallest commutative diagram. The analogue of Erdøos's Even Circuit Theorem for directed graphs allows to establish upper bound on the size of directed graphs with a fixed girth indicator. Size of members of infinite family of directed regular graphs of high girth is close to an upper bound.

Finite automata related to members of such a family of algebraic graphs over chosen commutative ring can be used effectively for the design of cryptographical algorithm for different problems of data security (stream ciphers, data base encryption, public key mode an digital signatures).

The explicit construction of infinite family of algebraic graphs of high girth defined over the arbitrarily chosen ring is given. Some results on their properties, based on theoretical studies or software implementations are given.

Keywords: Extremal graph theory, directed graphs of large girth, algebraic graphs over commutative rings, graph based cryptography, coding theory

1. Introduction

One of the important direction in the classical extremal graph theory is studies of the greatest number of edges $\mathrm{ex}(v,d) = ex(v, C_3, \dots C_{2d})$ of graphs on v vertices without cycles C_t of length $t = 3, 4, \dots, d$. It is known that $\mathrm{ex}(v, C_3, \dots, C_{2d}) \leq O(v^{1+1/d})$ (see [3]). Similar problem for directed graphs (roughly, finite automata) has been motivated by applications to cryptography and other areas of computer science.

We use term *binary relation graph* for the graph Γ of irreflexive binary relation ϕ over finite set V such that for each $v \in V$ sets $\{x|(x,v) \in \phi\}$ and $\{x|(v,x) \in \phi\}$ have same cardinality.

We say that the pair of passes $a = x_0 \to x_1 \to \cdots \to x_s = b$, $s \geq 1$ and $a = y_0 \to y_1 \to \cdots \to y_t = b$, $t \geq 1$ form an (s,t)-commutative diagram

$O_{s,t}$ if $x_i \neq y_j$ for $0 < i < s$, $0 < j < t$. Without loss of generality we assume $s \geq t$ and refer to the number s as the rank of $O_{s,t}$. The directed cycle with s arrows we denote as $O_{s,0}$. The minimal parameter $s = \max(s,t)$ of the commutative diagram $O_{s,t}$ with $s+t \geq 3$ in the binary relation graph Γ we call the *girth indicator* of the Γ and denote it as $\mathrm{gi}(\Gamma)$.

Let $\mathrm{E} = \mathrm{E}_d(v) = \mathrm{Ex}(v, O_{s,t}, s+t \geq 3 | 2 \leq s \leq d)$ be the maximal size (number of arrows) of the binary relation graphs with the girth indicator $> d$.

Notice , that the size of symmetric irreflexive relation is the double of the size of corresponding simple graph. because undirected edge of the simple graph corresponds to two arrows of $O_{2,0}$. In [27] the following bound has been obtained

$$\mathrm{E}_d(v) \leq v^{1+1/d} + O(v) \tag{1}$$

Via explicit constructions we find out that for $d = 2, 3, 4, 5$ and 6 the bound (1) is sharp up to magnitude.

It indicates that studies of extremal properties of graphs of binary relations with the high girth indicator and studies of $\mathrm{ex}(v, C_3, \ldots, C_n)$ are far from being equivalent. Really, the sharpness of the $\mathrm{ex}(v,n)$ for $n = 8$ and $n = 12$ are old open problems (similar to cases of cycles C_8 and C_{12} in Erdös' Even Circuit Theorem).

The girth of the simple graph is the minimal length of its minimal cycles. The infinite family of k-regular graphs Γ_i of fixed degree k is the *family of graphs of large girth* if the size of its members is close to $\mathrm{ex}v, C_3, c_4, \ldots, C_n$, i.e. girth of Γ_i of order v_i is $c\log_{k-1}(v_i)$, where c is independent on i constant. They turned out to be very useful in networking (see [2]).

The idea to use simple graphs of large girth in cryptography had been widely explored, in particular see [10], [19], [20], [22]-[26], [28]-[29].

The definitions of family of graphs of large girth for the class of irreflexive binary relation graphs formulated in [28], where more general encryption scheme for the "potentially infinite" text based on the graphs of binary relations with special "rainbow-like" coloring of arrows has been proposed (see section 2 of current paper for all details). In fact, a family of k-regular binary relation graphs Γ_i, $i = 1, \ldots$ is a family of graphs of large girth if the size of its members is close to the bound (1).

For the encryption purpose we identify the vertex of the graph with the plaintext, encryption procedure corresponds to the chain of adjacent vertices starting from the plaintext, the information on such chain is given by the sequence of colors (passwords). We assume that the end of the chain is the ciphertext.

The important feature of such encryption is the resistance to attacks, when adversary intercepts the pair plaintext - ciphertext. It is true because the best algorithm of finding the pass between given vertices (by Dijkstra, see [6] and latest modifications) has complexity $n\ln n$ where n is the order of the graph, i.e. the size of the plainspace. The situation is similar to the checking of the primality of Fermat's numbers $2^{2^m}+1$: if the input given by the string of binary digits, then the problem is polynomial, but if the input is given by just a parameter m, then the task is NP-complete.

We have an encryption scheme with the flexible length of the password (length of the chain). If graphs are connected then we can convert each potentially infinite plaintext into the chosen string "as fast as it is possible".

Finally, in the case of "algebraic graphs" (see [1]) with the special "rainbow-like" coloring (symbolic rainbow-like graphs of section 3) there is an option to use symbolic computations in the implementation of graph based algorithm. We can create public rules symbolically and use the above algorithm as public key tool (for the example of implementation look at [24]).

The first explicit examples of families with large girth with arbitrary large valency were given by Margulis. The constructions were Cayley graphs $X^{p,q}$ of group $SL_2(Z_q)$ with respect to special sets of $q+1$ generators, p and q are primes congruent to 1 mod 4. The family of $X^{p,q}$ is not a family of algebraic graphs because the neighborhood of each vertex is not an algebraic variety over F_q. For each p, graphs $X^{p,q}$, where q is running via appropriate primes, form a family of small world graph of unbounded diameter (see [15]-[17]).

The first family of connected algebraic graphs over F_q of large girth and arbitrarily large degree had been constructed in [13]. These graphs $CD(k,q)$, k is an integer ≥ 2 and q is odd prime power had been constructed as connected component of graphs $D(k,q)$ defined earlier. For each q graphs $CD(k,q)$, $k \geq 2$ form a family of large girth with $\gamma = 4/3\log_{q-1}q$.

Some new examples of simple algebraic graphs of large girth and arbitrary large degree the reader can find in [29].

2. Binary relations, related rainbow-like graphs and algorithms

2.1. *Binary relations and special colorings*

Let Φ be an irreflexive binary relation over the set V, i.e. $\Phi \in V \times V$ and for each v pair (v, v) is not the element of Φ.

We say that u is the neighbor of v if $(v, u) \in \Phi$. Recall, that we use term *binary relation graph* for the graph Γ of irreflexive binary relation ϕ over finite set V such that for each $v \in V$ sets $\{x|(x, v) \in \phi\}$ and $\{x|(v, x) \in \phi\}$ have the same cardinality. It is a directed graph without loops and multiple edges.

Let Γ be the graph of binary relation. The *pass* between vertices a and b is the sequence $a = x_0 \to x_1 \to \ldots x_s = b$ of length s, where x_i, $i = 0, 1, \ldots s$ are distinct vertices.

We shall use a term *the family of algebraic graphs* for the family of graphs $\Gamma(K)$, where K belongs to some infinite class F of commutative rings, such that the neighborhood of each vertex of $\Gamma(K)$ and the vertex set itself are quasi-projective varieties over K of dimension ≥ 1 (see [1]).

Such a family can be treated as special Turing machine with the internal and external alphabet K.

We say that the graph Γ of binary relation Φ has a rainbow-like coloring over the set of colors C if for each v, $v \in V$ we have a coloring function ρ_v, which is a bijection from the neighborhood $St(v)$ of v onto C, such that the operator $N_c(v)$ of taking the neighbor of v with color c is the bijection of V onto V.

We say that the rainbow like coloring ρ is invertible if there is a rainbow-like coloring of Φ^{-1} over C' such that $N_c{}^{-1} = N'_{c'}$ for some color $c' \in C'$.

Example 2.1. (Cayley graphs)

Let G be the group and S be subset of distinct generators, then the binary relation $\phi = \{(g_1, g_2)|g_i \in G, i = 1, 2, g_1 g_2{}^{-1} \in S\}$ admits the rainbow like coloring $\rho(g_1, g_2) = g_1 g_2{}^{-1}$

This rainbow like coloring is invertible because the inverse graph $\phi^{-1} = \{(g_2, g_1)|g_1 g_2{}^{-1} \in S\}$ admits the rainbow-like coloring $\rho'(g_2, g_1) = g_2 g_1{}^{-1} \in S^{-1}$.

Example 2.2. (Parallelotopic graphs and latin squares)

Let G be the graph with the coloring $\mu : V(G) \to C$ of the set of vertices $V(G)$ into colors from C such that the neighborhood of each vertex looks like rainbow, i.e. consists of $|C|$ vertices of different colors. In case of pair

(G, μ) we shall refer to G as *parallelotopic graph* with the local projection μ (see [20], [22] and further references).

It is obvious that parallelotopic graphs are k-regular with $k = |C|$. If C' is a subset of C, then induced subgraph $G^{C'}$ of G which consists of all vertices with colors from C' is also a parallelotopic graph. It is clear that connected component of the parallelotopic graph is also a parallelotopic graph.

The *arc* of the graph G is a sequence of vertices $v_1, \ldots, v_k$ such that $v_i I v_{i+1}$ for $i = 1, \ldots, k-1$ and $v_i \neq v_{i+2}$ for $i = 1, \ldots, k-2$. If $v_1, \ldots, v_k$ is an arc of the parallelotopic graph (G, μ) then $\mu(v_i) \neq \mu(v_{i+2})$ for $i = 1, \ldots, k-2$.

Let $+$ be the latin square defined on the set of colors C. Let us assume that $\rho(u, v) = \mu(u) - \mu(v)$. The operator $N_c(u)$ of taking the neighbor of the color is invertible, ${N_c}^{-1} = N_{-c}$, where $-c$ is the opposite for c element in the latin square. It means that ρ is invertible rainbow like coloring.

Example 2.3. The class of sparse parallelotopic bipartite graphs can be given by the following incidence structure defined over finite field F_q in [27].

Let $P = \{(x_1, \ldots, x_n) | x_i \in F_q\}$ and $L = \{[y_1, \ldots, y_n] | y_i \in F_q\}$ be the sets of points and lines. The point $(x_1, \ldots, x_n)$ is incident to the line $[y_1, y_2, \ldots, y_n] | y_i \in F_q]$ if and only if $x_i - y_i = x_{k}(i) y_{l}(i)$, $i = 2, 3, \ldots, n$, where parameters $k(i) < i$, $l(i) < i$ are chosen for each value of i. We can define the coloring $\mu((x_1, \ldots x_n) = x_1$, $\mu([y_1, \ldots, y_n]) = y_1$ and obtain the parallelotopic graph. The choice of the field addition $+$ as an appropriate latin square allows us to define an effective finite automaton: the operator of taking the neighbor of chosen color require $2n-1$ field operations.

2.2. *General symmetric algorithm*

Let us consider the encryption algorithm corresponding to the graph Γ with the chosen invertible rainbow-like coloring of edges. Let $\rho(u, v)$ be the color of arrow $u \to v$, C is the totality of colors and $N_c(u)$ is the operator of taking the neighbor of u with the color c.

The password is the string of colors $(c_1, c_2, \ldots, c_s)$ and the encryption procedure is the composition $N_{c_1} \times N_{c_2} \ldots N_{c_s}$ of bijective maps $N_{c_i} : V(\Gamma) \to V(\Gamma)$. So if the plaintext $v \in V(\Gamma)$ is given, then the encryption procedure corresponds to the following chain: $x_0 = v \to x_1 = N_{c_1}(x_0) \to x_2 = N_{c_2}(x_1) \to \cdots \to x_s = N_{c_s}(x_{s-1}) = u$ in the graph. The vertex u is the ciphertext.

Let $N'_{c'}(N_c(v)) = v$ for each $v \in V(\Gamma)$. The decryption procedure corresponds to the composition of maps $N'_{c'_s}, N'_{c'_{s-1}}, \ldots, N'_{c'_1}$. The above scheme gives a symmetric encryption algorithm with flexible length of the password (key). Let $A(\Gamma, \rho, s)$ be the above encryption scheme.

Examples 1 and 2 demonstrate that each known infinite family of graphs of large girth of unbounded degree can be used for the development of the encryption algorithm according to the above scheme; see [28] or [29] for the details.

2.3. *Symbolic computations and public keys*

Let K be the commutative ring. Recall that graph Γ is the algebraic graph over K if the set of vertices $V(\Gamma)$ and the neighborhood of each vertex u are algebraic quasi-projective varieties over the ring K; see [1].

In the case of *symbolic invertible rainbow-like graph* (Γ, ρ, ρ'), the vertex set $V(\Gamma)$ and the neighborhoods of each vertex are open algebraic varieties in Zariski topology as well as the color set C, maps $N(c, v) = N_c(u)$ and $N'(c, v) = N'_c(u)$ are polynomial maps from $C \times V(\Gamma)$ onto $V(\Gamma)$.

In the case of symbolic rainbow-like graph the encryption as above with the key $(t_1, t_2, \ldots, t_k)$ is given by some polynomial map from $C^k \times V(\Gamma) \to V(\Gamma)$. We can treat t_i, $i = 1, \ldots, k$ as symbolic variables.

The specializations $t_i = \alpha_i \in K$ give the public key map $P : V(\Gamma) \to V(\Gamma)$. Like in the known example of polynomial encryption proposed by Imai and Matsumoto (see [16]) and its modifications by J. Patarin (see [12]) we can combine P with two invertible affine transformations T_1 and T_2 (bijective polynomial maps of degree 1) and work with the public map $Q = T_1 P T_2$.

Let us use the characters Alice and Bob from books on Cryptography (see, for instance [11], [12]), where Bob is a public user and Alice is a key holder. So she knows the string $t_1, \ldots, , t_s$, the graph and affine transformations T_1 and T_2. She can decrypt via consecutive applications of $T_2{}^{-1}$, $N'_{t'_k}, N'_{t'_k - 1}, \ldots N'_{t_1}$ and $T_1{}^{-1}$.

The public user Bob has the encryption map Q only. He can encrypt, but the decryption is hard task because (1) Q is the polynomial map of degree ≥ 2 from many variables. (2) Even in the case, when Bob knows T_1, T_2 and the graph Γ. The problem of finding the pass between the plaintext vertex and the ciphertext vertex has complexity $n \ln n$, where $n = |V(\Gamma)|$ is the size of plainspace. So Bob is not able to decrypt if the plainspace is large enough.

2.4. *Coding theory, other applications*

The theory of distance transitive graphs is the theoretical basis for coding theory problems dealing with the problem of error detection and error correction (see [4], [5]). Some applications (not only in Coding Theory, but in Complexity Studies and Parallel Computing) require the expansion properties of the graphs (see [2], [8] and further references). For instance, error correcting codes by Tanner [18] use expansion properties of finite generalized polygons, which are both distance-regular and expanding graphs. In the paper [7] Tanner's idea (see [13]) had been implemented to graphs $CD(k, q)$ which are not distance regular, but have good expansion properties. We suggest to use the encryption based on the graphs $X(p, q)$ of fixed degree $q + 1$. They form the family of graphs of large girth (girth $= 4/3\log_q(v)$), family of small world graphs (diameter $= 4/3\log_q(v) + 2$, family of expanding graphs (the second largest eigenvalue is bounded by $2\sqrt{q}$ [14], so it is the Ramanujan case).

3. The incidence structures defined over commutative rings

We define the family of graphs $D(k, K)$, where $k > 2$ is positive integer and K is a commutative ring (see [20], [29]), such graphs have been considered in [13] for the case $K = F_q$. Let P and L be two copies of Cartesian power K^N, where K is the commutative ring and N is the set of positive integer numbers. Elements of P will be called *points* and those of L *lines*.

To distinguish points from lines we use parentheses and brackets, for $x \in V$, we write $(x) \in P$ or $[x] \in L$. It will also be advantageous to adopt the notation for co-ordinates of points and lines introduced in [20] for the case of general commutative ring K:

$$(p) = (p_{0,1}, p_{1,1}, p_{1,2}, p_{2,1}, p_{2,2}, p'_{2,2}, p_{2,3}, \ldots, p_{i,i}, p'_{i,i}, p_{i,i+1}, p_{i+1,i}, \ldots),$$
$$[l] = [l_{1,0}, l_{1,1}, l_{1,2}, l_{2,1}, l_{2,2}, l'_{2,2}, l_{2,3}, \ldots, l_{i,i}, l'_{i,i}, l_{i,i+1}, l_{i+1,i}, \ldots].$$

The elements of P and L can be thought as infinite ordered tuples of elements from K, such that only finite number of components are different from zero.

We now define an incidence structure (P, L, I) as follows. We say that the point (p) is incident with the line $[l]$, and we write $(p)I[l]$, if the following relations between their co-ordinates hold:

$$\begin{aligned} l_{i,i} - p_{i,i} &= l_{1,0}p_{i-1,i} \\ l'_{i,i} - p'_{i,i} &= l_{i,i-1}p_{0,1} \\ l_{i,i+1} - p_{i,i+1} &= l_{i,i}p_{0,1} \\ l_{i+1,i} - p_{i+1,i} &= l_{1,0}p'_{i,i} \end{aligned} \tag{2}$$

These four relations are defined for $i \geq 1$, $p'_{1,1} = p_{1,1}$, $l'_{1,1} = l_{1,1}$). This incidence structure (P, L, I) we denote by $D(K)$ and identify it with the bipartite *incidence graph* of (P, L, I), which has the vertex set $P \cup L$ and edge set consisting of all pairs $\{(p), [l]\}$ for which $(p)I[l]$.

For each positive integer $k \geq 2$ we obtain an incidence structure (P_k, L_k, I_k) as follows. First, P_k and L_k are obtained from P and L, respectively, by simply projecting each vector onto its k initial coordinates with respect to the above order. The incidence I_k is then defined by imposing the first $k-1$ incidence equations and ignoring all others. The incidence graph corresponding to the structure (P_k, L_k, I_k) is denoted by $D(k, K)$.

To facilitate notation in future results, it will be convenient for us to define $p_{-1,0} = l_{0,-1} = p_{1,0} = l_{0,1} = 0$, $p_{0,0} = l_{0,0} = -1$, $p'_{0,0} = l'_{0,0} = -1$, and to assume that (6) are defined for $i \geq 0$.

Notice that for $i = 0$, the four conditions (1) are satisfied by every point and line, and, for $i = 1$, the first two equations coincide and give $l_{1,1} - p_{1,1} = l_{1,0}p_{0,1}$.

The incidence relation is motivated by the linear interpretation of Lie geometries in terms of their Lie algebras [22] . Let us define the "root subgroups" U_α, where the "root" α belongs to the root system Root $= \{(1,0), (0,1), (1,1), (1,2), (2,1), (2,2), (2,2)' \ldots, (i,i), (i,i)', (i,i+1), (i+1,i) \ldots\}$. The "root system above" contains all real and imaginary roots of the Kac-Moody Lie Algebra $\tilde{A}_1$ with the symmetric Cartan matrix (see [9]). We just double the imaginary roots (i,i) by introducing $(i,i)'$.

Remark 3.1. For $K = F_q$ the following statement had been formulated in [13]. Let $k \geq 6$, $t = [\frac{k+2}{4}]$, and let

$$u = (u_\alpha, u_{11}, \cdots, u_{tt}, u'_{tt}, u_{t,t+1}, u_{t+1,t}, \cdots)$$

be a vertex of $D(k, K)$ ($\alpha \in \{(1,0), (0,1)\}$, it does not matter whether u is a point or a line). For every r, $2 \leq r \leq t$, let

$$a_r = a_r(u) = \sum_{i=0,r} (u_{ii}u'_{r-i,r-i} - u_{i,i+1}u_{r-i,r-i-1}),$$

and $a = a(u) = (a_2, a_3, \cdots, a_t)$.

Proposition 3.1. *(i) The classes of equivalence relation $\tau = \{(u,v)|a(u) = a(v)\}$ are connected components of graph $D(n,K)$, where $n \geq 2$ and K be the ring with unity of odd characteristic.*

(ii) For any $t-1$ ring elements $x_i \in K)$, $2 \leq t \geq [(k+2)/4]$, there exists a vertex v of $\mathrm{D}(k,K)$ *for which*

$$a(v) = (x_2, \ldots, x_t) = (x).$$

(3i) The equivalence class C for the equivalence relation τ on the set $K^n \cup K^n$ is isomorphic to the affine variety $K^t \cup K^t$, $t = [4/3n] + 1$ for $n = 0, 2, 3 \bmod 4$, $t = [4/3n] + 2$ for $n = 1 \bmod 4$.

Remark 3.2. Let K be the general commutative ring and C be the equivalence class on τ on the vertex set $\mathrm{D}(K)$ ($\mathrm{D}(n,K)$, then the induced subgraph, with the vertex set C is the union of several connected components of $\mathrm{D}(K)$ ($\mathrm{D}(n,K)$).

Without loss of generality we may assume that the vertex v of $C(n,K)$ satisfies to conditions $a_2(v) = 0, \ldots a_t(v) = 0$. We can find the values of components $v'_{i,i)}$ from this system of equations and eliminate them. Thus we can identify P and L with elements of K^t, where $t = [3/4n] + 1$ for $n = 0, 2, 3 \bmod 4$, and $t = [3/4n] + 2$ for $n = 1 \bmod 4$.

We shall use notation $C(t,K)$ ($C(K)$) for the induced subgraph of $D(n,K)$ with the vertex set C.

Remark 3.3. If $K = F_q$, q is odd, then the graph $C(t,k)$ coincides with the connected component $CD(n,q)$ of the graph $D(n,q)$ (see [29] and further references), graph $C(F_q)$ is a q-regular tree. In other cases the question on the connectivity of $C(t,K)$ is open. It is clear that $g(C(t,F_q))$ is $\geq 2[2t/3] + 4$.

Proposition 3.2. *Projective limit of graphs $D(n,K)$ (graphs $C(t,K)$, $CD(n,K)$) with respect to standard morphisms of $D(n+1,K)$ onto $D(n,K)$ (their restrictions on induced subgraphs) equals to $D(K)$ ($C(K)$.*

If K is an integrity domain, then $D(K)$ and $CD(K)$ are forests. Let C be the connected component, i.e a tree.

We define the parallelotopic coloring of the graphs $C(t,K)$), $D(n,K)$, $C(K)$ and $D(K)$ by formulae $\mu(p_{1,0}, p_{1,1}, \ldots) = p_{1,0}$, $\mu([l_{0,1}, l_{1,1}, \ldots]) = l_{0,1}$.

Let us consider the directed flag graphs $F(t,K)$ and $E(n,K)$ of the tactical configurations $C(t,K)$ and $D(n,K)$, respectively. The vertex set of $F(t,K)$ ($E(n,K)$) is a totality of flags $f = (([l],(p))$, where $(p)I[l]$ in the

$C(t,K)$ ($D((n,K)$, respectively), we have $f_1 \to f_2$ for $f_1 = ([l^1],(p^1))$, $f_2 = ([l^2],(p^2))$ if $[l_2]I(p_1)$,$[l^1] \neq [l^2]$, $(p^1) \neq (p^2)$. We can consider the symbolic invertible rainbow-like coloring $\rho(f_1, f_2)$ of $F(t,K)$ ($E(t,K)$) defined on the color set $K^* \times K^*$ by the following rule:

Let $f_1 = ([l^1],(p^1))$, $f_2 = ([l^2],(p^2))$ form the arrow in $F(t,K)$ $(E(t,K))$. So, $[l^2]I(p^1)$. We assume that $\rho(f_1, f_2) = (l^1_{1,0} - l^2_{1,0}, p^1_{0,1} - l^2_{0,1})$.

If K is finite, then the cardinality of the color set is $(|K|-1)^2$. Let Reg$K$ be the totality of regular elements, i.e. not zero divisors of the ring K. Let us delete all arrows with color (x,y), where one of the elements x and y is not a zero divisor for $F(t,K)$ and $E(t,K)$. New graph $RF(t,K)$ and $RE(t,K)$ are the symbolic rainbow-like graphs over the set of colors $(\mathrm{Reg}K)^2$. The following statement can be found in [28].

Theorem 3.1. *The girth indicator* gi *of the symbolic rainbow like graph* $RF(t,k)$ *are* $g \geq 1/3t$.

Corollary 3.1. *Let* K *be a finite such that* $k = |\mathrm{Reg}K| \geq 2$. *Then graphs* $RF(t,K)$, $t = 1,2,\ldots$ *form the family of symbolic rainbow-like graphs of large girth of degree* k^2.

4. Symmetric encryption, algorithms related to graphs $RF(n,K)$

We can apply the general scheme of symmetric encryption to the parallelotopic graphs $RF(t,K)$ or $RE(n,K)$ from the previous section. Other options are based on the fact that $RE(n,K)$ and $E(n,K)$ are the enveloping graph for $RF(t,K)$ and the description of "connectivity invariants" $a_i(u)$, $i \geq 2$ of $D(n,K)$.

The information on the vertex $f = \{(p),[l]\}$, $(p)I[l]$ can be given by the list of coordinates $(p_{1,0}, p_{1,1},\ldots)$ of the point (p) and the parallelotopic color $l_{0,1}$ of the line $[l]$. Obviously, (p) and $[l]$ are in the same connected component of the graph $D(n,K)$. So we can think of $a_i((p))$, $i \geq 2$ as connectivity invariants of the graph $RE(n,K)$.

Let Root be the list of all roots related to $D(n,K)$ and $\Omega = \mathrm{Root} - \{(0,1)\}$ be the list of the indexes of components of the tuple (p). Let $a_i(p)$, $i = 2,3,\ldots,t$ be the list of connectivity invariants. We choose two subsets $J = \{i_1, i_2, \ldots i_l\}$ and $J' = \{j_1,\ldots,J_m\}$, $|J \cap J'| = 0$ of the set $\{2,3,\ldots,t\}$, $l + m \leq t-1$. So we have 3^{t-1} options to make a choice of (J,J'). Let

$$F_j(x_1,x_2,\ldots,x_d),\ d = t-l-m-1,\ j = 1,2$$

be the polynomial maps from K^d into K. The pair (J, J') and functions F_i, $i = 1, 2$ form the "internal key" of our encryption algorithm.

Let us choose the "external key" in the form of the pair (b), $= (b_1, \ldots, b_d)$, c, where b $\in K^d$ and c $\in \mathrm{Reg}(K)^s$ for some even integer $1 \leq s < \mathrm{gi} - 1$. If the ring K is finite, then we have $|K|^d|\mathrm{Reg}K|^s$ options to form the external password for chosen parameter s.

4.1. *The encryption algorithm*

Let

$$\begin{aligned} \Delta(J) &= \{(i,i)|i \in J\} \\ \Delta(J') &= \{(i,i)|i \in J'\} \\ \mathrm{Root}' &= \mathrm{Root} - J \cup J' \end{aligned} \tag{3}$$

The plainspace is the totality of functions $f : \mathrm{Root}' \to K$. So the plainspace is the string of characters from the alphabet K.

Step 1. We will form the vertex of the graph $E(n, K)$ by the following rule: form the point (p) such that $p_\alpha = f(\alpha)$ for $\alpha \in \mathrm{Root}'$. For the $\alpha \in \Delta(J) \cup \Delta(J')$ values p_α ($\alpha = (i,i)$ or $\alpha = (i,i)'$) will be computed consequently from the equations $a_i(\mathrm{p}) = b_i$, $i \in J \cup J'$, [l] be the neighboring line for (p) with the parallelotopic color $f((0,1))$. We form the vertex $\mathrm{v} = ((p), [l])$ of the graph $E(n, K)$.

Step 2. Let $R_{t_1,t_2}(\mathrm{u})$ be the operator of taking the neighbor of the vertex u $=$ (p), [l]) of the parallelotopic color $(p_{1,0} + t_1, l_{0,1} + t_2)$, where $t_i \in K$, $i = 1, 2$. Let $s_1, s_2, \ldots, s_d$ be the list of elements of the complement for $J \cup J'$. We compute $R_{F_1(a_{s_1}(\mathrm{p}),\ldots,a_{s_d})(\mathrm{p}),F_2(a_{s_1}(\mathrm{p}),\ldots,a_{s_d})(\mathrm{p})} = \mathrm{v}_0$.

Step 3. Apply the composition of operators $R_{c_1,c_2}, \ldots, R_{c_{s-1},c_s}$ to the vertex v_0. In fact, we use here the general encryption scheme for the graph $RE(n, K)$ in case of the plaintext v_0). Let ((h), [g]) be the resulting vertex. Assume that the information on this pair is given by function $z : \mathrm{Root} \to K$, such that $z((0,1)) = g_{0,1}$ and $z_\alpha = h_\alpha$ for $\alpha \neq (0,1)$.

Step 4. The ciphertext is the restriction z' of the function z onto $\mathrm{Root} - \Delta(J) \cup \Delta(J')$.

Step 5. We combine polynomial map $\tau : f \to z'$ as above with two invertible sparse affine transformations A and B by taking the composition $A\tau B$.

4.2. *Decryption procedure*

Step 1. Let u be the ciphertext. We form the vertex $((p'), [l'])$ of the graph $RE(n, K)$ from the function $y = B^{-1}(U)$ by the following rule: $p'_\alpha = y(\alpha)$ for $\alpha \in Root'$, for the $\alpha \in \Delta(J) \cup \Delta(J')$ values p'_α ($\alpha = (i, i)$ or $\alpha = (i, i)'$) will be computed consequently from the equations

$$a_i(\mathrm{p}') = \mathrm{b_i},, \mathrm{i} \in \mathrm{J} \cup \mathrm{J}'.$$

The line $[\mathrm{l}']$ be the neighboring line for (p') with the parallelotopic color $y((0.1)$. We form the vertex

vertex $\mathrm{v}' = ((p'), [l'])$ of the graph $RE(n, K)^{-1}$.

Step 2. Let $R'_{t_1,t_2}(\mathrm{u})$ be the operator of taking the neighbor of the vertex $\mathrm{u} = ((\mathrm{p}), [\mathrm{l}])$ in the graph $E(n, K)^{-1}$ of the parallelotopic color $(p_{1,0} = t_1, l_{0,1} + t_2)$, where $t_i \in \mathrm{RegK}$, $i = 1, 2$. Let $s_1, s_2, \ldots s_d$ be the list of elements of the complement for $J \cup J'$. We compute $R_{-F_1(a_{s_1}(\mathrm{p}'),\ldots,a_{s_d}(\mathrm{p}')),-F_2(a_{s_1}(\mathrm{p}'),\ldots,a_{s_d}(\mathrm{p}'))}(\mathrm{v}') = \mathrm{v}'_0$.

Step 3. Apply the composition of operators $R'_{-c_{s-1},-c_{c_s}}, \ldots, -R_{c_1,c_2}$ to the vertex v'_0. In fact, we use here the general decryption scheme for the graph $RE(n, K)$ in case of the ciphertext v'_0).

Step 4. Let $((\mathrm{h}'), [\mathrm{g}'])$ be the resulting vertex. Assume that the information on this pair is given by function z_1 : Root $\to K$, such that $z((0, 1)) = g'_{0,1}$ and $z_\alpha = h'_\alpha$ for $\alpha \neq (0, 1)$. We take the restriction z'_1 of the function z_1 onto Root $- \Delta(J) \cup \Delta(J')$.

Step 5. Compute the plaintext $A^{-1}(z'_1)$.

Let us denote the above symmetric encryption algorithm as $\mathrm{Alg}(F_1, F_2, A, B)$.

Proposition 4.1. *(i) Let us keep the internal password fixed. Then different external passwords correspond to distinct ciphertext.*

(ii) If the values of $F_1(a_{s_1}(\mathrm{p}), \ldots, a_{s_d}(\mathrm{p}))$ and $F_2(a_{s_1}(\mathrm{p}), \ldots, a_{s_d}(\mathrm{p}))$

(Step 2) are regular elements and $B = A^{-1}$ of the ciphertext is always different from the plaintext.

Proof. Let us consider the transformation $\mathrm{Alg}(E, E, F_1, F_2)$, where E is the identity map. Steps 1 and 2 do not depend on the external alphabet. The encryption procedure corresponds to the directed pass in the graph $RE(n, K)$ of the length less than the girth indicator. So the property (i) holds. The affine transformations A and B are bijections, so the property (i) is true for the $\mathrm{Alg}(A, B, F_1, F_2)$.

If the condition of (ii) holds then steps 2 and 3 of $\mathrm{Alg}(E, E, F_1, F_2)$ correspond to the directed pass in the graph $RE(n, K)$ of the length less then the girth indicator. So this transformation have no fixed points. The map $\mathrm{Alg}(A, A^{-1}, F_1, F_2)$ is conjugate with $\mathrm{Alg}(E, E, F_1, F_2)$. □

Remark 4.1. In fact we can use conditions $F_1(a_{s_1}(\mathrm{p}), \ldots, a_{s_d}(\mathrm{p})) \times F_2(a_{s_1}(\mathrm{p}), \ldots, a_{s_d}(\mathrm{p})) \neq 0$ or $F_i(a_{s_1}(\mathrm{p}), \ldots, a_{s_d}(\mathrm{p})) = 0$ for some i instead of condition of (ii) for the above statement.

We say that functions A, B, F_i, $i = 1, 2$ are sparse if their computation requires $O(n)$ operations of the ring K.

Proposition 4.2. *(i) Let the complexity of transformations A, B, F_i, $i = 1, 2$. If $\max(|J|, |J'|)$ is bounded by independent on n constant, then the symmetric encryption as above requires $O(n)$ ring operations.*

(ii) For each positive integer m there is a choice of functions F_i, $i = 1, 2$, such that the degree of polynomial encryption map is $\geq m$.

(iii) If functions F_1 and F_2 are constants, then the encryption transformation of $\mathrm{Alg}(E, E, F_1, F_2)$ sending $(l_{0,1}, p_{1,0}, p_{1,1}, \ldots, p_\alpha, \ldots)$ into $(l'_{0,1}, p'_{1,0}, p'_{1,1}, \ldots, p'_\alpha, \ldots)$ ' is a triangular map of kind $l'_{0,1} \to l'_{0,1} + c$, $p'_\alpha \to p_\alpha + f_\alpha(l_{0,1}, p_{1,0}, p_{1,1}, \ldots, p_\alpha)$. So the value of p'_{alpha} depends components p_β such that $\beta < \alpha$ according to the natural order on the root set.

Remark 4.2. The properties (i) and (iii) of the statement above allow to use $\mathrm{Alg}(E, E, c_1, c_2$ as a stream cipher for "changing data on the fly" (telecommunications, encryption CD's with movies and etc). If we keep the external password fixed, then change of the single character p_α of the plaintext lead to change of characters p'_β of ciphertext with $\beta \geq \alpha$.

Remark 4.3. In the case of the root set with the highest roots $(l, l), (l, l)'$ such that l is not an element of J and J', $F_i(a_{s_1}(\mathrm{p}), \ldots, a_{s_d}(\mathrm{p}))$, $i = 1, 2$ are linear combinations of $a_{s_1}(\mathrm{p}), \ldots, a_{s_d}(\mathrm{p})$ containing terms $k_i a_i(\mathrm{p})$, $k_i \neq 0$ change of single character of the ciphertext lead to change with the probability close to 1 each character of the ciphertext. It justifies use of nontrivial

F_1 and F_2 for the self-coding in case of encryption of large file (data bases, Geological Information Systems, etc).

4.3. *Examples*

1) In the practically important case of the ring $K = Z_{2^n}$ (sizes of the ASKEE and binary alphabets are 2^7 and 2^8, respectively) the RegK is the totality of odd residues modulo 2^n, all values of functions of kind $2f(x_1, x_2, \ldots, x_n) + 1$ are regular elements. So if F_i, $i = 1, 2$ belong to this set, then the ciphertext for Alg(A, A^{-1}, F_1, F_2) is always different from each plaintext.

2) In the case of $K = z_{2k+1}$, the value of Euler function $\phi(2k+1)$ giving us the number of regular elements of the ring is $\geq k$, because of all even residues are regular, values of functions of kind $2f(x_1, \ldots, x_n)$ either are regular or zero, So if F_i, $i = 1, 2$ belong to this set, then the ciphertext for Alg(A, A^{-1}, F_1, F_2) is always different from each plaintext.

3) In the case of the integer domain RegK coincides with the $K - \{0\}$ any transformation Alg(A, A^{-1}, F_1, F_2) has no fixed points.

5. Public keys

We can use the following modification of the encryption Alg(A, B, F_1, F_2) with subsets $J = \{i_1, i_2, \ldots, i_l\}$ and $J' = \{j_1, \ldots, j_m\}$, $|J \cap J'| = 0$ of the set $\{2, 3, \ldots, t\}$, $l + m \leq t - 1$. Recall that F_i, $i = 1, 2$ depend on variables $(x_1, x_2, \ldots, x_d)$, $d = t - l - m - 1$.

Let us choose the "dynamical external key" in the form of the pair $\mathrm{b} = (b_1, \ldots b_d) \in K^d$ and $\mathrm{c} = (f_1(x_1, x_2, \ldots, x_d), \ldots, f_s(x_1, \ldots, x_d))$, where s is even integer and f_i are polynomial maps from K^d into K.

We have to complete Step 1 and 2 without any changes. After computation of "numerical password" $\mathrm{c}' = (f_1(a_{s_1}(\mathrm{p}), \ldots, a_{s_d}(\mathrm{p})) = (c'_1, \ldots, c'_d)$ and complete modified Step 3 i. e. apply the composition of operators $R_{c'_1, c'_2}, \ldots, R_{c'_{s-1}, c'_s}$ to the vertex v_0. In fact, we use here the general encryption scheme for the graph $E(n, K)$ in case of the plaintext v_0).

After we conduct remaining steps 4 and 5 without any changes.

The new algorithm Alg$(E(n, k), A, B, F_1, F_2, f_1, \ldots, f_s)$ defines the polynomial map of the free module K^r, $r = |\mathrm{Root} - J \cup J|$ into itself. We have to create (say with "Mathematica" or "Maple") the public rule:

$$y_1 = P_1(x_1, \ldots, x_r), \ldots, y_r = P_r(x_1, \ldots, x_r).$$

The public user Bob can use it for the encryption procedure. If parameter s and degrees of polynomials $F_1, F_2, f_1, \dots, f_s$ are sufficiently large then he is not able to find the inverse map.

The key holder Alice use information on the graph $E(n, K)$, matrices A, B, sets J, J', functions F_1, F_2 and external key i. e. string b of elements of the ring K and sequences of functions $f_1, \dots, f_s$. That is why she can use modified decryption scheme of $\mathrm{Alg}(A, B, F_1, F_2)$, where the only modification is the computation of "numerical password" $\mathrm{c} = (f_1(a_{s_1}(\mathrm{p}), \dots, a_{s_d}(\mathrm{p})) = (c_1, \dots, c_d)$ for the restriction p of the output of Step 2 onto the set $\mathrm{Root}\{(0,1)\}$.

If values functions $F_1, F_2, f_1, \dots, f_s$ are regular elements then the plaintext and the ciphertext of $\mathrm{Alg}(E(n, k), E, E, F_1, F_2, f_1, \dots, f_s)$ are at the distance $s/2 + 1$ in the graph $RE(n, K)$. If $s/2 + 1$ is $\leq$ gi, then the ciphertext is always different from the plaintext.

6. Other algebraic parallelotopic graphs

The algorithm $\mathrm{Alg}(E, E, F_1, F_2)$ with empty sets $J$and J can be easily generalized on the arbitrary algebraic parallelotopic graph $G(K)$ over ring K with the color set M and coloring function μ. We can assume that $M = M(K)$ is the open quasi-algebraic variety over the commutative ring K. Let $V(K)$ be the vertex set of G and $g : V(K) \to K^l$ be the connectivity invariant of the graph i. e. function which is constant on vertices from the same connected component. Let $R_a(\mathrm{v})$ be the operator of taking the neighbor of the vertex $\mathrm{v} \in V(G)$ of the color $a \in M$. Let $h_i(x, y)$, $x \in M$, $y \in M$, $i = 1, \dots, s$ be polynomial maps from $M \times K^l \to M$ such that each equation of kind $h_i(x, b) = c$ has unique solution in variable x. We define the following invertible procedure τ: $\mathrm{v} \to \mathrm{v}_1 = R_{h_1(\mu(\mathrm{v}), g(\mathrm{v})}(\mathrm{v}) \to R_{h_2(\mu(\mathrm{v}_1), g(\mathrm{v})} \to R_{h_s(\mu(\mathrm{v}_{s-1}), g(\mathrm{v})}$. The transformation $A\tau B$, where A and B are sparse invertible polynomial automorphisms of $V(G)$, can be used for various cryptographical problems.

If $L(x, y)$ be the latin squire on M, which is a polynomial map from $M \times M \to M$, then we can take $h_i = L(\phi_i(x), y)$, where ϕ_i are polynomial automorphisms of open variety M.

The reader can find examples of parallelotopic graphs over the set of colors K^m of high girth for each commutative ring K and each positive integer $m \geq 2$.

We can consider more general graphs $RF^{2s}(t, K)$ ($RE^{2s}(t, K)$), which vertices are chains $(p^1), [l^1], \dots, (p^s), [l^s]$ of old graph $C_t(K)$ ($D(n, K)$, respectively) such that $\rho(p^i) - \rho(p^{i-1})$, $i = 2, 3, \dots, s$ and $\rho(l^i) -$

$\rho(l^{i-1})$, $i = 2, 3, \ldots, s$. Two vertices $(p^1), [l^1], \ldots, (p^s), \ldots, [l^s]$ and $(x^1), [y^1], \ldots, (x^s), [y^s]$ are in binary relation $RF^s(t, s)$ ($RE^s(t, s)$, respectively) if $[l^s]I(x^1)$ and $\rho((x^1)) - \rho((p^s))$ and $\rho([y^1]) - \rho([l^s])$ are regular elements of the ring K.

We consider as well the directed bipartite graphs $RF^{2s+1}(t, K)$ with the point-set

$$\{((p^1), [l^1], \ldots, (p^s), [l^s], (p_{s+1}) | (p^i)I[l^i]I(p^{i+1}), i = 1, \ldots, s\}$$

and line-set

$$\{([y^1], (x^1) \ldots, [y^s], (x^s), [y^{s+1}]) | [l^i]I(p^i)I[l^{i+1}], i = 1, \ldots, s\}.$$

We have,

$$((p^1), l[l^1], \ldots, (p^s), [l^s], (p_{s+1}) \to ([y^1], (x^1) \ldots, [y^s], (x^s), [y^{s+1}])$$

if

$$(p^{s+1})I[y^1], \quad \textit{and} \quad \rho([y^1]) - \rho([l^s]),\ \rho((x^1)) - \rho(p^{s+1}$$

are regular elements of the ring K. Analogously,

$$([y^1], (x^1) \ldots, [y^s], (x^s), [y^{s+1}]) \to ((p^1), [l^1], \ldots, (p^s), [l^s], (p_{s+1})$$

if

$$(p^1)I[y^{s+1}], \quad \textit{and} \quad \rho((p^1)) - \rho((x^s)),\ \rho([l^1]) - \rho([y^{s+1}])$$

are regular elements.

Proposition 6.1. *The map π given by the close formula*

$$\mathrm{p}^\pi = [p_{10}, -p_{11}, p_{21}, p_{12}, -p'_{22}, -p_{22}, \ldots, -p'_{ii}, -p_{ii}, p_{i+1,i}, p_{i,i+1}, \ldots],$$
$$\mathrm{l}^\pi = (l_{01}, -l_{11}, l_{21}, l_{12}, -l'_{22}, -l_{22}, \ldots, -l'_{ii}, -l_{ii}, l_{i+1,i}, l_{i,i+1}, \cdots)$$

is the color preserving automorphism of $D(K)$ of order two. It preserves blocks of the equivalence relation τ. Its restriction on $V(D(2n, k))$ and $V(CD(2n, K))$ are color preserving graph automorphism of order two.

We define the polarity graph $RF_\pi^{2s+1}(t, K)$ with the vertex set

$$\{((p^1), [l^1], \ldots, (p^s), [l^s], (p_{s+1}) | (p^i)I[l^i]I(p^{i+1}), i = 1, \ldots, s\}$$

by declaring

$$((p^1), [l^1], \ldots, (p^s), [l^s], (p_{s+1})) \to ((x^1), [y^1], \ldots, (x^s), [ly^s], (x_{s+1}))$$

in the case when

$$((p^1), [l^1], \ldots, (p^s), [l^s], (p_{s+1})) \to (\pi((x^1)), \pi([y^1]), \ldots, \pi((x^s)), \pi([ly^s]), \pi((x^{s+1}))$$

in the directed graph $RF^{2s+1}(t,K)$. The following statement is immediate corollary from Theorem 3.

Corollary 6.1. *Let K be a finite such that $k = |\mathrm{Reg}K| \geq 2$. Then graphs $RF^s(t,K)$ $(RF_\pi^{2s+1}(t,K))$, $t = 1,2,\ldots$ form the family of symbolic rainbow-like graphs of large girth of degree k^s (k^{2s+1}, respectively).*

Password length m	3000	6000	9000
5	915	1760	2606
10	1830	3520	5211
15	2745	5280	7815
20	3666	7053	10440

7. Remarks on implementation

In our package for symmetric encryption we used the rings Z_{2^8} and F_{2^8}, same field as in the new U.S. Advanced Encryption Standard (AES). In fact, our crypto-system works primarily with bytes (8 bits), represented from the right as: b7b6b5b4b3b2b1b0. The 8-bit elements of the field are regarded as polynomials with coefficients in the field F_2: b7x7 + b6x6 + b5x5 + b4x4 + b3x3 + b2x2 + b1x1 + b0. The field elements will be denoted by their sequence of bits, using two hexadecimal digits. . We eight irreducible polynomial (a polynomial that cannot be factored into the product of two simpler polynomials). As for the AES, we use the following irreducible polynomial: $m(x) = x^8 + x^4 + x^3 + x + 1 = 0x11b$(hex). The intermediate product of the two polynomial we first generate a multiplication table for the 256 elements and once any multiplication of elements in the F_{2^8}, we use the loaded multiplication table for efficiency reasons.

To evaluate the performance of our algorithm (F_{2^8}), we use measure the encryption time for our method for different size of data files and using different password lengths of keys (in bytes). It shows also that for password of length 10 our algorithm is capable to encrypt at speed as fast as 2 kilo-bytes per millisecond. Together with the traditional table below, we just write the close formula $T = 2nk$, T - time, n, k are dimensions of the plainspace and the keyspace as vector spaces over the chosen field, respectively. Our program is written in Java and it runs on Pentium 4, 1GHZ. The better performance on better computers we get via C or C++ version.

Our algorithm uses binary code, it may encrypt any data type. We have developed a prototype software written in Java. We hope that our software can be a very attractive tool for reliable security of virtual organization (e-learning, e-business, etc ...). We used families of graphs $RE(k, K)$, $K \in \{Z_{2^8}, F_{2^8}\}$, $D(k, F_{2^8})$. Because of the use of loaded multiplication table the speed of computation does not depend on the choice of the family.

Remark. Case of rings Z_{2^s}. Let us assume that the password and the plaintext are numbers written base 2 (binary code). To encrypt we making two steps: first is the conversion of the plaintext and the key into strings of residues mod Z_{2^s} (numbers n and k base 2^s), second is our algorithm on symmetric mode with the numerical key. The complexity of first step $O(\log^2(n)) + O\log^2(k)$ does not depend on parameter s (see [11]). The complexity of second step is approximately $2nk2^{-2s}$ because the key (plaintext) is the string of length $k2^{-s}$ (n^{-2s}, respectively. So the algorithm in case of $s = 32$ ($s = 64$) works 16 times (64 times, respectively) faster then in case $s = 8$ evaluated by table below.

References

[1] N. Biggs, *Algebraic Graph Theory* (2nd ed), Cambridge, University Press, 1993.

[2] F. Bien, *Constructions of telephone networks by group representations*, Notices Amer. Mah. Soc., 36 (1989), 5-22.

[3] B. Bollobás, *Extremal Graph Theory*, Academic Press, London, 1978.

[4] A. Brower, A. Cohen, A. Nuemaier, *Distance regular graphs*, Springer, Berlin, 1989.

[5] P. J. Cameron and J.H. van Lint, *Graphs, Codes and Designs*, London. Math. Soc. Lecture Notes, 43, Cambridge (1980).

[6] E. Dijkstra, *A note on two problems in connection with graphs*, Num. Math., 1 (1959), 269-271.

[7] P. S. Guinard and J.Lodge, *Tanner Type Codes Arizing from Large Girth Graphs*, Communications Research Centre, Canada , Reprint GUI94, 2006.

[8] S. Hoory, N. Linial, and A.Wigderson, *Expander graphs and their applications*, Bulletin (New Series) of AMS, volume 43, N4, 439-461,

[9] V. Kac. *Infinite dimensional Lie algebras*, Birkhauser, Boston, 1983.

[10] Yu. Khmelevsky , V. A. Ustimenko, *Practical aspects of the Informational Systems reengineering*, The South Pacific Journal of Natural Science, volume 21, 2003, www.usp.ac.fj(spjns).

[11] N. Koblitz, *A Course in Number Theory and Cryptography*, Second Edition, Springer, 1994, 237 p.

[12] N. Koblitz, Algebraic aspects of Cryptography, in Algorithms and Computations in Mathematics, v. 3, Springer, 1998.

[13] F. Lazebnik, V. Ustimenko and A.J.Woldar, *A new series of dense graphs of high girth*, Bulletin of the AMS 32 (1) (1995), 73-79.

[14] A. Lubotsky, R. Philips, P. Sarnak, *Ramanujan graphs*, J. Comb. Theory, 115, N 2., (1989), 62-89.

[15] G. A. Margulis, *Explicit construction of graphs without short cycles and low density codes*, Combinatorica, 2, (1982), 71-78.

[16] G. Margulis, *Explicit group-theoretical constructions of combinatorial schemes and their application to desighn of expanders and concentrators*, Probl. Peredachi Informatsii, 24, N1, 51-60. English translation publ. Journal of Problems of Information transmission (1988), 39-46.

[17] M. Margulis, *Arithmetic groups and graphs without short cycles*, 6th Intern. Symp. on Information Theory, Tashkent, abstracts, vol. 1, 1984, pp. 123-125 (in Russion).

[18] R.Michael Tanner. *A recursive approach to low complexity codes*, IEEE Trans on Info.Th., IT, 27(5):533-547, Sept. 1981.

[19] A. Tousene, V. Ustimenko, *Graph Based Private Key Crypto System*, International Journal on Computer Research, Nova Science Publisher, volume 13 (2006), issue 4, 12p.

[20] V. A. Ustimenko, *Coordinatisation of regular tree and its quotients*, in "Voronoi's impact on modern science", eds P. Engel and H. Syta, book 2, National Acad. of Sci, Institute of Matematics, 1998, 228p.

[21] V. A. Ustimenko, *On the varieties of parabolic subgroups, their generalizations and combinatorial applications*, Acta Applicandae Mathematicae, 52 (1998), 223-238.

[22] V. Ustimenko, *Graphs with Special Arcs and Cryptography*, Acta Applicandae Mathematicae, 2002, vol. 74, N2, 117-153.

[23] V. Ustimenko, *CRYPTIM: Graphs as tools for symmetric encryption*, In Lecture Notes in Comput. Sci., 2227, Springer, New York, 2001.

[24] V. Ustimenko, *Maximality of affine group and hidden graph cryptsystems*, Journal of Algebra and Discrete Mathematics, October, 2004, v.10, pp. 51-65.

[25] V. A. Ustimenko, D. Sharma, *CRYPTIM: system to encrypt text and image data*, Proceedings of International ICSC Congress on Intelligent Systems 2000, Wollongong, 2001, 11pp.

[26] V. Ustimenko, A. Touzene, *CRYPTALL:system to encrypt all types of data*, Notices of the Kiev-Mohyla Academy, v 23, June , 2004, pp. 12-15.

[27] V. Ustimenko, *On the extremal binary relation graphs of high girth*, Proceedings of the Conference on infinite particle systems, Kazimerz- Dolny, 2006, World Scientific Publ.(to appear).

[28] V. Ustimenko, *On the graph based cryptography and symbolic computations*, Serdica Journal of Computing, Proceedings of the International Conference, ACA 2006, Warna, Bugaria (to appear).

[29] V. Ustimenko. *On linguistic Dynamical Systems, Graphs of Large Girth and Cryptography*, Journal of Mathematical Sciences, Springer, vol.140, N3 (2007) pp. 412-434.

Fast arithmetic on hyperelliptic curves via continued fraction expansions

M. J. Jacobson, Jr.

Department of Computer Science, University of Calgary,
2500 University Drive NW, Calgary, Alberta, Canada T2N 1N4
E-mail: jacobs@cpsc.ucalgary.ca

R. Scheidler*

Department of Mathematics and Statistics, University of Calgary,
2500 University Drive NW, Calgary, Alberta, Canada T2N 1N4
E-mail: rscheidl@math.ucalgary.ca

A. Stein

Department of Mathematics, University of Wyoming
1000 E. University Avenue, Laramie, WY 82071-3036, USA
Email: astein@uwyo.edu

In this paper, we present a new algorithm for computing the reduced sum of two divisors of an arbitrary hyperelliptic curve. Our formulas and algorithms are generalizations of Shanks's NUCOMP algorithm, which was suggested earlier for composing and reducing positive definite binary quadratic forms. Our formulation of NUCOMP is derived by approximating the irrational continued fraction expansion used to reduce a divisor by a rational continued fraction expansion, resulting in a relatively simple and efficient presentation of the algorithm as compared to previous versions. We describe a novel, unified framework for divisor reduction on an arbitrary hyperelliptic curve using the theory of continued fractions, and derive our formulation of NUCOMP based on these results. We present numerical data demonstrating that our version of NUCOMP is more efficient than Cantor's algorithm for most hyperelliptic curves, except those of very small genus defined over small finite fields.

Keywords: Hyperelliptic curve, reduced divisor, continued fraction expansion, infrastructure, Cantor's algorithm, NUCOMP

*The research of the first two authors is supported by NSERC of Canada.

1. Introduction and Motivation

Divisor addition and reduction is one of the fundamental operations required for a number of problems and applications related to hyperelliptic curves. The group law of the Jacobian can be realized by this operation, and as such, applications ranging from computing the structure of the divisor class group to cryptographic protocols depend on it. Furthermore, the speed of algorithms for solving discrete logarithm problems on hyperelliptic curves, particularly of medium and large size genus, depend on a fast computation of the group law. There has been a great deal of work on finding efficient algorithms for this operation (see for instance [5]).

Cantor's algorithm [2] is a generic algorithm that allows this operation to be explicitly computed. It works by first adding the two divisors and subsequently reducing the sum. One drawback of this approach, and most algorithms derived from it, is that one has to deal with intermediate operands of double size. That is, while the basis polynomials of the two starting divisors and the final reduced divisor have degree at most g, where g is the genus of the curve, the divisor sum has a basis consisting of two polynomials whose degree is usually as large as $2g$, and reduction only gradually reduces the degrees back down to g. This greatly reduces the speed of the operation, and it is highly desirable to be able to perform divisor addition and reduction without having to compute with quantities of double size.

The group operation of the class group of positive definite binary quadratic forms, composition and reduction, suffers from the same problem of large intermediate operands. In 1988, Shanks [13] devised a solution to this problem, an algorithm he called NUCOMP. The idea behind this algorithm is to stop the composition process before completion and apply a type of intermediate reduction before computing the composed form. Instead of using the rather expensive continued fraction algorithm that produces the aforementioned intermediate operands of double size, the reduction is performed using the much less costly extended Euclidean Algorithm. The coefficients are only computed once the form is reduced or almost reduced. As a result, the sizes of the intermediate operands are significantly smaller, and the binary quadratic form produced by NUCOMP is very close to being reduced.

In [11], van der Poorten generalized NUCOMP to computing with ideals in the infrastructure of a real quadratic number field by showing how the relative generator corresponding to the output can be recovered. Jacobson and van der Poorten [6] presented numerical evidence for the efficiency of their version of NUCOMP. They also sketched an adaptation of this method to arithmetic in the class group and infrastructure of a hyperelliptic curve. Their computational results indicated that their version of NUCOMP was more efficient than Cantor's algorithm for moderately small genera (between genus 5 and 10), and that the relative efficiency improved as both the genus and size of the ground field increase. However, a formal analysis and description of NUCOMP in the hyperelliptic curve setting was not provided.

Shanks's formulation of NUCOMP, as well as the treatments in [11] and [6], are based on the arithmetic of binary quadratic forms. In [8], the authors described NUCOMP in terms of ideal arithmetic in real quadratic number fields. They provided a clear and complete description of NUCOMP in terms of continued fraction expansions of real quadratic irrationalities and, in addition, showed how to optimize the formulas in this context.

In this paper, we provide a unified description of NUCOMP for divisor arithmetic on the three different possible models of a hyperelliptic curve: imaginary, real, and unusual [3]. We generalize the results in [8], describing and deriving NUCOMP in terms of continued fraction expansions in all three settings. Furthermore, we explain NUCOMP purely in terms of divisor arithmetic, also incorporating the infrastructure arithmetic of a real hyperelliptic curve. Our formulation of NUCOMP is complete and somewhat simpler than that in [6], and its relation to Cantor's algorithm is more clear. In addition, we prove its correctness and a number of related results, including the fact that the output is in most cases reduced, and is in the worst case only one step away from being reduced. The end result, supported by computational results, is that our improved formulation of NUCOMP offers performance improvements over Cantor's algorithm for even smaller genera than indicated in [6].

We begin in Sec. 2 with an overview of continued fractions, and explain divisor arithmetic on hyperelliptic curves and its connection to continued fractions in Sec. 3–Sec. 5. Based on this foundation, we describe divisor addition and reduction as well as NUCOMP in Sec. 6–Sec. 10. We conclude with numerical results in Sec. 11, including a discussion of the efficiency of our two different versions of NUCOMP as given in Sec. 9.

2. Continued Fraction Expansions

For brevity, we write the symbolic expression

$$s_0 + \cfrac{1}{s_1 + \cfrac{1}{\ddots \cfrac{1}{s_n + \cfrac{1}{\alpha_{n+1}}}}}$$

as $[s_0, s_1, \ldots, s_n, \alpha_{n+1}]$. If we wish to leave the end of the expression undetermined, we simply write $[s_0, s_1, \ldots]$.

Let k be any field, $k[t]$ the ring of polynomials in the indeterminate t with coefficients in k, and $k(t)$ the field of rational functions in t with coefficients in k. It is well-known that the completion of $k(t)$ with respect to the place at infinity of $k(t)$ (corresponding to the discrete valuation "denominator degree minus numerator degree") is the field $k\langle t^{-1}\rangle$ of Puiseux series in t^{-1}; that is, any non-zero element in $k\langle t^{-1}\rangle$ is of the form

$$\alpha = \sum_{i=-\infty}^{d} a_i t^i \ ,$$

where $d \in \mathbb{Z}$, $a_i \in k$ for $i \leq d$, and $a_d \neq 0$. Define

$$\lfloor \alpha \rfloor = \sum_{i=0}^{d} a_i t^i \ , \qquad \mathrm{sgn}(\alpha) = a_d \ , \qquad \deg(\alpha) = d \ . \tag{2.1}$$

Also, define $\lfloor 0 \rfloor = 0$ and $\deg(0) = -\infty$.

Let $n \geq 0$, $s_0, s_1, \ldots, s_n$ a sequence of polynomials in $k[t]$, and $\alpha \in k\langle t^{-1}\rangle$ non-zero. Then the expression

$$\alpha = [s_0, s_1, \ldots, s_n, \alpha_{n+1}] \tag{2.2}$$

is referred to as the (*ordinary*) *continued fraction expansion* of α with *partial quotients* $s_0, s_1, \ldots, s_n$. It uniquely defines a Puiseux series $\alpha_{n+1} \in k\langle t^{-1}\rangle$ where $\alpha_0 = \alpha$ and $\alpha_{i+1} = (\alpha_i - s_i)^{-1}$ for $0 \leq i \leq n$. If we set

$$\begin{aligned} A_{-2} = 0 \ , \ A_{-1} = 1 \ , \ A_i = s_i A_{i-1} + A_{i-2} \ , \\ B_{-2} = 1 \ , \ B_{-1} = 0 \ , \ B_i = s_i B_{i-1} + B_{i-2} \ , \end{aligned} \tag{2.3}$$

for $0 \leq i \leq n$, then $A_i/B_i = [s_0, s_1, \ldots, s_i]$ for $0 \leq i \leq n-1$. Since $A_i B_{i-1} - A_{i-1} B_i = (-1)^{i-1}$ for $-1 \leq i \leq n$, A_i and B_i are coprime for $-2 \leq i \leq n$.

If $s_i = q_i$ with $q_i = \lfloor \alpha_i \rfloor$ for $i \geq 0$, then Eq. (2.2) is the well-known *regular* continued fraction expansion of α. Here, the partial quotients $q_0, q_1, \ldots$

are uniquely determined by α, and $\deg(q_i) \geq 1$ for all $i \in \mathbb{N}$. The rational function $A_i/B_i = [q_0, q_1, \ldots, q_i]$ is the *i-th convergent* of α. This term is motivated by the well-known inequalities

$$\deg\left(\alpha - \frac{A_i}{B_i}\right) \leq -\deg(B_i B_{i+1}) < -2\deg(B_i) \tag{2.4}$$

for all $i \geq 0$. The following result is also well-known:

Lemma 2.1. *Let $\alpha \in k\langle t^{-1}\rangle$, $E, F \in k[t]$ with $\alpha F \neq 0$ and $\gcd(E, F) = 1$. If*

$$\deg\left(\alpha - \frac{E}{F}\right) < -2\deg(F) ,$$

then E/F is a convergent in the regular continued fraction expansion of α.

Throughout this paper, we reserve the symbols q_i and $\hat{q}_i$ for the quotients of a regular continued fraction expansion; for arbitrary partial quotients, we use the symbol s_i. To distinguish expansions of rational functions from those of Puiseux series, we henceforth use the convention that partial quotients and convergents relating to expansions of rational functions are equipped with a "ˆ" symbol, whereas quantities pertaining to expansions of Puiseux series do not have this symbol.

One of the main ideas underlying NUCOMP is to approximate the regular continued fraction expansion of a Puiseux series by that of a rational function "close" to it. We then expect the convergents, and hence the two expansions, to agree up to a certain point:

Theorem 2.1. *Let $\alpha \in k\langle t^{-1}\rangle$ and $\hat{\alpha} \in k(t)$ be non-zero, and write $\hat{\alpha} = E/F$ with $E, F \in k[t]$. Let $\hat{q}_i$ $(0 \leq i \leq m)$ and $\hat{r}_i$ $(-1 \leq i \leq m)$ be the sequences of quotients and remainders, respectively, obtained by applying the Euclidean Algorithm to $\hat{\alpha}$; that is, $\hat{r}_{-2} = E$, $\hat{r}_{-1} = F$, $\hat{r}_{i-2} = \hat{q}_i\hat{r}_{i-1} + \hat{r}_i$ with $\hat{q}_i = \lfloor r_{i-2}/r_{i-1} \rfloor$ for $0 \leq i \leq m$, so $\hat{r}_{m-1} = \gcd(E, F)$ and $\hat{r}_m = 0$. If there exists $n \in \mathbb{Z}$, $-1 \leq n \leq m-1$, such that $2\deg(\hat{r}_n) > \deg(F^2(\alpha - \hat{\alpha}))$, then the first $n+2$ partial quotients in the regular continued fraction expansions of α and $\hat{\alpha}$ are equal.*

Proof. Let $\alpha = [q_0, q_1, \ldots, q_m, \ldots]$ be the regular continued fraction expansion of α. The regular continued fraction expansion of $\hat{\alpha}$ is obviously $\hat{\alpha} = [\hat{q}_0, \hat{q}_1, \ldots, \hat{q}_m]$. Then $A_i/B_i = [q_0, q_1, \ldots, q_i]$ and $\hat{A}_i/\hat{B}_i = [\hat{q}_0, \hat{q}_1, \ldots, \hat{q}_i]$ are the i-th convergents of α and $\hat{\alpha}$, respectively. We wish to prove that $q_i = \hat{q}_i$ for $0 \leq i \leq n+1$.

Suppose n as in the statement exists. If $n = -1$, then $2\deg(\hat{r}_{-1}) = 2\deg(F) > \deg(F^2(\alpha - \hat{\alpha}))$ implies $\deg(\alpha - \hat{\alpha}) < 0$, so $q_0 = \lfloor \alpha \rfloor = \lfloor \hat{\alpha} \rfloor = \hat{q}_0$.

Assume now inductively that $2\deg(\hat{r}_{n-1}) > \deg(F^2(\alpha - \hat{\alpha}))$ implies $q_i = \hat{q}_i$ for $0 \leq i \leq n$ and suppose that $2\deg(\hat{r}_n) > \deg(F^2(\alpha - \hat{\alpha}))$. Since the r_i are decreasing in degree for $-1 \leq i \leq m$, we have $2\deg(\hat{r}_{n-1}) > 2\deg(\hat{r}_n) > \deg(F^2(\alpha - \hat{\alpha}))$, so $q_i = \hat{q}_i$ for $0 \leq i \leq n$ by induction hypothesis, and we only need to show $q_{n+1} = \hat{q}_{n+1}$.

A simple induction argument yields $\hat{r}_i = (-1)^{i-1}(\hat{A}_i F - \hat{B}_i E)$ for $-2 \leq i \leq m$, so by assumption and Eq. (2.4),

$$\deg(\alpha - \hat{\alpha}) < 2\deg\left(\frac{\hat{r}_n}{F}\right) = 2\deg(\hat{A}_n - \hat{B}_n\hat{\alpha}) \leq -2\deg(\hat{B}_{n+1}) \ .$$

It follows again from Eq. (2.4) that

$$\deg\left(\alpha - \frac{\hat{A}_{n+1}}{\hat{B}_{n+1}}\right) \leq \max\left\{\deg(\alpha - \hat{\alpha}), \deg\left(\hat{\alpha} - \frac{\hat{A}_{n+1}}{\hat{B}_{n+1}}\right)\right\} < -2\deg(\hat{B}_{n+1}) \ .$$

Since $\gcd(\hat{A}_{n+1}, \hat{B}_{n+1}) = 1$, Lemma 2.1 implies that $\hat{A}_{n+1}/\hat{B}_{n+1} = A_j/B_j$ for some $j \geq 0$. If $j < n+1$, then $[q_{j+1}, \ldots, q_{n+1}] = 0$ which is a contradiction. If $j > n+1$, then similarly $[\hat{q}_{n+2}, \ldots, \hat{q}_j] = 0$, again a contradiction. Thus, $\hat{A}_{n+1}/\hat{B}_{n+1} = A_{n+1}/B_{n+1}$, and hence $q_{n+1} = \hat{q}_{n+1}$. □

Let $E, F \in k[t]$ be non-zero, and assume that $\deg(E) > \deg(F)$. Consider again the regular continued fraction expansion of the rational function $E/F = [\hat{q}_0, \hat{q}_1, \ldots, \hat{q}_m]$, where $m \geq 0$ is again minimal with that property. Set $\hat{\phi}_0 = E/F$ and $\hat{\phi}_{i+1} = (\hat{\phi}_i - \hat{q}_i)^{-1}$, so $\hat{q}_i = \lfloor \hat{\phi}_i \rfloor$ for $i \geq 0$. This continued fraction expansion corresponds to the Euclidean algorithm applied to E and F. We define

$$\begin{aligned} b_{-1} &= E \ , \ b_0 = F \ , \quad b_{i+1} = b_{i-1} - \hat{q}_i b_i \ , \\ a_{-1} &= 0 \ , \ a_0 = -1 \ , \ a_{i+1} = a_{i-1} - \hat{q}_i a_i \ , \end{aligned} \tag{2.5}$$

so $\hat{q}_i = \lfloor b_{i-1}/b_i \rfloor$, for $0 \leq i \leq m$. Then $\hat{q}_i$ and b_{i+1} are the quotients and remainders, respectively, when dividing b_{i-1} by b_i. We have

$$b_{i-1} = \hat{q}_i\, b_i + b_{i+1} \quad , \quad \deg(b_{i+1}) < \deg(b_i) \quad (-1 \leq i \leq m) \ , \tag{2.6}$$

and the b_i strictly decrease in degree for $-1 \leq i \leq m+1$. Then m is minimal such that $b_{m+1} = 0$, so $b_m = \gcd(E, F)$.

As before, denote by $\hat{A}_i/\hat{B}_i = [\hat{q}_0, \hat{q}_1, \ldots, \hat{q}_i]$ the i-th convergents of $\hat{\phi}_0$ for $0 \leq i \leq m$. The quantities $\hat{A}_i, \hat{B}_i$ can be computed recursively by

$$\begin{aligned} \hat{A}_{-2} &= 0, \ \hat{A}_{-1} = 1, \ \hat{A}_i = \hat{q}_i \hat{A}_{i-1} + \hat{A}_{i-2} \quad (0 \leq i \leq m) \ , \\ \hat{B}_{-2} &= 1, \ \hat{B}_{-1} = 0, \ \hat{B}_i = \hat{q}_i \hat{B}_{i-1} + \hat{B}_{i-2} \quad (0 \leq i \leq m) \ . \end{aligned} \tag{2.7}$$

Then induction yields $a_i = (-1)^{i-1}\hat{A}_{i-1}$ for $-1 \le i \le m+1$; in particular, we see that the a_i increase in degree for $-1 \le i \le m+1$. We also obtain

$$b_{-1} = (-1)^i(a_{i-1}b_i - a_ib_{i-1}) \qquad (0 \le i \le m+1) \; . \tag{2.8}$$

We require the following basic degree properties later on:

Lemma 2.2.

(a) $\deg(b_i) = \deg(b_{i-1}) - \deg(\hat{q}_i) \le \deg(b_{i-1}) - 1 \qquad (0 \le i \le m)$.
(b) $\deg(a_i) = \deg(a_{i-1}) + \deg(\hat{q}_{i-1}) \ge \deg(a_{i-1}) + 1 \qquad (1 \le i \le m+1)$.
(c) $\deg(b_i) \le \deg(b_{-1}) - i - 1 \qquad (-1 \le i \le m+1)$.
(d) $\deg(a_i) \ge i \qquad (0 \le i \le m+1)$.
(e) $\deg(a_i) + \deg(b_{i-1}) = \deg(b_{-1}) \qquad (0 \le i \le m+1)$.

Proof. Since $\deg(\hat{q}_i) \ge 1$ for $0 \le i \le m$ by Eq. (2.6), (a) and (b) follow from Eq. (2.5). Parts (c) and (d) can then be obtained from (a) and (b), respectively, using induction. Finally, since $\deg(a_ib_{i-1}) > \deg(a_{i-1}b_i)$ by (a) and (b), (e) now follows from Eq. (2.8). □

3. Hyperelliptic Curves

We employ an algebraic framework of hyperelliptic curves based on the treatments of function fields given in [12], [17], and [4], as opposed to a more geometric treatment. Let k be a finite field of order q. Following [3], we define a *hyperelliptic function field of genus* $g \in \mathbb{N}$ to be a quadratic extension of genus g over the rational function field $k(u)$, and a *hyperelliptic curve of genus* g *over* k to be a plane, smooth[a], absolutely irreducible, affine curve C over k whose function field $k(C)$ is hyperelliptic of genus g. The curve C and its function field are called *imaginary*, *unusual*, or *real*, if the place at infinity of $k(u)$ is ramified, inert, or split in $k(C)$, respectively. Then C is of the form

$$C : v^2 + h(u)v = f(u) \; , \tag{3.1}$$

where $f, h \in k[u]$, $h = 0$ if k has odd characteristic, h is monic if k has even characteristic, and every irreducible factor in $k[u]$ of h is a simple factor of f; in particular, f is squarefree if k has odd characteristic. Then the function field of C is $k(C) = k(u, v)$ and its *maximal order* is the integral domain $k[C] = k[u, v]$, the coordinate ring of C over k. The different signatures at infinity can easily be distinguished as follows:

[a]A hyperelliptic curve does have singularities at infinity if it is not elliptic, i.e. $g \ge 2$

(1) C is imaginary if $\deg(f) = 2g+1$, and if $\deg(h) \leq g$ if k has characteristic 2;

(2) C is unusual if the following holds: if k has odd characteristic, then $\deg(f) = 2g+2$ and $\operatorname{sgn}(f)$ is a non-square in k, whereas if k has characteristic 2, then $\deg(h) = g+1$, $\deg(f) = 2g+2$ and the leading coefficient of f is not of the form $e^2 + e$ for any $e \in k^*$.

(3) C is real if the following holds: if k has odd characteristic, then $\deg(f) = 2g+2$ and $\operatorname{sgn}(f)$ is a square in k, whereas if k has characteristic 2, then $\deg(h) = g+1$, and either $\deg(f) \leq 2g+1$, or $\deg(f) = 2g+2$ and the leading coefficient of f is of the form $e^2 + e$ for some $e \in k^*$.

In some literature sources, unusual curves are counted among the imaginary ones, as there is a unique place in $k(C)$ lying above the place at infinity of $k(u)$ for both models. Note also that an unusual curve over k is real over a quadratic extension of k; whence the term "unusual".

It is well-known that the places of $k(u)$ are given by the monic irreducible polynomials in $k[u]$ together with the place at infinity of $k(u)$. Define S to be the set of places of $k(C)$ lying above the place at infinity of $k(u)$, and write $S = \{\infty\}$ if C is imaginary or unusual, and $S = \{\infty_1, \infty_2\}$ if C is real. Then the places of $k(C)$ are the prime ideals lying above the places of $k(u)$ (the *finite* places) together with the elements of S (the *infinite* places). To every place $\mathfrak{p}$ of $k(C)$ corresponds a normalized additive valuation $\nu_{\mathfrak{p}}$ on $k(C)$ and a discrete valuation ring $\mathcal{O}_{\mathfrak{p}} = \{\alpha \in k(C) \mid \nu_{\mathfrak{p}}(\alpha) \geq 0\}$; for brevity, we write $\nu_i = \nu_{\infty_i}$ $(i = 1, 2)$ if C is real. The *degree* $\deg(\mathfrak{p})$ of a place $\mathfrak{p}$ is the field extension degree $\deg(\mathfrak{p}) = [\mathcal{O}_{\mathfrak{p}}/\mathfrak{p} : k]$. Note that $\deg(\infty) = 1$ if C is imaginary, $\deg(\infty) = 2$ if C is unusual, and $\deg(\infty_1) = \deg(\infty_2) = 1$ if C is real. The *norm* of a finite place $\mathfrak{p}$ is the polynomial $N(\mathfrak{p}) = \mathfrak{P}^{\deg(\mathfrak{p})} \in k[u]$, where $\mathfrak{P}$ is the unique place of $k(u)$ lying below $\mathfrak{p}$.

For any place $\mathfrak{p}$ of $k(C)$, denote by $k(C)_{\mathfrak{p}}$ the completion of $k(C)$ with respect to $\mathfrak{p}$. Then it is easy to see that the completions $k(C)_S$ of $k(C)$ with respect to the places in S are, respectively,

$$k(C)_S = \begin{cases} k(C)_\infty &= k\langle u^{-1/2}\rangle & \text{if } C \text{ is imaginary },\\ k(C)_\infty &= k'\langle u^{-1}\rangle & \text{if } C \text{ is unusual },\\ k(C)_{\infty_1} &= k(C)_{\infty_2} = k\langle u^{-1}\rangle & \text{if } C \text{ is real },\end{cases}$$

where $k' = k(\operatorname{sgn}(v))$ is a quadratic extension of k. For C imaginary or unusual, the embedding of $k(C)$ into $k(C)_S$ is unique, whereas for the real case, we have two embeddings of $k(C)$ into $k\langle u^{-1}\rangle$. Here, we number the indices so that $\nu_1(v) \leq \nu_2(v)$, and choose the embedding with $\deg(\alpha) = -\nu_1(\alpha)$ for all $\alpha \in k(C)$.

To unify our discussion over all hyperelliptic models, we henceforth interpret elements in $k(C)$ as series in powers of u^{-1}, where in the imaginary case, the exponents of these powers are half integers. All degrees of function field elements are then taken with respect to u; more exactly, we set

$$\deg(\alpha) = \deg_u(\alpha) = \begin{cases} -\nu_\infty(\alpha)/2 & \text{if } C \text{ is imaginary} \ , \\ -\nu_\infty(\alpha) & \text{if } C \text{ is unusual} \ , \\ -\nu_1(\alpha) \ = \ -\nu_2(\overline{\alpha}) & \text{if } C \text{ is real} \ , \end{cases}$$

for $\alpha \in k(C)$. Here, if $\alpha = a + bv \in k(C)$ with $a, b \in k(u)$, then $\overline{\alpha} = a - b(v + h)$ is the *conjugate* of α. Note that for imaginary curves, $\deg(\alpha)$ can be a half integer. The following properties are easily seen:

Lemma 3.1.

(a) *If C is imaginary, then* $\deg(v) = \deg(v + h) = g + 1/2$.
(b) *If C is unusual or real with* $\deg(f) = 2g+2$, *then* $\deg(v) = \deg(v+h) = g + 1$.
(c) *If C is real and* $\deg(f) \leq 2g + 1$, *then* $\deg(v) = g + 1$ *and* $\deg(v + h) = \deg(f) - (g + 1) \leq g$.

A *divisor*[b] is a formal sum $D = \sum_{\mathfrak{p}} \nu_{\mathfrak{p}}(D)\mathfrak{p}$ where $\mathfrak{p}$ runs through all the places of $k(C)$ and $\nu_{\mathfrak{p}}(D) = 0$ for all but finitely many places $\mathfrak{p}$. The *support* $\mathrm{supp}(D)$ of D is the set of places for which $\nu_{\mathfrak{p}}(D) \neq 0$, and the *degree* of D is $\deg(D) = \sum_{\mathfrak{p}} \nu_{\mathfrak{p}}(D) \deg(\mathfrak{p})$; this agrees with the notion of degree of a place. A divisor whose support is disjoint from S is a *finite* divisor. Every divisor D of $k(C)$ can be written uniquely as a sum of two divisors

$$D = D_S + D^S \quad \text{where } D_S \text{ is finite and } \mathrm{supp}(D) \subseteq S \ .$$

The norm map extends naturally to all finite divisors D_S via $\mathbb{Z}$-linearity, and we can now define the norm of any divisor D to be $N(D) = N(D_S)$.

For two divisors D_1 and D_2 of $k(C)$, we write $D_1 \geq D_2$ if $\nu_{\mathfrak{p}}(D_1) \geq \nu_{\mathfrak{p}}(D_2)$ for all places $\mathfrak{p}$ of $k(C)$. With this notation, we see that $k[C]$ is the set of all $\alpha \in k(C)$ with $\mathrm{div}(\alpha)_S \geq 0$ and its unit group $k[C]^*$ consists of exactly those $\alpha \in k(C)$ with $\mathrm{div}(\alpha)_S = 0$.

---

[b] An equivalent geometric definition of a divisor (defined over k) that is frequently used in the literature on hyperelliptic curves is as follows: it is a formal sum $D = \sum_P \nu_P(D)P$ that is invariant under the Galois action of k, where P runs through all the points on C with coordinates in some algebraic closure of k. The degree of D is then simply $\sum_P \nu_P(D)$.

Let $\mathcal{D}$ denote the group of divisors of $k(C)$, $\mathcal{D}^0$ the subgroup of $\mathcal{D}$ of degree 0 divisors of $k(C)$, and $\mathcal{P}$ the subgroup of $\mathcal{D}^0$ of principal divisors of $k(C)$. Then the *degree 0 divisor class group* $\mathrm{Pic}^0 = \mathcal{D}^0/\mathcal{P}$ of $k(C)$ is a finite Abelian group whose order h is the *(degree 0 divisor) class number* of C.

Recall that the conjugation map on $k(C)$, arising from the hyperelliptic involution on C, maps each element $\alpha = a + bv \in k(C)$ with $a, b \in k(u)$ to $\overline{\alpha} = a - b(v + h)$. This map thus acts on all the finite places of $k(C)$ as well as on S via $\overline{\infty} = \infty$ if C is imaginary or unusual and $\overline{\infty}_1 = \infty_2$ if C is real. This action extends naturally to the groups $\mathcal{D}$, $\mathcal{D}^0$, $\mathcal{P}$, and hence to Pic^0. Note that $N(D) = N(\overline{D})$ and $D + \overline{D} = \mathrm{div}(N(D))$ for any degree 0 divisor D.

Define $\mathcal{D}_S = \{D_S \mid D \in \mathcal{D}\}$, $\mathcal{D}^S = \{D^S \mid D \in \mathcal{D}\}$, $\mathcal{P}_S = \mathcal{P} \cap \mathcal{D}_S$, and $\mathcal{P}^S = \mathcal{P} \cap \mathcal{D}^S$. By Proposition 14.1, p. 243, of [12], there are exact sequences

$$(0) \to k^* \to k[C]^* \to \mathcal{P}^S \to (0) \ , \tag{3.2}$$

$$(0) \to (\mathcal{D}^S \cap D^0)/\mathcal{P}^S \to \mathrm{Pic}^0 \to \mathcal{D}_S/\mathcal{P}_S \to \mathbb{Z}/f\mathbb{Z} \to (0) \ , \tag{3.3}$$

where $f = \gcd\{\deg(\mathfrak{p}) \mid \mathfrak{p} \in S\}$, so $f = 2$ if C is unusual and $f = 1$ otherwise. If C is imaginary or unusual, then $\mathcal{D}^S \cap D^0 = \mathcal{P}^S = 0$, whereas if C is real, then $\mathcal{D}^S \cap D^0 = \langle \infty_1 - \infty_2 \rangle$ and $\mathcal{P}^S = \langle R(\infty_1 - \infty_2) \rangle$, where R is the order of the divisor class of $\infty_1 - \infty_2$ in Pic^0 and is called the *regulator* of C. The principal divisor $R(\infty_1 - \infty_2)$ is the divisor of a fundamental unit of $k(C)$, i.e. a generator of the infinite cyclic group $k[C]^*/k^*$. For completeness, if C is imaginary or unusual, simply define the regulator of C to be $R = 1$.

A *fractional* $k[C]$*-ideal* is a subset $\mathfrak{f}$ of $k(C)$ such that $d\mathfrak{f}$ is a $k[C]$-ideal for some non-zero $d \in k[u]$. Let $\mathcal{I}$ denote the group of non-zero fractional $k[C]$-ideals, $\mathcal{H}$ the subgroup of $\mathcal{I}$ of non-zero principal fractional $k[C]$-ideals (which we write as (α) for $\alpha \in k(C)^*$), $\mathcal{C} = \mathcal{I}/\mathcal{H}$ the ideal class group of $k(C)$, and $h' = |\mathcal{C}|$ the ideal class number of $k(C)$. There is a natural isomorphism

$$\Phi : \mathcal{D}_S \to \mathcal{I}, \quad D_S \mapsto \{\alpha \in k(C)^* \mid \mathrm{div}(\alpha)_S \geq D_S\} \tag{3.4}$$

with inverse

$$\Phi^{-1}: \quad \mathcal{I} \to \mathcal{D}_S$$

$$\mathfrak{f} \mapsto D_S = \sum_{\mathfrak{p} \notin S} m_{\mathfrak{p}} \mathfrak{p} \quad \text{where } m_{\mathfrak{p}} = \min\{\nu_{\mathfrak{p}}(\alpha) \mid \alpha \in \mathfrak{f} \text{ non-zero}\} \ .$$

The conjugate $\overline{\mathfrak{f}}$ of a fractional ideal $\mathfrak{f}$ is the image of $\mathfrak{f}$ under the conjugation map. If $\mathfrak{f}$ is non-zero, then the *norm* $N(\mathfrak{f})$ of $\mathfrak{f}$ is simply $N(\Phi^{-1}(\mathfrak{f}))$, the norm of the finite divisor corresponding to $\mathfrak{f}$ under Φ^{-1}, with Φ given by Eq. (3.4). Note that $\mathfrak{f}\bar{\mathfrak{f}}$ is the principal fractional ideal generated by $N(\mathfrak{f})$.

The isomorphism Φ extends to an isomorphism from the factor group $\mathcal{D}_S/\mathcal{P}_S$ onto the ideal class group $\mathcal{C}$ (see p. 401 of [4] and Theorem 14.5, p. 247, of [12]). Thus, we have $h = Rh'/f$ by Eq. (3.3). The Hasse-Weil bounds $(\sqrt{q}-1)^{2g} \leq h \leq (\sqrt{q}+1)^{2g}$ imply $h \sim q^g$, and for real curves, we generally expect that h' is small and hence $R \approx h$. The isomorphism Φ in Eq. (3.4) can further be extended to the group $\mathcal{D}^0$, or a subgroup thereof, as follows.

3.1. *Imaginary Curves*

Since $\deg(\infty) = 1$ in this case, every degree 0 divisor of $k(C)$ can be written uniquely in the form $D = D_S - \deg(D_S)\infty$. Hence, every degree 0 divisor D is uniquely determined by D_S, and the isomorphism in Eq. (3.4) extends naturally to an isomorphism $\mathcal{D}^0 \to \mathcal{I}$.

3.2. *Unusual Curves*

Here, $\deg(\infty) = 2$, so every degree 0 divisor D of $k(C)$ can be written as $D = D_S - (\deg(D_S)/2)\infty$ and must have $\deg(D_S)$ even. Again, every degree 0 divisor D is uniquely determined by D_S. Thus, Φ as given in Eq. (3.4) extends to an isomorphism from $\mathcal{D}^0$ onto the group of fractional ideals whose norm have even degree.

3.3. *Real Curves*

If C is real, then $\deg(\infty_1) = \deg(\infty_2) = 1$, so every degree 0 divisor of $k(C)$ can be uniquely written in the form

$$D = D_S - \deg(D_S)\infty_2 + \nu_1(D)(\infty_1 - \infty_2) \ .$$

Hence, every degree 0 divisor D is uniquely determined by D_S and $\nu_1(D)$. Here, Φ extends to an isomorphism from the subgroup of $\mathcal{D}^0$ of degree 0 divisors D with $\nu_1(D) = 0$ onto $\mathcal{I}$.

We conclude this section with the observation that the choice of the transcendental element u determines the signature at infinity (ramified, inert, or split) and hence the set S of places lying above infinity. So the ideal class group $\mathcal{C}$, its order h', and the regulator R depend on the model of C (imaginary, unusual, or real), whereas the genus g, the divisor groups $\mathcal{D}$, $\mathcal{D}^0$ and $\mathcal{P}$, as well as the degree 0 divisor class group Pic^0 and its order h are model-independent.

4. Reduced Ideals and Divisors

Some of the material in this and the next section can be found in [2], [5], and [7]. As before, let $C : v^2 + h(u)v = f(u)$ be a hyperelliptic curve of genus g over a finite field k. The maximal order $k[C]$ of $k(C)$ is an integral domain and a $k[u]$-module of rank 2 with $k[u]$-basis $\{1, v\}$. The non-zero integral ideals in $k[C]$ are exactly the $k[u]$-modules of the form $\mathfrak{a} = k[u]\, SQ + k[u]\, S(P+v)$ where $P, Q, S \in k[u]$ and Q divides $f + hP - P^2$. Here, S and Q are unique up to factors in k^* and P is unique modulo Q. For brevity, write $\mathfrak{a} = S(Q, P)$. An ideal $\mathfrak{a} = S(Q, P)$ is *primitive* if $S \in k^*$, in which case we simply take $S = 1$ and write $\mathfrak{a} = (Q, P)$. A primitive ideal $\mathfrak{a}$ is *reduced* if $\deg Q \leq g$. The basis Q, P of a primitive ideal $\mathfrak{a} = (Q, P)$ is *adapted* if $\deg(P) < \deg(Q)$ and *reduced* if C is real and $\deg(P - h - v) < \deg(Q) < \deg(P + v)$; the latter is only possible if C is real. In practice, it is common to have reduced divisors given in adapted form for imaginary and unusual curves and in reduced (or possibly adapted) form for real curves.

A divisor D of $k(C)$ is *effective* if $D \geq 0$. An effective finite divisor D_S is *semi-reduced* [c] if there does not exist any subset $U \subseteq \text{supp}(D_S)$ such that $\sum_{\mathfrak{p} \in U} \nu_{\mathfrak{p}}(D_S)\mathfrak{p}$ is the divisor of a polynomial in $k[u]$, and *reduced* if in addition $\deg(D_S) \leq g$. Under the isomorphism in Eq. (3.4), effective finite divisors of $k(C)$ map to integral $k[C]$-ideals, semi-reduced divisors to primitive ideals, and reduced divisors to reduced ideals. Analogous to the ideal notation, we write $D_S = (Q, P)$ for the semi-reduced divisor of $k(C)$ corresponding to the primitive $k[C]$-ideal $\mathfrak{a} = (Q, P)$ under Φ, and refer to the polynomials Q and P as a *basis* of D_S; note that $N(D_S) = N(\mathfrak{a}) = \text{sgn}(Q)^{-1}Q$. It is easy to see that the conjugation map of $k(C)$ acts on semi-reduced and reduced divisors $D_S = (Q, P)$ via $\overline{D}_S = (Q, -P - h)$.

Up to now, we have only defined the notions of reduced and semi-reduced for finite divisors. We simply extend this notion to arbitrary degree 0 divisors of $k(C)$ by declaring a degree 0 divisor D to be (semi-)reduced if D_S is (semi-)reduced. We then say that a semi-reduced divisor D is in *adapted* or *reduced* form if D_S is given by an adapted or reduced basis, respectively.

We would like to represent degree 0 divisor classes via reduced divisors. In the imaginary case, this is well-known, but we repeat it briefly here for completeness; for the other two hyperelliptic curve models, it is less simple. In particular, for the unusual case, reduced divisors need not exist in some

[c] For geometric and ideal-independent definitions of the notions of semi-reduced and reduced divisors, see for example [2] or [5].

divisor classes, so we will have to allow divisors D with $\deg(D_S) = g+1$ when representing elements in Pic^0. For simplicity, we will say that a degree 0 divisor D in a given class $\mathbf{C} \in \mathrm{Pic}^0$ has *minimal norm* if D is semi-reduced and $\deg(N(E)) \geq \deg(N(D))$ for every semi-reduced divisor $E \in \mathbf{C}$. We will see that if C is imaginary, unusual with g even, or real, then D will always be reduced, otherwise (C unusual and g odd), we have $\deg(N(D)) \leq g+1$.

4.1. *Imaginary Curves*

Here, it is well-known that reduced divisors are pairwise inequivalent (see [2]), and every degree 0 divisor class in Pic^0, and hence every ideal class in $\mathcal{C}$, has exactly one reduced representative.

4.2. *Unusual Curves*

Again, reduced degree 0 divisors are pairwise inequivalent, and every degree 0 divisor class contains at most one reduced divisor. Those classes that do not contain any reduced divisor contain exactly $q+1$ pairwise equivalent semi-reduced divisors D with $\deg(D_S) = g+1$ (see p. 183 of [1]). Note that this can only occur if g is odd, so in this case, the norm of a reduced divisor must have degree $\leq g-1$. Hence if g is even, then in complete analogy to the imaginary case, every divisor class does in fact have a unique representative. In order to represent divisor classes without reduced divisors, i.e. with $q+1$ pairwise equivalent divisors of minimal norm of degree $g+1$, for g odd, a fast equivalence test or a systematic efficient way to identity a distinguished divisor of minimal norm in a given degree 0 divisor class are required.

4.3. *Real Curves*

By Proposition 4.1 of [10], every degree 0 divisor class of $k(C)$ contains a unique[d] reduced divisor D such that $0 \leq \deg(D_S) + \nu_1(D) \leq g$, or equivalently, $-g \leq \nu_2(D) \leq 0$. Using these reduced representatives for arithmetic in Pic^0 is somewhat slower that for imaginary curves, so we concentrate instead on reduced divisors $D = D_S - \deg(D_S)\infty_2$ with $\nu_1(D) = 0$. By the Paulus-Rück result cited above, these divisors are pairwise inequivalent, so every degree 0 divisor class of $k(C)$ contains at most one such reduced divisor.

[d] The proposition as stated in [10] reads "$0 \leq \nu_1(D) \leq g - \deg(D_S)$". The correct statement is "$0 \leq \deg(D_S) + \nu_1(D) \leq g$".

Rather than examining degree 0 divisor classes, we now consider ideal classes of $k(C)$. Recall that the isomorphism Φ defined in Eq. (3.4) can be extend to an isomorphism from the set $\{D \in \mathcal{D}^0 \mid \nu_1(D) = 0\}$ onto $\mathcal{I}$. For any non-zero fractional ideal $\mathfrak{f}$, set $D(\mathfrak{f}) = \Phi^{-1}(\mathfrak{f})$ to be the divisor with no support at ∞_1 corresponding to $\mathfrak{f}$; note that $\mathfrak{f}$ is reduced if and only if $D(\mathfrak{f})$ is reduced. Let $\mathbf{C}$ be any ideal class of $k(C)$, and define the set

$$\mathcal{R}_{\mathbf{C}} = \{D(\mathfrak{a}) \mid \mathfrak{a} \in \mathbf{C} \text{ reduced}\} \ .$$

By our above remarks, all the divisors in $\mathcal{R}_{\mathbf{C}}$ are reduced and pairwise inequivalent even though the corresponding ideals are all equivalent. Since the basis polynomials of a reduced divisor or ideal have bounded degree, $\mathcal{R}_{\mathbf{C}}$ is a finite set.

We now fix any reduced ideal $\mathfrak{a} \in \mathbf{C}$; for example, if $\mathbf{C}$ is the principal ideal class, then we always chose $\mathfrak{a} = (1)$ to be the trivial ideal. Then for every $\mathfrak{b} \in \mathbf{C}$, there exists $\alpha \in k(C)^*$ with $\mathfrak{b} = (\alpha)\mathfrak{a}$; if $\mathfrak{a} = (1)$, then α is in fact a generator of $\mathfrak{b}$. By multiplying α with a suitable power of a fundamental unit of $k(C)$, or equivalently, adding a suitable multiple of $R(\infty_1 - \infty_2)$ to its divisor, we may assume that $-R < \nu_1(\alpha) \leq 0$, or equivalently, $0 \leq \deg(\alpha) < R$. Then we define the *distance* of the divisor $D(\mathfrak{b})$ (with respect to $D(\mathfrak{a})$) to be $\delta(D(\mathfrak{b})) = \deg(\alpha)$. It follows that the set $\mathcal{R}_{\mathbf{C}}$ is ordered by distance, and if we set $D_1 = D(\mathfrak{a})$ and $r_{\mathbf{C}} = |\mathcal{R}_{\mathbf{C}}|$, then we can write

$$\mathcal{R}_{\mathbf{C}} = \{D_1, D_2, \ldots, D_{r_{\mathbf{C}}}\}$$

and $\delta_i = \delta(D_i)$, with $0 = \delta_1 < \delta_2 < \cdots < \delta_{r_{\mathbf{C}}} < R$. The set $\mathcal{R}_{\mathbf{C}}$ is called the *infrastructure* of $\mathbf{C}$; we will motivate this term later on. Note that if $\mathbf{C}$ is the principal class and $\mathfrak{b} \in \mathbf{C}$, $D(\mathfrak{b})$ and $D(\overline{\mathfrak{b}})$ both belong to $\mathcal{R}_{\mathbf{C}}$, and $\delta(D(\overline{\mathfrak{b}})) = R + \deg(D(\mathfrak{b})_S) - \delta(D(\mathfrak{b}))$ if $\mathfrak{b}$ is nontrivial.

5. Reduction and Baby Steps

We continue to assume that we have a hyperelliptic curve C given by Eq. (3.1). Our goal is to develop a unified framework for reduction on all hyperelliptic curves. We begin with the standard approach for reduction on imaginary curves — which we however apply to any hyperelliptic curve — and then link this technique to the traditional continued fractions method for real curves.

Starting with polynomials R_0, S_0 such that $\deg(R_0) < \deg(S_0)$ and S_0

dividing $f + hR_0 - R_0^2$, $\deg(S_0)$ even if C is unusual, the recursion

$$S_{i+1} = \frac{f + hR_i - R_i^2}{S_i} \; , \quad R_{i+1} = h - R_i + \left\lfloor \frac{R_i - h}{S_{i+1}} \right\rfloor S_{i+1} \; , \tag{5.1}$$

produces a sequence of semi-reduced, pairwise equivalent divisors $E_i = (S_{i-1}, R_{i-1})$, $i \in \mathbb{N}$. To avoid the costly full division in the expression for S_{i+1}, we can rewrite Eq. (5.1) as follows. Given S_0 and R_0, generate S_1 and R_1 using Eq. (5.1) and $s_1 = \lfloor (R_0 - h)/S_1 \rfloor$. Then for $i \in \mathbb{N}$:

$$\begin{aligned} S_{i+1} &= S_{i-1} + s_i(R_{i-1} - R_i) \; , \quad s_{i+1} = \left\lfloor \frac{R_i - h}{S_{i+1}} \right\rfloor \; , \\ R_{i+1} &= h - R_i + s_{i+1}S_{i+1} \equiv h - R_i \pmod{S_{i+1}} \; . \end{aligned} \tag{5.2}$$

Note that s_{i+1} and R_{i+1} are simply obtained by applying the division algorithm, i.e. $R_i - h = s_{i+1}S_{i+1} + (-R_{i+1})$ and $\deg(-R_{i+1}) < \deg(S_{i+1})$. Similar to [2] and [15], we derive the following properties.

Lemma 5.1.

(a) $\deg(R_i) < \deg(S_i)$ *for all* $i \geq 0$, *so all the* E_i *are in adapted form.*
(b) If $\deg(S_i) \geq g + 2$, *then* $\deg(S_{i+1}) \leq \deg(S_i) - 2$.
(c) If $\deg(S_i) = g + 1$, *then* $\deg(S_{i+1}) \leq g$ *if* C *is imaginary and* $\deg(S_{i+1}) = g + 1$ *if* C *is unusual or real. Hence, unless* C *is real,* E_{i+2} *has minimal norm.*
(d) There is a minimal index j *such that* $\deg(S_j) \leq \deg(v) < \deg(S_{j-1})$, *so unless* C *is real,* E_{j+1} *is the first of divisor of minimal norm. We have* $j \leq \lceil (\deg(S_0) - g)/2 \rceil$ *if* $\deg(S_j) \leq g$ *and* $j \leq \lceil (\deg(S_0) - g - 1)/2 \rceil$ *if* $\deg(S_j) = g + 1$.
(e) If C *is unusual, then* $\deg(S_i)$ *is even for all* $i \geq 0$.

Proof. (a) is obvious from Eq. (5.1). Since $\deg(h) \leq g + 1$, Eq. (5.1) and (a) imply

$$\begin{aligned} \deg(S_{i+1}) &= \deg(f + hR_i - R_i^2) - \deg(S_i) \\ &\leq \max\{\deg(f), \deg(S_i) + g, 2\deg(S_i) - 2\} - \deg(S_i) \; , \end{aligned} \tag{5.3}$$

yielding (b) and (c). Now (d) can easily be derived from (b) and (c). To see (e), note that if $\deg(R_i) \geq g + 1$, then $\deg(S_i) \geq g + 2$, so by Eq. (5.3), $\deg(S_{i+1}) = 2\deg(R_i) - \deg(S_i)$ (note that by the assumptions on $\mathrm{sgn}(f)$, there can never be cancellation in the numerator of S_{i+1} in the case where $\deg(R_i) = g + 1$), and if $\deg(R_i) \leq g$, then $\deg(S_{i+1}) = 2g + 2 - \deg(S_i)$. In either case, $\deg(S_{i+1})$ has the same parity as $\deg(S_i)$, so (e) is obtained by induction, since $\deg(S_0)$ was assumed to be even if C is unusual. □

Suppose $\deg(S_j) \leq \deg(v) < \deg(S_{j-1})$ as in part (d) of Lemma 5.1. If C is imaginary, or C is unusual with $\deg(S_j) \leq g$, then E_{j+1} is the unique reduced divisor in the class of D_1. If C is unusual and $\deg(S_j) = g+1$, (g odd), then the other q semi-reduced divisors equivalent to E_{j+1} whose norm have degree $g+1$ can be obtained from E_{j+1} as follows (see also [1] for the case where q is odd).

Proposition 5.1. *Let C given by Eq. (3.1) be unusual of odd genus g and $E = (S, R)$ a semi-reduced divisor with* $\deg(R) \leq \deg(S) = g+1$. *Then the $q+1$ divisors in the divisor class of E whose norm have degree $g+1$ are given by E and $E_a = (S_a, R_a)$ for $a \in \mathbb{F}_q$ where*

$$R_a = h - R + aS, \qquad S_a = \frac{f + hR_a - R_a^2}{S} \ . \tag{5.4}$$

Proof. Since $E_a = E + \mathrm{div}((R_a + v)/S)$ for all $a \in \mathbb{F}_q$, all E_a are equivalent to E. Furthermore, $\deg(R_a) \leq g+1$ and hence $\deg(S_a) = g+1$, since the conditions on $\mathrm{sgn}(f)$ prevent cancellation of leading terms in the numerator of S_a. So it only remains to show that E and and all the E_a are pairwise distinct. To that end, we prove that equality among any two of these $q+1$ divisors leads to a sequence of divisibility conditions that yield a singular point on C.

So fix $a \in \mathbb{F}_q$ and suppose that $E_a = E$ or $E_a = E_b$ for some $b \in \mathbb{F}_q \backslash \{a\}$. We first claim that

$$S_a \text{ and } S \text{ differ by a constant factor in } \mathbb{F}_q \ . \tag{5.5}$$

This is clear if $E_a = E$, so suppose $E_a = E_b$ with $b \in \mathbb{F}_q$, $b \neq a$. Then S_a and S_b differ by a constant factor, and $R_a \equiv R_b \pmod{S_a}$. By Eq. (5.4), $R_a \equiv R_b \pmod{S}$, so since $\deg(R_a - R_b) = \deg(S_a) = \deg(S) = g+1$, we see that S_a and S must also differ by a constant factor in $\mathbb{F}_q$.

Next, we claim that

$$S \text{ divides } 2R - h \ . \tag{5.6}$$

If $E_a = E$, then $R \equiv R_a \pmod{S}$. On the other hand, $R_a \equiv h - R \pmod{S}$ by Eq. (5.4), so $R \equiv h - R \pmod{S}$, proving Eq. (5.6). Suppose now that $E_a = E_b$ for some $b \in \mathbb{F}_q$ distinct from a. Then S_a and S_b differ by a constant factor, so by Eq. (5.5), both differ from S by a constant factor. Now a simple calculation yields $S_a - S_b = (a-b)(2R - h - (a+b)S)$. Since $a \neq b$ and S divides the left hand side of this equality, S must again divide $2R - h$.

Our next assertion is that

$$S^2 \text{ divides } f + hR - R^2 \ . \tag{5.7}$$

By Eq. (5.4) and Eq. (5.6), $R_a \equiv h - R \equiv R \pmod{S}$. Since $\deg(R_a - R) \leq g + 1 = \deg(S)$, there exists $c_a \in \mathbb{F}_q$ with $R_a = R + c_a S$. Substituting into Eq. (5.4) yields $SS_a = f + hR - R^2 + c_a S(h - 2R - c_a S)$. By Eq. (5.5), S^2 divides the left hand side of this equality. Invoking Eq. (5.6), we obtain Eq. (5.7).

Our fourth and final claim is that

$$S \text{ divides } f' + hR' \ , \tag{5.8}$$

where f' denotes the derivative of f with respect to u; similarly for R'. To prove this claim, we simply observe that taking derivatives in Eq. (5.7) implies that S divides $f' + h'R + hR' - 2RR' = f' + hR' + R'(h - 2R)$, so Eq. (5.8) now follows from Eq. (5.6).

Now let r be a root of S in some algebraic closure of k. Then Eq. (5.6)–Eq. (5.8) easily imply that $(r, -R(r))$ is a singular point on C, a contradiction. So no two among the divisors E and E_a $(a \in \mathbb{F}_q)$ can be be equal, proving the proposition. □

We now relate Eq. (5.1) to a regular continued fraction expansion, which is the usual approach to reduction on real curves. Let $P, Q \in k[u]$ with Q non-zero and Q dividing $f + hP - P^2$, and let $s_0, s_1, \ldots$ be a sequence of polynomials in $k[u]$. Set $P_0 = P$, $Q_0 = Q$, and

$$P_{i+1} = h - P_i + s_i Q_i, \quad Q_{i+1} = \frac{f + hP_{i+1} - P_{i+1}^2}{Q_i} \ , \tag{5.9}$$

for $i \geq 0$. If we set $\phi_i = (P_i + v)/Q_i$, then $\phi_{i+1} = (\phi_i - s_i)^{-1}$, so $\phi_0 = [s_0, s_1, \ldots, s_i, \phi_{i+1}]$ for all $i \geq 0$. Thus, Eq. (5.9) determines a continued fraction expansion of ϕ_0 in the completion $k(C)_S$. It is clear that Eq. (5.9) defines a sequence $D_i = (Q_{i-1}, P_{i-1})$ of semi-reduced divisors with corresponding primitive ideals $\mathfrak{a}_i$. The operation $D_i \to D_{i+1}$ is referred to as a *baby* or *reduction step*[e].

Set $\theta_1 = 1$ and $\theta_i = \prod_{j=1}^{i-1} \phi_j^{-1}$ for $i \geq 2$. Since $\phi_i \overline{\phi}_i = -Q_{i-1}/Q_i$, it is easy to see that $Q_0 \theta_i \overline{\theta}_i = (-1)^{i-1} Q_{i-1}$. Thus

$$\overline{\theta}_i = \prod_{j=1}^{i-1} \overline{\phi}_j^{-1} = (-1)^{i-1} \frac{Q_{i-1}}{Q_0 \theta_i} = (-1)^{i-1} \frac{Q_{i-1}}{Q_0} \prod_{j=1}^{i-1} \phi_j \ . \tag{5.10}$$

[e]Note that Eq. (5.4) is a special case of Eq. (5.9), with $s_i = a \in \mathbb{F}_q$. However, in this case, the recursion only alternates between E and E_a.

Then $\mathfrak{a}_{i+1} = (\overline{\phi_i^{-1}})\mathfrak{a}_i$ and hence $\mathfrak{a}_i = (\overline{\theta}_i)\mathfrak{a}_1$, for $i \in \mathbb{N}$. Therefore, the ideals $\mathfrak{a}_i$ are all equivalent, so baby steps preserve ideal equivalence.

If we choose s_i in Eq. (5.9) to be $s_i = q_i = \lfloor \phi_i \rfloor$, i.e. the quotient in the regular continued fraction expansion of ϕ_0 in $k(C)_S$, then we have the baby steps

$$q_i = \left\lfloor \frac{P_i + v}{Q_i} \right\rfloor , \; P_{i+1} = h - P_i + q_i Q_i, \; Q_{i+1} = \frac{f + hP_{i+1} - P_{i+1}^2}{Q_i} \; . \quad (5.11)$$

If $\deg(Q_i) > \deg(v)$, then $q_i = \lfloor P_i/Q_i \rfloor$. It is now easy to deduce that if j is as in part (d) of Lemma 5.1 and S_i, R_i are defined as in Eq. (5.1), then

$$q_i = \lfloor P_i/Q_i \rfloor \in k[u], \quad P_{i+1} = h - R_i, \quad Q_{i+1} = S_{i+1} \; , \quad (5.12)$$

for $0 \le i < j$. Therefore, for this range of indices, Eq. (5.11) is equivalent to Eq. (5.1) and hence produces the same sequence of divisors. For imaginary and unusual curves, we will only consider baby steps as in Eq. (5.11) in the range $0 \le i < j$. For C real, baby steps as in Eq. (5.11) can be performed beyond that range as well. However, for $i \ge j$, $q_j \ne \lfloor P_j/Q_j \rfloor$, so Eq. (5.12) is false. Here, if we use Eq. (5.11) to compute the sequence $D_{i+1} = (Q_i, P_i)$, starting with $i = j$, then D_{i+1} is reduced for $i > j$. We have $\deg(P_{j+1} - h - v) \le g$, $\deg(P_{j+1} + v) = g + 1$, and for $i \ge j + 2$, $D_{i+1} = (Q_i, P_i)$ is in reduced form.

We now see that for all hyperelliptic curves, there exists an index $l \ge 0$ such that Eq. (5.11) repeatedly applied to $D_1 = (Q_0, P_0)$ produces a reduced divisor D_{l+1}, if one exists, after $l \le \lceil (\deg(Q_0) - g)/2 \rceil$ steps. If C is unusual, g is odd, and the class of D_1 contains no reduced divisor, then Eq. (5.11) produces a divisor D_{l+1} whose norm has degree $g + 1$ after $l \le \lceil (\deg(Q_0) - g - 1)/2 \rceil$ steps. In the imaginary and unusual scenarios, we have $l = j$ with j as in part (d) of Lemma 5.1; for C real, we have $l = j + 1$. For $0 \le i < l$, Eq. (5.11) is equivalent to

$$\begin{aligned} q_i &= \left\lfloor \frac{P_i + e_i v}{Q_i} \right\rfloor , \quad e_i = \begin{cases} 1 \text{ if } C \text{ real, } \deg(Q_i) = g + 1, \\ 0 \text{ otherwise,} \end{cases} \\ P_{i+1} &= h - P_i + q_i Q_i, \quad Q_{i+1} = \frac{f + hP_{i+1} - P_{i+1}^2}{Q_i} \; . \end{aligned} \quad (5.13)$$

Again, the recursion in Eq. (5.13) can be made more efficient for $i \ge 1$, i.e. for all but the first baby step. Given Q_0 and P_0, we compute Q_1 and P_1 using Eq. (5.13). Then for $i \in \mathbb{N}$:

$$\begin{aligned} q_i &= \left\lfloor \frac{P_i + \lfloor e_i v \rfloor}{Q_i} \right\rfloor , \quad r_i \equiv P_i + \lfloor e_i v \rfloor \pmod{Q_i} \; , \\ P_{i+1} &= h + \lfloor e_i v \rfloor - r_i \; , \quad Q_{i+1} = Q_{i-1} + q_i (r_i - r_{i-1}) \; . \end{aligned} \quad (5.14)$$

As before, the first line in Eq. (5.14) is equivalent to applying the division algorithm in order to compute polynomials q_i and r_i such that $P_i + \lfloor e_i v \rfloor = q_i Q_i + r_i$ and $\deg(r_i) < \deg(Q_i)$.

Suppose now that C is real. If we repeatedly apply Eq. (5.11), or equivalently, Eq. (5.14), starting with a reduced divisor $D_1 = D(\mathfrak{a})$ for some reduced ideal $\mathfrak{a}$ of $k(C)$, then we can generate the entire infrastructure $\mathcal{R}_{\mathbf{C}} = \{D_i \mid 1 \leq i \leq r_{\mathbf{C}}\}$ of the ideal class $\mathbf{C}$ containing $\mathfrak{a}$. Here, $D_i = D(\mathfrak{a}_i)$ where $\mathfrak{a}_i = (\overline{\theta}_i)\mathfrak{a}$ with $\overline{\theta}_i$ as in Eq. (5.10), so the distance of D_i is $\delta_i = \deg(\overline{\theta}_i)$. In particular, $\overline{\theta}_{r_{\mathbf{C}}}$ is a fundamental unit of $k(C)$ of positive degree, and $\deg(\overline{\theta}_{r_{\mathbf{C}}}) = R$ is the regulator of $k(C)$.

We conclude this section by showing how to compute the distances $\delta_i = \delta(D_i)$. By Eq. (5.10), the distance satisfies

$$\delta_i = \deg(\overline{\theta}_i) = \deg(Q_{i-1}) - \deg(Q_0) + \sum_{j=1}^{i-1} \deg(q_j) \tag{5.15}$$

for $i \in \mathbb{N}$. Since $\overline{\phi}_i = (P_i - h - v)/Q_i = -Q_{i-1}/(P_i + v)$ and $\delta_{i+1} - \delta_i = -\deg(\overline{\phi}_i) = \deg(P_i + v) - \deg(Q_{i-1}) = g + 1 - \deg(Q_{i-1})$ by Eq. (5.10), we have $1 \leq \delta_{i+1} - \delta_i \leq g$ if D_i is non-zero, and $\delta_{i+1} = g + 1$ if $D_i = 0$, in which case $\mathbf{C}$ is the principal class.

6. Giant Steps and the Idea of NUCOMP

As before, let C be given by Eq. (3.1), and let $D' = (Q', P')$, $D'' = (Q'', P'')$ be two semi-reduced divisors of $k(C)$. Then it is well-known that there exists a semi-reduced divisor $D = (Q, P)$ in the divisor class of the sum $D' + D''$ that can be computed as follows.

$$\begin{aligned} S &= \gcd(Q', Q'', P' + P'' - h) = VQ' + WQ'' + X(P' + P'' - h) \ , \\ Q &= \frac{Q'Q''}{S^2} \ , \\ P &= P'' + U\,\frac{Q''}{S} \quad \text{with } U \equiv W(P' - P'') + XR'' \pmod{Q'/S} \ , \end{aligned} \tag{6.1}$$

where $U, V, W, X \in k[u]$, $\deg(U) < \deg(Q'/S)$, and $R'' = (f + hP'' - P''^2)/Q''$. Note that D is in adapted form if $\deg(P'') < \deg(Q)$.

Since S tends to have very small degree (usually $S = 1$), we expect $\deg Q \approx \deg Q' + \deg Q''$; in particular, even if D' and D'' have minimal norm, then D will generally not have minimal norm. We now apply repeated baby steps as in Eq. (5.14) to $P_0 = P$ and $Q_0 = Q$ until we obtain a divisor of minimal norm. The first divisor thus obtained is defined to be $D' \oplus D''$. The operation $(D', D'') \to D' \oplus D''$ is called a *giant step*.

6.1. *Imaginary Curves*

Here, $D' \oplus D''$ is the unique reduced divisor in the class of $D' + D''$, and the algorithm above is Cantor's algorithm [2]. Thus, the group operation on Pic^0 can be performed efficiently via reduced representatives.

6.2. *Unusual Curves*

In this case, if g is even, then everything is completely analogous to the imaginary setting. However, if g is odd, then $D' \oplus D''$ may or may not be reduced, so the set of reduced divisors is no longer closed under the operation $\oplus$. However, as mentioned earlier, if we could either perform fast equivalence testing, or efficiently and systematically identify a distinguished divisor D with $\deg(D_S) = g+1$ in every divisor class that contains no reduced divisor, then we could perform arithmetic in Pic^0 via these distinguished representatives plus reduced representatives if they exist.

6.3. *Real Curves*

Suppose D' and D'' are reduced, and $D' \in \mathcal{R}_{\mathbf{C}'}$, $D'' \in \mathcal{R}_{\mathbf{C}''}$ for suitable ideal classes $\mathbf{C}', \mathbf{C}''$ of $k(C)$. Then $D' \oplus D'' \in \mathcal{R}_{\mathbf{C}'\mathbf{C}''}$. In particular, if $\mathbf{C}''$ is the principal ideal class, then $D' \oplus D'' \in \mathcal{R}_{\mathbf{C}'}$, and we have

$$\delta(D' \oplus D'') = \delta(D') + \delta(D'') - \delta \quad \text{with} \quad 0 \le \delta \le 2g \ . \tag{6.2}$$

Here, distances in the principal class are taken with respect to $D_1 = 0$, and distances in $\mathbf{C}'$ with respect to some some suitable first divisor. The "error term" δ in Eq. (6.2) is linear in g and hence very small compared to the two distances $\delta(D')$ and $\delta(D'')$. The quantity δ in Eq. (6.2) can be efficiently computed as part of the giant step.

Suppose now that $D' = (Q', P')$ and $D'' = (Q'', P'')$ are two divisors of minimal norm. A giant step as described above finds the divisor $D' \oplus D''$ in two steps. First set $D_1 = (Q, P)$ with P and Q given by Eq. (6.1); Q and P have degree approximately $2g$, i.e. double size. Then apply repeated baby steps as in Eq. (5.14) to D_1 until the first divisor $D_{l+1} = D' \oplus D''$ of minimal norm is obtained; by Lemma 5.1, we have $l \le \lceil g/2 \rceil$ for all three curve models, so this takes at most $\lceil g/2 \rceil$ such steps. The reduction process produces a sequence of semi-reduced divisors $D_{i+1} = (Q_i, P_i)$, $0 \le i \le l$, via the continued fraction expansion of $\phi = (P+v)/Q = [q_0, q_1, \ldots, q_l, \phi_{l+1}]$. It slowly shrinks the degrees of the Q_i and P_i again to original size, reducing them by about 2 in each step by Lemma 5.1. The obvious disadvantage of this method is that the polynomials Q_i, P_i have large degree while i is small, and are costly to compute.

NUCOMP is an algorithm for computing $D' \oplus D''$ that eliminates these costly baby steps on large operands. The idea of NUCOMP is to perform arithmetic on polynomials of much smaller degree. Instead of computing Q as well as the Q_i and P_i explicitly via the continued fraction expansion of ϕ, one computes sequences of polynomials a_i, b_i, c_i, and d_i such that

$$\begin{aligned} Q_i &= (-1)^i(b_{i-1}c_{i-1} - a_{i-1}d_{i-1}) \\ P_i &= (-1)^i(b_{i-2}c_{i-1} - a_{i-1}d_{i-2}) + P'' \ . \end{aligned}$$

Only two basis coefficients Q_{n+2} and P_{n+2} are evaluated at the end in order to obtain a divisor D_{n+3}. Here, the value of n is determined by the property that a_n, b_n, c_n, and d_n have approximately equal degree of about $g/2$. More exactly, we will have $l = n + 2$ or $n + 3$, i.e. $D' \oplus D'' = D_{n+3}$ or D_{n+4}.

The key observation is that $\hat{\phi} = U/(Q'/S)$, with U as given in Eq. (6.1), is a very good rational approximation of $\phi = (P+v)/Q$, and that the continued fraction expansion of $\hat{\phi}^{-1}$ is given by $Q'/(SU) = [q_1, q_2, \ldots, q_{n+1}, \ldots]$. Note that $\deg(U) < \deg(Q'/S) \leq g$ (or possibly $g + 1$), so all quantities involved are of small degree. The polynomials a_i, b_i, c_i, and d_i are computed recursively along with the continued fraction expansion of $Q'/(SU)$ which is basically the extended Euclidean algorithm applied to Q'/S and U; in fact, the b_i are the remainders obtained in this Euclidean division process. Alternatively, only the a_i and b_i are computed recursively, and c_{n-1}, d_{n-1}, and d_n are then obtained from these two sequences; this approach turns out to employ polynomials of smaller degree (as c_0 and d_0 have large degree), but requires an extra full division by Q'/S. We describe the details of NUCOMP in the next two sections.

7. NUCOMP

Let $D' = (Q', P')$ and $D'' = (Q'', P'')$ be two divisors of minimal norm, and let P, Q, S, U be defined as in Eq. (6.1). We assume that

$$\deg(P'') \leq g + 1 < \deg(Q) \ . \tag{7.1}$$

The first inequality in Eq. (7.1) is equivalent to $\deg(P'' + v) \leq g + 1$, and holds if D'' is given in adapted or reduced form. While it can always be achieved by reducing P modulo Q, for example, we will see that this will generally not be necessary, i.e. usually NUCOMP outputs a divisor

$\hat{D} = (\hat{Q}, \hat{P})$ that again satisfies[f] $\deg(\hat{P}) \leq g + 1$.

The second inequality in Eq. (7.1) is no great restriction, since if $\deg(Q) \leq g + 1$, then $D = (Q, P)$ is at most one baby step away from having minimal norm, so one would simply compute $D' \oplus D''$ using one of the recursions in Section 5 and not use NUCOMP in this case. We now define

$$M = \max\{g \, , \deg(P'' + v)\} \in \frac{1}{2}\mathbb{Z} \, . \tag{7.2}$$

Note that $M \in \{g + 1/2, g + 1\}$ if C is imaginary, $M = \deg(P'' + v) = g + 1$ if C is unusual (since $\mathrm{sgn}(P'') \in k$ and $\mathrm{sgn}(v) \notin k$ can never cancel each other), and $M \in \{g, g + 1\}$ if C is real. Furthermore, if D'' is given in adapted or reduced form, then $M = \deg(P'' + v)$.

The quantity

$$N = \frac{1}{2}(\deg(Q') - \deg(Q'') + M) \in \frac{1}{4}\mathbb{Z} \tag{7.3}$$

will play a crucial role in our discussion. Since D'' is of minimal norm, we have $\deg(Q'') \leq M$ for all hyperelliptic curve models, so $N \geq \deg(Q')/2 > 0$. Furthermore, $N < \deg(Q'/S)$ by the second inequality in Eq. (7.1), so $N < g + 1$. Usually, we expect N to be of magnitude $g/2$.

Let $Q'/SU = [\hat{q}_0, \hat{q}_1, \ldots, \hat{q}_m]$ be the regular continued fraction expansion of Q'/SU, where as usual, $m \geq 0$ is minimal. Setting $E = Q'/S$ and $F = U$, Eq. (2.5) defines sequences a_i, b_i for $-1 \leq i \leq m$, i.e.

$$\begin{aligned} b_{-1} &= Q'/S \ , \ b_0 = U \ , \quad b_{i+1} = b_{i-1} - \hat{q}_i b_i \ , \\ a_{-1} &= 0 \ , \qquad a_0 = -1 \ , \ a_{i+1} = a_{i-1} - \hat{q}_i a_i \ . \end{aligned} \tag{7.4}$$

If we put $b_{-2} = U$ and $\hat{q}_{-1} = 0$, then for $i \geq -1$, the remainder sequence of the Euclidean algorithm applied to $\hat{\phi} = SU/Q'$ is the same as the one applied to $\hat{\phi}^{-1} = Q'/SU$ since $\deg(U) < \deg(Q'/S)$. The first step then simply reads $U = b_{-2} = 0 \cdot b_{-1} + b_0$. Since $Q'/SU = [\hat{q}_0, \hat{q}_1, \ldots, \hat{q}_m]$, we then see that the continued fraction expansion of $\hat{\phi}$ is $\hat{\phi} = [0, \hat{q}_0, \hat{q}_1, \ldots, \hat{q}_m]$.

Set $\hat{P}_0 = P$, $\hat{Q}_0 = Q$, and recall that $\hat{q}_{-1} = 0$. We investigate the sequence of semi-reduced divisors $\hat{D}_i = (\hat{Q}_{i-1}, \hat{P}_{i-1})$, $1 \leq i \leq m + 3$, obtained by choosing $s_i = \hat{q}_{i-1}$ in Eq. (5.9). That is

$$\hat{P}_{i+1} = h - \hat{P}_i + \hat{q}_{i-1}\hat{Q}_i, \quad \hat{Q}_{i+1} = \frac{f + h\hat{P}_{i+1} - \hat{P}_{i+1}^2}{\hat{Q}_i} \ , \tag{7.5}$$

[f] If C is unusual, g is odd, and $\deg(\hat{Q}) = g + 1$, then we expect $\deg(\hat{P}) \leq g + 2$. However, in this situation, it suffices to assume $\deg(P'') \leq g + 2$ as well. In order to avoid having to distinguish between too many different cases, we will henceforth ignore this scenario.

for $0 \le i \le m+1$. To facilitate the computation of $\hat{P}_i, \hat{Q}_i$, we proceed as in [8] and introduce two more sequences of polynomials c_i, d_i, $-1 \le i \le m+1$ as follows.

$$\begin{aligned} c_{-1} &= \frac{Q''}{S}, & c_0 &= \frac{P-P'}{b_{-1}}, & c_{i+1} &= c_{i-1} - \hat{q}_i c_i, \\ d_{-1} &= P' + P'' - h, & d_0 &= \frac{d_{-1}b_0 - SR''}{b_{-1}}, & d_{i+1} &= d_{i-1} - \hat{q}_i d_i, \end{aligned} \tag{7.6}$$

for $0 \le i \le m$. We point out an interesting symmetry between the sequences b_i and c_i, $-1 \le i \le m+1$; namely, reversing the roles of D' and D'' in Eq. (6.1) results in a swap of these two sequences. An easy induction yields

$$c_i = \frac{1}{b_{-1}}\left(b_i\frac{Q''}{S} + a_i(P' - P'')\right) , \tag{7.7}$$

$$d_i = \frac{1}{b_{-1}}\left(b_i(P' + P'' - h) + a_i SR''\right) , \tag{7.8}$$

for $-1 \le i \le m+1$. Using induction simultaneously on both formulas, we obtain

$$\hat{Q}_i = (-1)^i(b_{i-1}c_{i-1} - a_{i-1}d_{i-1}) , \tag{7.9}$$

$$\hat{P}_i = (-1)^i(b_{i-2}c_{i-1} - a_{i-1}d_{i-2}) + P'' , \tag{7.10}$$

for $0 \le i \le m+2$.

As outlined above, we wish to determine a point up to which the divisors $D_{i+1} = (Q_i, P_i)$ with $P_0 = P$, $Q_0 = Q$, and P_i, Q_i given by Eq. (5.13) or equivalently, by Eq. (5.11) or Eq. (5.14) are identical to the divisors $\hat{D}_{i+1} = (\hat{Q}_i, \hat{P}_i)$ with $\hat{P}_i, \hat{Q}_i$ given by Eq. (7.5) or equivalently, by Eq. (7.9) and Eq. (7.10). Clearly, $\hat{D}_1 = D_1$ by definition, so our goal is to find a maximal index $n \ge -1$ that guarantees $Q_i = \hat{Q}_i$ and $P_i = \hat{P}_i$, and hence $D_{i+1} = \hat{D}_{i+1}$, for $0 \le i \le n+2$ (see Theorem 7.1). Such an index will have to satisfy $n \le m$ to ensure that the polynomials $\hat{Q}_i, \hat{P}_i$ as given in Eq. (7.5) are in fact defined. Our next task will then be to see how many baby steps if any we need to apply to the last divisor $D_{n+3} = (Q_{n+2}, P_{n+2})$ to obtain the divisor $D' \oplus D''$.

Theorem 7.1. *Let $D' = (Q', P')$, $D'' = (Q'', P'')$ be two divisors, and let P and Q be given by Eq. (6.1). Set $P_0 = \hat{P}_0 = P$, $Q_0 = \hat{Q}_0 = Q$, and define P_i, Q_i ($i \in \mathbb{N}$) by Eq. (5.11), $\hat{P}_i, \hat{Q}_i$ ($1 \le i \le m+2$) by Eq. (7.5), and b_i ($-1 \le i \le m+1$) by Eq. (7.4). Then there exists $n \in \mathbb{Z}$, $-1 \le n \le m$, such* *that* $\deg(b_n) > N$, *with* N *as in Eq. (7.3). Furthermore,*

$$\begin{aligned} q_i &= \hat{q}_{i-1} \quad (0 \le i \le n+1) \ , \\ P_i &= \hat{P}_i \quad (0 \le i \le n+2) \ , \\ Q_i &= \hat{Q}_i \quad (0 \le i \le n+2) \ . \end{aligned}$$

Proof. We already observed that $\deg(b_{-1}) = \deg(Q'/S) > N$, so since $\deg(b_i)$ decreases as i increases, there must exist $n \ge -1$ with $\deg(b_n) > N$. Since $\deg(b_{m+1}) = -\infty < N$, we must have $n \le m$. So n as specified above exists and all the quantities $\hat{q}_{i-1}, \hat{P}_i, \hat{Q}_i$ above are in fact well-defined.

Set $\phi = (P+v)/Q$ and $\hat{\phi} = SU/Q'$. Then $\phi = [q_0, q_1, \ldots]$ with $q_i = (P_i + v)/Q_i$ is the continued fraction expansion of ϕ in a suitable field of Puiseux series; also, recall that $\hat{\phi} = [\hat{q}_{-1}, \hat{q}_0, \ldots, \hat{q}_m]$ where $\hat{q}_{-1} = 0$. We wish to apply[g] Theorem 2.1 to ϕ and $\hat{\phi}$. Since $\phi - \hat{\phi} = (P'' + v)/Q$, we have $b_{-1}^2(\phi - \hat{\phi}) = Q'(P'' + v)/Q''$. The definition of N implies $2N \ge \deg(Q'(P'' + v)/Q'')$, so

$$2\deg(b_n) > 2N \ge \deg\left(b_{-1}^2(\phi - \hat{\phi})\right) \ . \tag{7.11}$$

Set[h] $\hat{r}_{-2} = U$, $\hat{r}_{-1} = Q'/S$, and $\hat{r}_i = \hat{r}_{i-2} - \hat{q}_{i-1}\hat{r}_{i-1}$ for $0 \le i \le m+1$. Then $\hat{r}_i = b_i$ for $-1 \le i \le m+1$, so the $\hat{r}_i$ are the remainders when applying the Euclidean algorithm to $E = U$ and $F = Q'/S$. By Theorem 2.1, Eq. (7.11) implies that $q_i = \hat{q}_{i-1}$ for $0 \le i \le n+1$. Now $P_0 = \hat{P}_0$, $Q_0 = \hat{Q}_0$, and inductively by Eq. (5.11) and Eq. (7.5),

$$\begin{aligned} P_{i+1} &= h - P_i + q_i Q_i = h - \hat{P}_i + \hat{q}_{i-1}\hat{Q}_i = \hat{P}_{i+1} \ , \\ Q_{i+1} &= \frac{f + hP_{i+1} - P_{i+1}^2}{Q_i} = \frac{f + h\hat{P}_{i+1} - \hat{P}_{i+1}^2}{\hat{Q}_i} = \hat{Q}_{i+1} \ , \end{aligned}$$

for $0 \le i \le n+1$. □

Corollary 7.1. *With the notation of Theorem 7.1, we have* $D_i = \hat{D}_i$ *for* $1 \le i \le n+3$.

---

[g] Although the degrees in Theorem 2.1 are taken with respect to $u^{1/2}$ if C is imaginary, the statement still holds if degrees are taken with respect to u as is done here, since this only changes both sides of the degree inequality in Theorem 2.1 by a factor of 2.

[h] Note that the indices of the partial quotients $\hat{q}_i$ in the definition of the $\hat{r}_i$ are offset by 1 compared to the proof of Theorem 2.1 because here, the continued fraction in question is $\hat{\phi} = [\hat{q}_{-1}, \hat{q}_0, \hat{q}_1, \ldots, \hat{q}_m]$ (with $\hat{q}_{-1} = 0$), whereas in Theorem 2.1, it is $\hat{\phi} = [\hat{q}_0, \hat{q}_1, \ldots, \hat{q}_m]$.

Since $\deg(b_i)$ is a decreasing sequence for $-1 \le i \le m+1$, there exists a unique index n with $-1 \le n \le m$ such that

$$\deg(b_n) > N \ge \deg(b_{n+1}) \ , \tag{7.12}$$

with N as in Eq. (7.3). By Corollary 7.1, $D_i = \hat{D}_i$ for $1 \le i \le n+3$.

8. Giant Steps with NUCOMP

We now show that D_{n+3} is at most one baby step away from being reduced if C is imaginary or real, and always has minimal norm if C is unusual. Furthermore, D_{n+2} never has minimal norm. Note that this implies that if D_{n+3} actually has minimal norm, then $D_{n+3} = D' \oplus D''$.

Substituting Eq. (7.7) and Eq. (7.8) into Eq. (7.9) yields

$$\hat{Q}_i = \frac{(-1)^i}{b_{-1}} \left(\frac{Q''}{S} b_{i-1}^2 + (h - 2P'')a_{i-1}b_{i-1} - SR''a_{i-1}^2 \right) \tag{8.1}$$

for $0 \le i \le m+2$. For brevity, we define sequences of rational functions u_i, v_i, w_i via

$$u_i = \frac{Q''}{b_{-1}S} b_i^2 \ , \quad v_i = \frac{h - 2P''}{b_{-1}} a_i b_i \ , \quad w_i = \frac{SR''}{b_{-1}} a_i^2 \ , \tag{8.2}$$

for $-1 \le i \le m+1$, where as before, $R'' = (f + hP'' - P''^2)/Q''$. Then

$$(-1)^{i+1}\hat{Q}_{i+1} = u_i + v_i + w_i \qquad (1 \le i \le m+1) \ . \tag{8.3}$$

Note that u_i decreases and w_i increases in degree as i increases. Furthermore, u_i, v_i, w_i satisfy the following properties:

Lemma 8.1. *Let N and n be given by Eq. (7.3) and Eq. (7.12), respectively, and define*

$$\begin{aligned} L = \deg(Q''R'') &= \deg(f + hP'' - P''^2) \\ &= \deg(P'' + v) + \deg(P'' - h - v) \ . \end{aligned} \tag{8.4}$$

Then we have the following:

(a) $\deg(v_i) \le g$ *for* $-1 \le i \le m+1$.
(b) $\deg(w_i) = L - \deg(u_{i-1})$ *for* $0 \le i \le m+1$.
(c) $\deg(u_{n+1}) \le M < \deg(u_i)$ *for* $-1 \le i \le n$.
(d) $\deg(w_i) \le \deg(P'' - h - v) - 1 \le g$ *for* $-1 \le i \le n+1$.

Proof. Since $\deg(h - 2P'') \leq g + 1$, (a) can be derived using Lemma 2.2 (e) and (b), since

$$\begin{aligned}\deg(v_i) &= \deg(a_i) + \deg(b_i) + \deg(h - 2P'') - \deg(b_{-1}) \\ &= \deg(a_i) - \deg(a_{i+1}) + \deg(h - 2P'') \leq -1 + (g + 1) = g\end{aligned}$$

for $0 \leq i \leq m + 1$. The definition of u_{i-1} as well as Eq. (8.4) and part (e) of Lemma 2.2 imply

$$\begin{aligned}\deg(w_i) &= 2\deg(a_i) + \deg(S) + \deg(R'') - \deg(b_{-1}) \\ &= \deg(b_{-1}) - 2\deg(b_{i-1}) + \deg(S) + L - \deg(Q'') \\ &= L - \deg(u_{i-1})\end{aligned}$$

for $0 \leq i \leq m + 1$, whence follows (b). For (c), we note that

$$\begin{aligned}\deg(u_i) &= 2\deg(b_i) + \deg(Q''/S) - \deg(b_{-1}) \\ &= \deg(Q''/Q') + 2\deg(b_i) \\ &= M - 2N + 2\deg(b_i)\end{aligned}$$

for $-1 \leq i \leq m+1$. We then see from Eq. (7.2) and Eq. (7.3) that $\deg(u_i) \leq M$ if and only if $\deg(b_i) \leq N$. Part (c) now follows from Eq. (7.12). For (d), we note that $\deg(w_{-1}) = -\infty$, and for $0 \leq i \leq n + 1$, by Eq. (8.4), Eq. (7.2), and parts (b) and (c),

$$\begin{aligned}\deg(w_i) &= L - \deg(u_{i-1}) < L - M \\ &\leq L - \deg(P'' + v) = \deg(P'' - h - v) \ . \qquad \square\end{aligned}$$

Corollary 8.1. *Let N and n be given by Eq. (7.3) and Eq. (7.12), respectively. Then the following holds.*

(a) $\deg(Q_{i+1}) = \deg(u_i) \geq g + 2$ *for* $-1 \leq i \leq n$.
(b) $\deg(Q_{n+2}) \leq M + 1 \leq g + 1$.
(c) $\deg(Q_{n+2}) \leq g$ *if and only if* $\deg(b_{n+1}) < N$ *or* $M < g + 1$.

Proof. Parts (a) and (b) immediately follow from Eq. (8.3) as well as parts (a), (c), and (d) of Lemma 8.1. For part (c) of the Corollary, note that $\deg(u_{n+1}) = M - 2(N - \deg(b_{n+1}))$, so $\deg(Q_{n+2}) = g + 1$ if and only if $\deg(u_{n+1}) = g + 1$, which in turn holds if and only if $\deg(b_{n+1}) = N$ and $M = g + 1$. $\square$

We now determine how to obtain the divisor $D' \oplus D''$ using NUCOMP. First, we recall that Eq. (7.5), or equivalently, Eq. (7.10) and Eq. (7.9), define a sequence of divisors $\hat{D}_{i+1} = (\hat{Q}_i, \hat{P}_i)$ for $0 \leq i \leq n + 2$. If C

is imaginary or real and $\deg(\hat{Q}_{n+2}) = g + 1$, then we define the divisor $D_{n+4} = (Q_{n+3}, P_{n+3})$ where

$$q_{n+2} = \left\lfloor \frac{\hat{P}_{n+2} + e_{n+2}v}{\hat{Q}_{n+2}} \right\rfloor \quad \text{with } e_{n+2} = \begin{cases} 1 \text{ if } C \text{ is real} , \\ 0 \text{ if } C \text{ is imaginary} , \end{cases}$$
$$P_{n+3} = h - \hat{P}_{n+2} + q_{n+2}\hat{Q}_{n+2}, \quad Q_{n+3} = \frac{f + hP_{n+3} - P_{n+3}^2}{\hat{Q}_{n+2}} . \tag{8.5}$$

so P_{n+3} and Q_{n+3} are obtained by applying Eq. (5.13) to $P_{n+2} = \hat{P}_{n+2}$ and $Q_{n+2} = \hat{Q}_{n+2}$. For brevity, we define the integer

$$K = \deg(Q'') + \deg(Q') - g . \tag{8.6}$$

Then we can determine $D' \oplus D''$ as follows.

Proposition 8.1. *Let N, n, and K be given by Eq. (7.3), Eq. (7.12), and Eq. (8.6), respectively. Then the following holds.*

(a) If C is unusual, then $D' \oplus D'' = D_{n+3}$.
(b) If C is imaginary or real and $\deg(P''+v) < g+1$, then $D' \oplus D'' = \hat{D}_{n+3}$.
(c) If C is imaginary or real and $\deg(P'' + v) = g + 1$, then $D' \oplus D'' = \hat{D}_{n+3}$ if K is even. If K is odd, then $D' \oplus D'' = D_{n+3}$ and only if $\deg(Q_{n+2}) \leq g$, or equivalently, $\deg(b_{n+1}) < N$, otherwise $D' \oplus D'' = D_{n+4}$.

Proof. Note that $\deg(P''+v) < g+1$ if and only if $M < g+1$, and $\deg(P''+v) = g+1$ if and only if $M = g+1$. We now use the definition of $D' \oplus D''$ and invoke Corollary 7.1. Then parts (a) and (b) follow immediately from parts (a) and (c) of Corollary 8.1, respectively. For part (c) of the Proposition, we have $M = \deg(P'' + v) = g + 1$, so $D' \oplus D'' = D_{n+3}$ if and only if $\deg(Q_{n+2}) \leq g$, which by part (c) of Corollary 8.1 holds if and only if $\deg(b_{n+1}) < N$. Now if K is even, then $2N = K + 1 + 2(g - \deg(Q''))$ is an integer and odd, and $2\deg(b_n)$ is even, so we must have $\deg(b_{n+1}) < N$. If K is odd and $\deg(Q_{n+2}) = g + 1$, then D_{n+3} is not reduced, so it suffices to prove that D_{n+4} is reduced.

To that end, note that by Eq. (8.5), $\deg(P_{n+3} - h - e_{n+2}v) < \deg(Q_{n+2}) = g + 1$. If C is imaginary, then this implies $\deg(P_{n+3}) \leq g$, whereas if C is real, then $\deg(P_{n+3} - h - v) \leq g$. In either case, $\deg(Q_{n+3}) \leq 2g + 1 - \deg(Q_{n+2}) = g$ by Eq. (8.5), so D_{n+4} is reduced. □

Remark 8.1. We note that if C is imaginary or real, $\deg(P'' + v) = g + 1$, and K as given in Eq. (8.6) is odd, then we will almost always have

$D' \oplus D'' = D_{n+4}$, i.e. it is very unlikely that D_{n+3} is reduced. In fact, under these conditions, if D_{n+3} is reduced, then it is easy to show that $\deg(b_{n+1}) \le N - 1$ and $\deg(b_n) \ge N + 1$, so

$$\deg(b_n) - \deg(b_{n+1}) \ge 2 \ . \tag{8.7}$$

If $b_{n+1} = 0$, then $b_n = \gcd(Q'/S, U)$, so Eq. (8.7) would imply that Q'/S and U have a non-trivial common factor which is highly unlikely. If $b_{n+1} \neq 0$, then Eq. (8.7) implies $\deg(\hat{q}_{n+1}) \ge 2$. But all but the first partial quotient in a regular continued fraction expansion are expected to have degree 1 with very high probability.

To compute the relative distance $\delta = \delta(D') + \delta(D'') - \delta(D' \oplus D'')$ using NUCOMP in the case where C is real, let $\mathfrak{a}$, $\mathfrak{a}'$, $\mathfrak{a}''$ be the reduced ideals corresponding to the divisors $D' \oplus D''$, D', D'', respectively. Then $\mathfrak{a} = (S/\overline{\theta})\mathfrak{a}'\mathfrak{a}''$ where $\overline{\theta} = \overline{\theta}_i$ with $\mathfrak{a}_1 = \mathfrak{a}'\mathfrak{a}''$, $\mathfrak{a}_i = \mathfrak{a}$, and $i = n+3$ or $n+4$ by Proposition 8.1. Setting $d = \deg(S) - \deg(\overline{\theta}_{n+3})$, we obtain by Eq. (5.10), Eq. (6.1), and Theorem 7.1,

$$\begin{aligned} d &= \deg(S) - \left(\deg(Q_{n+2}) - \deg(Q_0) + \sum_{j=1}^{n+2} \deg(q_j) \right) \\ &= \deg(Q') + \deg(Q'') - \deg(S) - \deg(\hat{Q}_{n+2}) - \sum_{j=0}^{n} \deg(\hat{q}_j) - \deg(q_{n+2}) \ . \end{aligned}$$

If $D' \oplus D'' = D_{n+3}$, then $\delta = d$, and if $D' \oplus D'' = D_{n+4}$, then $\delta = d - \deg(q_{n+3})$ with $q_{n+3} = \lfloor (P_{n+3} + v)/Q_{n+3} \rfloor$, so $\deg(q_{n+3}) = g + 1 - \deg(Q_{n+3})$.

We now give upper bounds on the index n of Eq. (7.12).

Theorem 8.1. *Let N, n and K be defined by Eq. (7.3), Eq. (7.12) and Eq. (8.6), respectively. Then the following holds:*

(a) If K is even, then $n \le (K-4)/2$ and $D' \oplus D'' = D_{n+3}$ is reduced.
(b) If K is odd, then we have the following:

  (a) If C is unusual, then $n \le (K-5)/2$ and $D' \oplus D'' = D_{n+3}$.
  (b) If C is imaginary or real and $\deg(P'' + v) < g + 1$, then $n \le (K-3)/2$ and $D' \oplus D'' = D_{n+3}$.
  (c) If C is imaginary or real and $\deg(P'' + v) = g + 1$, then $n \le (K-5)/2$, and $D' \oplus D'' = D_{n+3}$ if and only if $\deg(b_{n+1}) < N$, otherwise $D' \oplus D'' = D_{n+4}$.

Proof. From Lemma 2.2 (c), Eq. (7.3), Eq. (7.12), and Eq. (8.6), we obtain

$$n \leq \deg(b_{-1}) - \deg(b_n) - 1 < \deg(Q') - N - 1 = \frac{1}{2}(K - M + g) - 1 \; .$$

If K is even, then as before, $\deg(b_{n+1}) < N$, which holds if and only if $\deg(Q_{n+2}) < M$, or equivalently, $\deg(Q_{n+2}) \leq g$. Thus, D_{n+3} is reduced, and we simply use $M \geq g$ to obtain $n < K/2 - 1$ and hence $n \leq (K-4)/2$.

Suppose now that K is odd. Then all the claims in Theorem 8.1 except for the bounds on n follow from Proposition 8.1. If $\deg(P'' + v) < g + 1$, then we again use $M \geq g$ to obtain $n \leq (K - 3)/2$. If $\deg(P'' + v) = g + 1$ then $M = g+1$, yielding $n \leq (K-5)/2$. Note that this includes the unusual scenario. □

Remark 8.2. The bounds in Theorem (8.1) can also be derived as follows. If $D' \oplus D'' = D_{l+1}$, then by our remarks just before Eq. (5.13), $l \leq \lceil K/2 \rceil$ if $\deg(Q_l) \leq g$, and $l \leq \lceil (K-1)/2 \rceil$ if $\deg(Q_l) = g + 1$ for C unusual and g odd. Now distinguish between the cases $l = n + 2$ and $l = n + 3$ using Proposition 8.1.

In lieu of Remark 8.1, we see that in the imaginary and real cases, $D' \oplus D''$ can usually be found in $(K-4)/2$ "NUCOMP steps" if K is even and in either $(K-3)/2$ NUCOMP steps or $(K-5)/2$ NUCOMP steps plus one reduction step if K is odd. Furthermore, if D' and D'' have minimal norm, then we expect that $\deg(Q') = \deg(Q'')$. This degree will generally be equal to g if C is imaginary, unusual with g even, or real, and tends to be equal to $g + 1$ if C is unusual and g odd. In the latter case, we expect that the norm of $D' \oplus D''$ again has degree $g + 1$. We thus obtain the following Corollary:

Corollary 8.2. *Let N, n and K be defined by Eq. (7.3), Eq. (7.12) and Eq. (8.6), respectively, and assume that*

- $M = \deg(P'' + v) = g + 1$.
- $\deg(Q') = \deg(Q'') = g$ *if C is imaginary, unusual with g even, or real.*
- $\deg(Q') = \deg(Q'') = g + 1$ *if C is unusual and g odd.*
- $\deg(b_n) - \deg(b_{n+1}) = 1$.

Then the following holds:

(a) If g is even, then $D' \oplus D'' = D_{n+3}$ is reduced and $n \leq (g-4)/2$.
(b) If g is odd and C is unusual, then $D' \oplus D'' = D_{n+3}$ and $n \leq (g-3)/2$.
(c) If g is odd and C is imaginary or real, then $D' \oplus D'' = D_{n+4}$ and $n \leq (g-5)/2$.

Proof. Since $\deg(Q') = \deg(Q'')$, g has the same parity as K. If g is even, or g is odd and C is imaginary or real, then $\deg(Q') = \deg(Q'') = g$, so $K = g$. The bounds on n for these cases now again follow immediately from Theorem 8.1. If g is odd and C is unusual, then $K = 2(g+1) - g = g+2$, so $(K-5)/2 = (g-3)/2$. □

In all three cases of Corollary 8.2, as pointed out in Sec. 6, $D' \oplus D''$ is reached after at most $\lceil g/2 \rceil$ steps; these are all NUCOMP steps except in case (c), where all but the last step are NUCOMP steps and the last step is a baby step.

Finally, recall our assumption Eq. (7.1) that $\deg(P'' + v) \leq g + 1$. We argue that if $D' \oplus D'' = (\hat{Q}, \hat{P})$, then we generally have $\deg(\hat{P}) \leq g + 1$ as well if C is imaginary or real. If $\hat{P} = P_{n+3}$, then we saw that $\deg(P_{n+3} - h - v) \leq g$, so $\deg(\hat{P}) \leq g$ if C is imaginary and $\deg(\hat{P}) \leq g+1$ if C is real. Suppose now that $\hat{P} = \hat{P}_{n+2}$, so $\deg(Q_{n+2}) \leq g$, implying $\deg(u_{n+1}) \leq g$ by Eq. (8.3) and Lemma 8.1. Since $\gcd(Q'/S, U)$ is very likely to have small degree (usually the gcd is 1), it is highly improbable that $b_{n+1} = 0$. Therefore, $\hat{q}_{n+1}$ is defined, and from part (a) of Lemma 2.2 and the definition of u_i, we see that

$$\deg(Q_{n+1}) = \deg(u_n) = 2\deg(\hat{q}_{n+1}) + \deg(u_{n+1}) \leq 2\deg(\hat{q}_{n+1}) + g \ .$$

It follows from Eq. (5.13) and part (a) of Corollary 8.1 that $P_{n+2} = h - P_{n+1} + \lfloor P_{n+1}/Q_{n+1} \rfloor Q_{n+1}$, so $\deg(P_{n+2}) \leq \deg(Q_{n+1}) - 1 \leq 2\deg(\hat{q}_{n+1}) + g - 1$. Since $\hat{q}_{n+1}$, as the partial quotient of a continued fraction expansion, is expected to have degree 1, we obtain $\deg(P_{n+2}) \leq g + 1$ with high probability.

Note that if C is unusual, then we may have $\deg(P_{n+2}) \leq g + 2$, but all the proofs in Sec. 8 can be easily adjusted to work for this case under the assumption $\deg(P'') \leq g + 2$. We omit the details of this reasoning.

If we impose stronger conditions than Eq. (7.1) on P'', then $\hat{P}$ need not satisfy the same conditions. For example, if D'' is given in adapted form, then $D' \oplus D''$ will usually not be in adapted form. Similarly, if C is real and D'' is in reduced form, then $D' \oplus D''$ will generally not be in reduced form. In this case, if the application requires the basis $\hat{Q}, \hat{P}$ to be of a particular form, then a suitable multiple of $\hat{Q}$ will need to be added to $\hat{P}$. However, we point out that in many applications, the above question does not even play a role. For example, if we apply NUCOMP repeatedly to a starting divisor $D'' = (Q'', P'')$, say to generate a "scalar product" $D'' \oplus D'' \oplus \cdots \oplus D''$ computed as part of a cryptographic protocol, then it is sufficient to ensure that $\deg(P'') \leq g + 1$ once at the beginning of the computation.

9. NUCOMP Algorithms

The basic strategy of the NUCOMP algorithm is as follows. Suppose we are given two divisors $D' = (Q', P')$ and $D'' = (Q'', P'')$ of minimal norm with $\deg(P'') \leq g + 1$; for reasons of efficiency, we will also input the polynomials $R' = (f + hP' - P'^2)/Q'$ and $R'' = (f + hP'' - P''^2)/Q''$. Begin by computing S, U as in Eq. (6.1). If $\deg(Q') + \deg(Q'') - 2\deg(S) \leq g + 1$, then the divisor $D = (Q, P)$ defined in Eq. (6.1) is at most one step away from having minimal norm, so simply compute Q and P as in Eq. (6.1) and, if necessary, apply one reduction step — Eq. (5.2) if C is imaginary or Eq. (5.14) otherwise — to $D = (Q, P)$ to obtain $D' \oplus D''$.

Suppose now that $\deg(Q') + \deg(Q'') - 2\deg(S) \geq g + 2$. Then we simultaneously compute the sequences b_i, a_i, c_i, d_i for $-1 \leq i \leq n + 1$; this is what we referred to as "NUCOMP steps" in the previous section. Finally, recover P_{n+2} and Q_{n+2} using Eq. (7.10) and Eq. (7.9) and, if necessary, apply one iteration of Eq. (5.14) to P_{n+2}, Q_{n+2} to obtain $D' \oplus D''$. We describe this method in algorithmic form below.

Algorithm 9.1. NUCOMP (original)

Input: *(Q', P', R'), (Q'', P'', R'') with $Q'R' = f + hP' - P'^2$ and $Q''R'' = f + hP'' - P''^2$, representing two semi-reduced divisors D' and D'' of minimal norm.*

Output: *$(\hat{Q}, \hat{P}, \hat{R})$ representing $D' \oplus D''$ with $\hat{Q}\hat{R} = f + h\hat{P} - \hat{P}^2$.*

(1) *// Compute $D' + D''$*

- (a) *Compute $S_1, W_1 \in \mathbb{F}[u]$ such that $S_1 = \gcd(Q', Q'') = V_1Q' + W_1Q''$.*
- (b) *IF $S_1 = 1$ THEN $S := S_1 = 1$, $X := 0$, $W := W_1$, GOTO (d).*
- (c) *Compute $S, W_2, X \in \mathbb{F}[u]$ such that $S = \gcd(S_1, P' + P'' - h) = W_2S_1 + X(P' + P'' - h)$. Put $W := W_1W_2$.*
- (d) *Put $b_{-1} := Q'/S$ and $U :\equiv W(P' - P'') + XR'' \pmod{b_{-1}}$.*

(2) *IF $\deg(Q') + \deg(Q'') - 2\deg(S) \leq g + 1$ THEN // at most one baby step*

- (a) *Put*

$$\hat{Q} := \frac{Q'Q''}{S^2}, \quad \hat{P} := P'' + U\frac{Q''}{S} \pmod{Q}, \quad \hat{R} := \frac{f + hP - P^2}{Q}.$$

- (b) *IF $\deg(\hat{Q}) = g + 1$ AND C is imaginary THEN*

$$\hat{Q} := \hat{R}\,, \quad \hat{P} := h - \hat{P} \pmod{\hat{Q}}\,, \quad \hat{R} := \frac{f + h\hat{P} - \hat{P}^2}{\hat{Q}}.$$

(c) *IF* $\deg(\hat{Q}) = g + 1$ *AND* C *is real THEN*

  (i) *Put* $\tilde{P} := \hat{P}$, $\tilde{Q} := \hat{Q}$

  (ii) $\tilde{q} := \lfloor (\tilde{P} + v)/\tilde{Q} \rfloor$.

  (iii) $\hat{P} := h - \tilde{P} + \tilde{q}\tilde{Q}$.

  (iv) $\hat{Q} := \hat{R} + \tilde{q}(\tilde{P} - \hat{P})$, $\hat{R} := \tilde{Q}$.

(d) *RETURN*(Q, P, R)

(3) *// Now apply NUCOMP*

  (a) $b_0 := U, \quad a_{-1} := 0, \quad a_0 := 1$.

  (b) $c_{-1} := Q''/S, \quad P = P'' + UQ''/S, \quad c_0 := (P - P')/b_{-1}$.

  (c) $d_{-1} := P' + P'' - h, \quad d_0 := (d_{-1}b_0 - SR'')/b_{-1}$.

  (d) $i := 0, \quad N := (\deg(Q') - \deg(Q'') + \max\{g, \deg(P'' + v)\})/2$.

(4) *While* $\deg(b_i) > N$ *do*

  (a) $\hat{q}_i := \lfloor b_{i-1}/b_i \rfloor$, $b_{i+1} := b_{i-1} \pmod{b_i}$. *// Division with remainder*

  (b) $a_{i+1} := a_{i-1} - \hat{q}_i a_i$.

  (c) $c_{i+1} := c_{i-1} - \hat{q}_i c_i$.

  (d) $d_{i+1} := d_{i-1} - \hat{q}_i d_i$.

  (e) $i := i + 1$.

(5) *// Now* $i = n + 1$, *so* $\deg(b_{n+1}) \leq N < \deg(b_n)$.

  (a) $Q_{i+1} := (-1)^{i+1}(b_i c_i - a_i d_i)$ *//* $Q_{i+1} = Q_{n+2}$.

  (b) $P_{i+1} := (-1)^{i+1}(b_{i-1}c_i - a_i d_{i-1}) + P''$ *//* $P_{i+1} = P_{n+2}$.

  (c) $R_{i+1} := (-1)^{i-1}(a_{i-1}d_{i-1} - b_{i-1}c_{i-1})$ *//* $R_{i+1} = R_{n+2} = Q_{n+1}$

  (d) *IF* C *is imaginary or real and* $\deg(Q_{i+1}) = g + 1$ *THEN*

    i. *IF* C *is imaginary,* $q_{i+1} := \lfloor P_{i+1}/Q_{i+1} \rfloor$
    ELSE $q_{i+1} := \lfloor (P_{i+1} + v)/Q_{i+1} \rfloor$

    ii. $P_{i+2} := h - P_{i+1} + q_{i+1}Q_{i+1}$.

    iii. $Q_{i+2} := R_{i+1} + q_{i+1}(P_{i+1} - P_{i+2})$.

    iv. $R_{i+2} := Q_{i+1}$.

    v. $i := i + 1$.

  (e) *put* $\hat{Q} := Q_{i+1}$, $\hat{P} := P_{i+1}$, $\hat{R} := R_{i+1}$.

  (f) *RETURN*$(\hat{Q}, \hat{P}, \hat{R})$.

There is an alternative version of this algorithm that is aimed at keeping the size of the intermediate operands low. In the context of binary quadratic forms, this idea is originally due to Atkin. Instead of computing all four sequences, we only compute b_i, a_i for $-1 \leq i \leq n + 1$. Then compute c_{n+1}, d_n and d_{n+1} using Eq. (7.7) and Eq. (7.8), and finally, P_{n+2} and Q_{n+2} using Eq. (7.10) and Eq. (7.9). Since $N \approx g/2$, we expect b_n and b_{n+1}

to have approximate degree $g/2$. By Lemma 2.2 (e), we thus also expect $\deg(a_{n+1}) \approx g/2$, and Eq. (7.7) and Eq. (7.8) show that c_{n+1}, d_n and d_{n+1} also have approximate degree $g/2$. So all operands have very small degree; only the numerators in Eq. (7.7) for $i = n+1$ and Eq. (7.8) for $i = n$ and $i = n+1$ have degree $\approx 3g/2$. These degrees are much smaller than those of the numerators of c_0 and d_0 which are roughly $2g$. On the other hand, the computation of c_{n+1}, d_n and d_{n+1} requires three divisions by b_{-1}, compared to only two such divisions required for computing c_0 and d_0. We again present this technique algorithmically below.

Algorithm 9.2. NUCOMP (small operands)
Input: (Q', P', R'), (Q'', P'', R'') *with* $Q'R' = f + hP' - P'^2$ *and* $Q''R'' = f + hP'' - P''^2$, *representing two semi-reduced divisors* D' *and* D'' *of minimal norm.*
Output: $(\hat{Q}, \hat{P}, \hat{R})$ *representing* $D' \oplus D''$ *with* $\hat{Q}\hat{R} = f + h\hat{P} - \hat{P}^2$.

(1) // Compute $D' + D''$

(a) Compute $S_1, W_1 \in \mathbb{F}[u]$ *such that* $S_1 = \gcd(Q', Q'') = V_1Q' + W_1Q''$.

(b) IF $S_1 = 1$ *THEN* $S := S_1 = 1$, $X := 0$, $W := W_1$, *GOTO (d).*

(c) Compute $S, W_2, X \in \mathbb{F}[u]$ *such that* $S = \gcd(S_1, P' + P'' - h) = W_2S_1 + X(P' + P'' - h)$. *Put* $W := W_1W_2$.

(d) Put $b_{-1} := Q'/S$ *and* $U :\equiv W(P' - P'') + XR'' \pmod{b_{-1}}$.

(2) IF $\deg(Q') + \deg(Q'') - 2\deg(S) \leq g + 1$ *THEN // at most one baby step*

(a) Put

$$\hat{Q} := \frac{Q'Q''}{S^2}, \quad \hat{P} := P'' + U\frac{Q''}{S} \pmod{Q}, \quad \hat{R} := \frac{f + hP - P^2}{Q}.$$

(b) IF $\deg(\hat{Q}) = g + 1$ *AND* C *is imaginary THEN*

$$\hat{Q} := \hat{R}, \quad \hat{P} := h - \hat{P} \pmod{\hat{Q}}, \quad \hat{R} := \frac{f + h\hat{P} - \hat{P}^2}{\hat{Q}}.$$

(c) IF $\deg(\hat{Q}) = g + 1$ *AND* C *is real THEN*

(i) Put $\tilde{P} := \hat{P}$, $\tilde{Q} := \hat{Q}$

(ii) $\tilde{q} := \lfloor (\tilde{P} + v)/\tilde{Q} \rfloor$.

(iii) $\hat{P} := h - \tilde{P} + \tilde{q}\tilde{Q}$.

(iv) $\hat{Q} := \hat{R} + \tilde{q}(\tilde{P} - P)$, $\hat{R} := \tilde{Q}$.

(d) RETURN$(\hat{Q}, \hat{P}, \hat{R})$

(3) // Now apply NUCOMP

  (a) $b_0 := U,\ \ a_{-1} := 0,\ \ a_0 := 1.$
  (b) $i := 0,\ \ N := (\deg(Q') - \deg(Q'') + \max\{g, \deg(P'' + v)\})/2.$

(4) *While* $\deg(b_i) > N$ *do*

  (a) $\hat{q}_i := \lfloor b_{i-1}/b_i \rfloor,\ b_{i+1} := b_{i-1} \pmod{b_i}.$ *// Division with remainder*
  (b) $a_{i+1} := a_{i-1} - \hat{q}_i a_i.$
  (c) $i := i + 1.$

(5) *// Now* $i = n + 1$, *so* $\deg(b_{n+1}) \leq N < \deg(b_n)$.

  (a) $c_i := (b_i Q''/S + a_i(P' - P''))/b_{-1}.$
  (b) $d_{i-1} := (b_{i-1}(P' + P'' - h) + a_{i-1}SR'')/b_{-1}.$
  (c) $X_1 := b_{i-1}c_i,\ c_{i-1} := (X_1 + (-1)^i(P' - P''))/b_i.$
  (d) $X_2 := (-1)^{i-1}a_i d_{i-1},\ d_i := ((P' + P'' - h) - X_2)/(-1)^{i-2}a_{i-1}.$
  (e) $Q_{i+1} := (-1)^{i+1}(b_i c_i - a_i d_i)$ *//* $Q_{i+1} = Q_{n+2}.$
  (f) $P_{i+1} := (-1)^{i+1}(X_2 - X_1) + P''$ *//* $P_{i+1} = P_{n+2}.$
  (g) $R_{i+1} := (-1)^{i-1}(a_{i-1}d_{i-1} - b_{i-1}c_{i-1})$ *//* $R_{i+1} = R_{n+2} = Q_{n+1}$
  (h) *IF* C *is imaginary or real and* $\deg(Q_{i+1}) = g + 1$ *THEN*

    i. *IF* C *is imaginary,* $q_{i+1} := \lfloor P_{i+1}/Q_{i+1} \rfloor$
    ELSE $q_{i+1} := \lfloor (P_{i+1} + v)/Q_{i+1} \rfloor$
    ii. $P_{i+2} := h - P_{i+1} + q_{i+1}Q_{i+1}.$
    iii. $Q_{i+2} := R_{i+1} + q_{i+1}(P_{i+1} - P_{i+2}).$
    iv. $R_{i+2} := Q_{i+1}.$
    v. $i := i + 1.$

  (i) *put* $\hat{Q} := Q_{i+1},\ \hat{P} := P_{i+1},\ \hat{R} := R_{i+1}.$
  (j) *RETURN*$(\hat{Q}, \hat{P}, \hat{R})$.

10. An Extra Reduced Divisor

For real curves, if D_{n+3} is not reduced, then one can compute an alternative reduced divisor different from D_{n+4} under certain circumstances. Let C be a real hyperelliptic curve, and $\deg(P'' - h - v) \leq g$; this is the case, for example, if D'' is given in reduced form. If L is as in Eq. (8.4), then $L \leq 2g+1$, and $L \leq g$ if D'' is in reduced form. Furthermore, $\deg(P''+v) = M = g + 1$, so by Proposition 8.1 (c), $D' \oplus D'' = D_{n+4}$ if and only if K as given in Eq. (8.6) is odd and $\deg(b_{n+1}) = N$; note that in this case, $b_{n+1} \neq 0$, so $\hat{q}_{n+1}$ and b_{n+2} are defined. So suppose that this is the case, and define a new divisor $\hat{D}_{n+4} = (\hat{Q}_{n+3}, \hat{P}_{n+3})$ as follows:

$$\hat{P}_{n+3} = h - \hat{P}_{n+2} + \hat{q}_{n+1}\hat{Q}_{n+2}, \quad \hat{Q}_{n+3} = \frac{f + h\hat{P}_{n+3} - \hat{P}_{n+3}^2}{\hat{Q}_i}, \tag{10.1}$$

i.e. $\hat{D}_{n+4}$ is obtained by applying Eq. (7.5) to $\hat{D}_{n+3} = (\hat{Q}_{n+2}, \hat{P}_{n+2})$ (or alternatively, by using Eq. (7.10) and Eq. (7.9) with $i = n+3$). We prove $\hat{D}_{n+4}$ is a reduced divisor that is almost always different from D_{n+4}.

Proposition 10.1. *Let C be real, $\deg(P'' - h - v) \le g$, $\hat{D}_{n+3}$ not reduced, and $\hat{D}_{n+4} = (\hat{Q}_{n+3}, \hat{P}_{n+3})$ be given by Eq. (10.1). Then $\hat{D}_{n+4}$ is reduced.*

Proof. We have $\deg(\hat{Q}_{n+2}) = \deg(u_{n+1}) = g+1$. Then $\deg(u_{n+2}) \le \deg(u_{n+1}) - 2 = g-1$ by Lemma 2.2 (a), $\deg(v_{n+2}) \le g$ by Lemma 8.1 (a), and $\deg(w_{n+2}) = L - \deg(u_{n+1}) \le g$ by Lemma 8.1 (b), since $L \le 2g+1$. Thus, $\deg(\hat{Q}_{n+3}) \le g$ by Eq. (8.3), so $\hat{D}_{n+4}$ is reduced. □

Before we can prove that $\hat{D}_{n+4} \ne D_{n+4}$ almost always, we first require a lemma.

Lemma 10.1. *Under the assumptions of Proposition 10.1, we have*

$$\deg(\hat{P}_{n+3} + v) \le g \ .$$

Proof. Analogous to Eq. (8.1), we can derive

$$(-1)^{i+1}(\hat{P}_{i+1} + P'' - h) = u_i' + v_i' + w_i'$$

where

$$u_i' = \frac{Q''}{b_{-1}S} b_{i-1} b_i \ , \quad v_i' = \frac{h - 2P''}{b_{-1}} a_{i-1} b_i \ , \quad w_i' = \frac{SR''}{b_{-1}} a_{i-1} a_i \ ,$$

for $0 \le i \le m+1$. Using Lemmas 2.2 and 8.1, we obtain

$$\begin{aligned} \deg(u_{n+2}') &\le \deg(u_{n+1}) - 1 = (g+1) - 1 = g \ , \\ \deg(v_{n+2}') &\le \deg(v_{n+1}) - 1 \le g - 1 \ , \\ \deg(w_{n+2}') &\le \deg(w_{n+2}) - 1 = L - \deg(u_{n+1}) - 1 = g - 1 \ . \end{aligned}$$

It follows that

$$\deg(\hat{P}_{n+3} + v) = \deg\Big((\hat{P}_{n+3} + P'' - h) - (P'' - h - v)\Big) \le g \ . \qquad \square$$

Proposition 10.2. *Under the assumptions of Proposition 10.1, and with D_{n+4} given by Eq. (8.5), we have $\hat{D}_{n+4} \ne D_{n+4}$, provided $D_{n+4} \ne 0$.*

Proof. Recall that Eq. (8.5) yielded $\deg(P_{n+3} - h - v) \le g$, so $\deg(P_{n+3} + v) = g+1$. Thus, by Lemma 10.1, $\deg(\hat{P}_{n+3} + v) \le g < \deg(P_{n+3} + v)$. It follows that $\deg(P_{n+3}) = \deg(\hat{P}_{n+3}) = g+1$ and

$\hat{P}_{n+3} \neq P_{n+3}$. Now $\hat{P}_{n+3} - P_{n+3} = s\hat{Q}_{n+2}$ with $s = \hat{q}_{n+1} - q_{n+2}$. Since $\deg(\hat{Q}_{n+2}) = g + 1$, we must have $s \in \mathbb{F}_q^*$.

By way of contradiction, assume that $\hat{D}_{n+4} = D_{n+4} \neq 0$. Then Q_{n+3} and $\hat{Q}_{n+3}$ differ by a factor in k^*, and Q_{n+3} divides $\hat{P}_{n+3} - P_{n+3} = s\hat{Q}_{n+2}$. Since $s \in \mathbb{F}_q^*$, we see that Q_{n+3} divides $\hat{Q}_{n+2}$. By Eq. (8.5) and Eq. (10.1), we have

$$\begin{aligned}\hat{Q}_{n+2}(\hat{Q}_{n+3} - Q_{n+3}) &= (f + h\hat{P}_{n+3} - \hat{P}_{n+3}^2) - (f + hP_{n+3} - P_{n+3}^2) \\ &= (\hat{P}_{n+3} - P_{n+3})(h - \hat{P}_{n+3} - P_{n+3}) \\ &= s\hat{Q}_{n+2}(h - 2P_{n+3} - s\hat{Q}_{n+2}) \ ,\end{aligned}$$

so Q_{n+3} divides $h - 2P_{n+3}$. Now $D_{n+4} \neq 0$ forces Q_{n+3} to be non-constant. Let r be a root of Q_{n+3} in some algebraic closure of k. Then we can use reasoning analogous to the proof of Proposition 5.1 to infer that $(r, -P_{n+3}(r))$ is a singular point on C, a contradiction. □

Remark 10.1. Let $\hat{\mathfrak{a}}_{n+3}$, $\mathfrak{a}_{n+4}$ and $\hat{\mathfrak{a}}_{n+4}$ be the reduced ideals corresponding to $\hat{D}_{n+3}$, D_{n+4}, and $\hat{D}_{n+4}$, respectively. Then $(\hat{Q}_{n+2})\mathfrak{a}_{n+4} = (P_{n+3} + v)\hat{\mathfrak{a}}_{n+3}$ and $(\hat{Q}_{n+2})\hat{\mathfrak{a}}_{n+4} = (\hat{P}_{n+3} + v)\hat{\mathfrak{a}}_{n+3}$. If we now take distances with respect to some starting divisor and set $\delta_{n+4} = \delta(D_{n+4})$ and $\hat{\delta}_{n+4} = \delta(\hat{D}_{n+4})$, then we have $\delta_{n+4} = \hat{\delta}_{n+4} + \delta$ with

$$\delta = \deg(P_{n+3} + v) - \deg(\hat{P}_{n+3} + v) \ .$$

Since $\deg(P_{n+3} + v) = g + 1 > \deg(\hat{P}_{n+3} + v)$, we have $\delta \geq 1$. Furthermore, since $\deg(P_{n+3} + v) = \deg(\hat{P}_{n+3} - h - v) = g + 1$,

$$\frac{P_{n+3} + v}{\hat{P}_{n+3} + v} = \frac{(P_{n+3} + v)(\hat{P}_{n+3} - h - v)}{\hat{Q}_{n+2}Q_{n+3}} \ ,$$

and $\deg(Q_{n+3}) \geq 1$, we have $\delta \leq 2(g+1) - (g+1) - 1 = g$. In summary, $1 \leq \delta \leq g$, so D_{n+4} and $\hat{D}_{n+4}$ are not far from each other in the infrastructure of the appropriate ideal class. In general, we expect $\deg(Q_{n+3}) = g$ and hence $\delta = 1$, so D_{n+4} and $\hat{D}_{n+4}$ are neighbors.

11. Numerical Results

The following numerical experiments were performed on a Pentium IV 2.4 GHz computer running Linux. We used the computer algebra library NTL [14] for finite field and polynomial arithmetic and the GNU `C++` compiler version 3.4.3.

11.1. *Binary Exponentiation*

In order to test the efficiency of our versions of NUCOMP, we implemented routines for binary exponentiation using Cantor's algorithm in Eq. (6.1), NUCOMP (Algorithm 9.1), and NUCOMP with small operands (Algorithm 9.2). All three algorithms were implemented using real, imaginary, and unusual curves defined over prime finite fields $\mathbb{F}_p$ and characteristic 2 finite fields $\mathbb{F}_{2^n}$.

Table 11.1-11.5 contain the ratio of runtimes for binary exponentiation using Algorithm 9.1 (NUCOMP using recurrences to compute c_i and d_i) divided by the runtime using Algorithm 9.2 (NUCOMP using formulas to compute the final values of c_i and d_i). For each genus and field size listed, 1000 binary exponentiations were performed with random 100-bit exponents. The same 1000 exponents were used for both algorithms and for all genera and finite field sizes. The divisors produced by NUCOMP were normalized; adapted basis was used for imaginary and unusual curves and reduced basis was used for real curves [5]. The data clearly show that Algorithm 9.1 is more efficient that Algorithm 9.2 for $g < 10$ approximately, but that Algorithm 9.2 is ultimately more efficient as g grows.

Table 11.1. Exponentiation ratios (Alg 9.1 / Alg 9.2) over $\mathbb{F}_p$, imaginary.

g	$\log_2 p$								
	2	4	8	16	32	64	128	256	512
2	0.9839	0.9012	0.8983	0.9037	0.9038	0.8909	0.9110	0.9140	0.8976
3	0.8703	0.9471	0.9289	0.8934	0.9523	0.9659	0.9503	0.9568	0.9591
4	0.9619	0.9342	0.9266	0.9662	0.9503	0.9514	0.9634	0.9644	0.9672
5	0.9693	0.9550	0.9518	0.9576	0.9567	0.9474	0.9327	0.9341	0.9318
6	0.9754	0.9548	0.9631	0.9624	0.9378	0.9467	0.9413	0.9442	0.9434
7	0.9407	0.9530	0.9608	0.9561	0.9518	0.9532	0.9559	0.9592	0.9613
8	0.9726	0.9663	0.9666	0.9600	0.9576	0.9668	0.9785	0.9641	0.9671
9	0.9751	0.9764	0.9840	0.9776	0.9645	0.9760	0.9947	0.9710	0.9784
10	0.9793	0.9708	0.9817	0.9724	0.9629	0.9746	0.9976	0.9775	0.9864
11	0.9853	0.9792	0.9854	0.9877	0.9705	0.9839	1.0067	0.9875	0.9974
12	0.9983	0.9969	0.9971	0.9875	0.9777	0.9907	0.9924	0.9917	1.0023
13	0.9851	1.0084	1.0000	0.9963	0.9874	0.9993	0.9986	1.0024	1.0102
14	1.0126	1.0039	1.0049	0.9988	0.9845	1.0010	1.0003	1.0038	1.0130
15	1.0143	1.0085	1.0102	1.0097	0.9913	1.0079	1.0076	1.0103	1.0204
20	1.0823	1.1033	1.1029	1.1017	1.0670	1.1102	1.0568	1.0710	1.0866
25	1.1003	1.1185	1.1137	1.1203	1.1103	1.1187	1.0718	1.0988	1.0896
30	1.0872	1.0908	1.0927	1.0895	1.1152	1.1107	1.0839	1.0946	1.1129

Table 11.6–11.10 contain the ratio of runtimes for binary exponentiation using Cantor's algorithm as compared to that using the faster of Algorithm 9.1 or Algorithm 9.2. Again, for each genus and field size listed, 1000 binary exponentiations were performed with random 100-bit exponents. The same 1000 exponents were used for both algorithms and for all genera and finite field sizes. The data clearly show that NUCOMP outperforms Cantor's algorithm except for very small genera and finite field sizes, and that its relative performance improves as both the genus and

Table 11.2. Exponentiation ratios (Alg 9.1 / Alg 9.2) over $\mathbb{F}_p$, real.

g	$\log_2 p$								
	2	4	8	16	32	64	128	256	512
2	0.8661	0.8743	0.9368	0.9557	0.8414	0.8766	0.8830	0.8859	0.8761
3	0.8579	0.9149	0.9163	0.9216	0.8967	0.8996	0.8924	0.8761	0.8851
4	0.9395	0.9647	0.9485	0.9582	0.9533	0.9545	0.9633	0.9648	0.9694
5	0.9294	0.9209	0.9335	0.9489	0.9629	0.9661	0.9652	0.9695	0.9726
6	0.9477	0.9397	0.9595	0.9523	0.9636	0.9566	0.9499	0.9535	0.9570
7	0.8635	0.9431	0.9466	0.9370	0.9644	0.9606	0.9580	0.9586	0.9595
8	0.9349	0.9667	0.9684	0.9860	0.9869	0.9793	1.0003	0.9783	0.9781
9	0.9549	0.9723	0.9683	0.9561	0.9859	0.9818	0.9997	0.9774	0.9788
10	0.9522	0.9963	0.9913	0.9820	0.9968	0.9942	1.0116	0.9857	0.9962
11	0.9540	0.9645	0.9854	0.9874	0.9966	0.9975	0.9957	0.9902	0.9992
12	0.9726	0.9872	0.9960	0.9809	1.0166	1.0098	1.0058	1.0011	1.0130
13	0.9806	0.9948	0.9926	0.9941	1.0191	1.0148	1.0078	1.0018	1.0105
14	0.9883	1.0135	1.0023	0.9989	1.0239	1.0197	1.0171	1.0139	1.0237
15	0.9807	0.9989	1.0117	1.0071	1.0229	1.0226	1.0168	1.0127	1.0209
20	1.0995	1.1180	1.1156	1.1109	1.1063	1.1291	1.0692	1.0856	1.0932
25	1.0968	1.1090	1.1164	1.1060	1.0847	1.1100	1.0745	1.0784	1.0989
30	1.0981	1.1088	1.1149	1.1068	1.0980	1.1066	1.0863	1.0979	1.1258

Table 11.3. Exponentiation ratios (Alg 9.1 / Alg 9.2) over $\mathbb{F}_p$, unusual.

g	$\log_2 p$								
	2	4	8	16	32	64	128	256	512
2	0.9108	0.8800	0.8571	0.8910	0.8969	0.8896	0.9069	0.9082	0.9019
3	0.9175	1.0081	1.0161	1.0109	0.9715	0.9466	0.9658	0.9583	0.9649
4	0.9504	1.0290	1.0311	0.9967	0.9552	0.9542	0.9614	0.9603	0.9660
5	0.9684	0.9690	0.9853	0.9844	0.9730	0.9486	0.9475	0.9439	0.9491
6	0.9649	0.9626	0.9862	0.9731	0.9584	0.9471	0.9418	0.9368	0.9389
7	0.9816	1.0212	1.0139	0.9620	0.9854	0.9705	0.9868	0.9672	0.9724
8	0.9929	0.9867	0.9911	0.9980	0.9775	0.9666	0.9782	0.9590	0.9629
9	0.9938	0.9981	1.0131	0.9832	1.0047	0.9870	1.0063	0.9792	0.9899
10	1.0000	0.9982	0.9964	1.0017	0.9959	0.9834	0.9993	0.9729	0.9854
11	1.0000	1.0235	1.0103	1.0072	1.0228	1.0015	1.0012	0.9924	1.0058
12	1.0048	1.0046	1.0085	1.0014	1.0163	0.9956	0.9953	0.9851	0.9975
13	1.0000	1.0077	1.0243	1.0024	1.0362	0.9985	1.0101	1.0058	1.0184
14	0.9960	1.0245	1.0037	1.0070	1.0313	1.0101	1.0034	0.9958	1.0099
15	1.0094	1.0301	1.0370	1.0321	1.0448	1.0145	1.0176	1.0184	1.0264
20	1.1394	1.1789	1.1526	1.1262	1.1024	1.1000	1.0621	1.0671	1.0884
25	1.1014	1.1103	1.1209	1.1168	1.0860	1.1069	1.0716	1.0799	1.0981
30	1.0932	1.1047	1.1064	1.1018	1.0939	1.1108	1.0799	1.0885	1.1068

Table 11.4. Exponentiation ratios (Alg 9.1 / Alg 9.2) over $\mathbb{F}_{2^n}$, imaginary.

g	$\log_2 p$								
	2	4	8	16	32	64	128	256	512
2	0.9308	0.9002	0.8891	0.8889	0.8976	0.8744	0.9006	0.8880	0.8871
3	0.9622	0.9511	0.9547	0.9514	0.9446	0.9440	0.9600	0.9571	0.9585
4	0.9507	0.9395	0.9480	0.9507	0.9528	0.9592	0.9663	0.9613	0.9610
5	0.9682	0.9436	0.9396	0.9557	0.9443	0.9440	0.9396	0.9356	0.9343
6	0.9661	0.9544	0.9468	0.9528	0.9519	0.9530	0.9469	0.9474	0.9458
7	0.9819	0.9620	0.9674	0.9662	0.9681	0.9669	0.9611	0.9644	0.9622
8	0.9881	0.9663	0.9653	0.9693	0.9725	0.9780	0.9693	0.9691	0.9688
9	1.0071	0.9929	0.9868	0.9853	0.9920	0.9890	0.9807	0.9830	0.9820
10	1.0026	1.0011	0.9864	0.9917	0.9918	0.9891	0.9872	0.9878	0.9876
11	1.0205	0.9981	1.0046	1.0010	0.9947	0.9960	0.9960	0.9986	0.9964
12	1.0272	1.0124	1.0193	1.0137	1.0019	0.9984	1.0016	1.0042	1.0022
13	1.0341	1.0191	1.0311	1.0249	1.0116	1.0092	1.0118	1.0060	1.0148
14	1.0441	1.0311	1.0322	1.0242	1.0145	1.0081	1.0148	1.0181	1.0187
15	1.0504	1.0311	1.0415	1.0324	1.0208	1.0133	1.0190	1.0221	1.0216
20	1.1072	1.1263	1.1350	1.1218	1.1051	1.0890	1.0923	1.0893	1.0885
25	1.1624	1.1662	1.1724	1.1556	1.1337	1.1104	1.1119	1.1203	1.1146
30	1.1869	1.1797	1.1930	1.1826	1.1419	1.1375	1.1335	1.1337	1.1309

finite field size increase. The findings are consistent with those presented in [6], but our improved versions of NUCOMP presented here out-perform Cantor's algorithm for even smaller genera and finite field sizes than in [6].

Table 11.5. Exponentiation ratios (Alg 9.1 / Alg 9.2) over $\mathbb{F}_{2^n}$, real

	$\log_2 p$								
g	2	4	8	16	32	64	128	256	512
2	0.9249	0.8800	0.8630	0.8604	0.8649	0.8603	0.8816	0.8737	0.8725
3	0.8406	0.8562	0.8682	0.8670	0.8710	0.8910	0.8613	0.8745	0.8723
4	0.9331	0.9424	0.9480	0.9561	0.9524	0.9526	0.9561	0.9614	0.9618
5	0.9217	0.9480	0.9562	0.9596	0.9600	0.9526	0.9614	0.9668	0.9665
6	0.9471	0.9655	0.9548	0.9628	0.9574	0.9711	0.9444	0.9503	0.9504
7	0.9557	0.9580	0.9531	0.9511	0.9588	0.9574	0.9462	0.9512	0.9506
8	0.9765	0.9776	0.9781	0.9750	0.9819	0.9800	0.9711	0.9737	0.9759
9	0.9761	0.9709	0.9752	0.9799	0.9729	0.9705	0.9701	0.9701	0.9676
10	0.9891	1.0057	1.0019	0.9996	0.9970	0.9857	0.9868	0.9892	0.9952
11	0.9810	0.9962	1.0070	0.9997	0.9920	0.9849	0.9866	0.9905	0.9910
12	1.0080	1.0064	1.0220	1.0158	1.0081	0.9958	1.0006	1.0082	1.0085
13	1.0029	1.0162	1.0208	1.0120	1.0041	0.9845	1.0009	1.0093	1.0082
14	1.0243	1.0326	1.0379	1.0284	1.0162	0.9981	1.0093	1.0214	1.0215
15	1.0228	1.0329	1.0327	1.0270	1.0175	1.0016	1.0111	1.0182	1.0176
20	1.1270	1.1450	1.1401	1.1737	1.1256	1.0984	1.0998	1.0968	1.0937
25	1.1456	1.1565	1.1748	1.1471	1.0596	1.1021	1.1083	1.1207	1.1049
30	1.1672	1.1757	1.1822	1.1820	1.1477	1.1239	1.1288	1.1374	1.1328

Table 11.6. Exponentiation ratios (NUCOMP / Cantor) over $\mathbb{F}_p$, imaginary.

	$\log_2 p$								
g	2	4	8	16	32	64	128	256	512
2	1.0991	1.0504	1.0743	1.0432	0.9308	0.9141	0.9242	0.8847	0.8447
3	1.0662	1.0707	1.0609	1.0140	0.9523	0.9419	0.9008	0.8865	0.8652
4	1.0632	1.0607	1.0390	1.0158	0.9309	0.9286	0.9068	0.8582	0.8540
5	1.0766	1.0376	1.0350	1.0194	0.9120	0.9046	0.8865	0.8571	0.8642
6	1.0931	1.0462	1.0150	1.0056	0.8888	0.8963	0.8452	0.8594	0.8573
7	1.0235	0.9865	0.9679	0.9583	0.8692	0.8755	0.8310	0.8558	0.8586
8	0.9924	0.9349	0.9414	0.9268	0.8532	0.8697	0.8237	0.8500	0.8424
9	0.9557	0.9212	0.9212	0.9273	0.8405	0.8588	0.8144	0.8472	0.8420
10	0.9423	0.8975	0.8961	0.8910	0.8275	0.8451	0.8078	0.8402	0.8334
11	0.9538	0.8968	0.8981	0.9046	0.8128	0.8407	0.7921	0.8371	0.8333
12	0.9441	0.9043	0.8991	0.8918	0.8075	0.8332	0.8047	0.8278	0.8235
13	0.9361	0.9320	0.9035	0.9063	0.8060	0.7857	0.7995	0.8184	0.8148
14	0.9308	0.9038	0.8971	0.8981	0.7926	0.7821	0.8035	0.8184	0.8071
15	0.9135	0.8747	0.8704	0.8694	0.7851	0.7715	0.8043	0.8130	0.8059
20	0.8255	0.7956	0.7861	0.7989	0.7536	0.7242	0.7910	0.7843	0.7769
25	0.7949	0.7662	0.7693	0.7727	0.7208	0.7398	0.7854	0.7943	0.7759
30	0.7921	0.7714	0.7730	0.7716	0.7157	0.7372	0.7743	0.7616	0.7588

Table 11.7. Exponentiation ratios (NUCOMP / Cantor) over $\mathbb{F}_p$, real.

	$\log_2 p$								
g	2	4	8	16	32	64	128	256	512
2	0.8943	1.1268	1.2192	1.2763	1.0659	1.0987	1.0872	1.0731	1.0835
3	1.0449	1.1497	1.1165	1.1330	1.0503	1.0515	1.0434	1.0376	1.0500
4	1.0745	1.1081	1.0932	1.0784	1.0169	1.0137	1.0150	0.9847	1.0060
5	1.0549	1.0659	1.0300	1.0570	0.9635	0.9771	0.9650	0.9664	0.9787
6	1.0507	1.0124	1.0350	1.0327	0.9444	0.9555	0.9243	0.9540	0.9569
7	0.9705	0.9525	0.9231	0.9209	0.9144	0.9309	0.8950	0.9289	0.9512
8	0.9724	0.9539	0.9426	0.9338	0.9094	0.9195	0.8816	0.9244	0.9254
9	0.9591	0.9179	0.9028	0.9023	0.8726	0.8876	0.8608	0.8913	0.9013
10	0.9105	0.9056	0.8818	0.8877	0.8625	0.8879	0.8642	0.8933	0.8955
11	0.9396	0.9043	0.9159	0.9145	0.8402	0.8596	0.8415	0.8836	0.8862
12	0.9668	0.9341	0.9149	0.9135	0.8356	0.8536	0.8512	0.8832	0.8745
13	0.9581	0.9128	0.8856	0.8942	0.8047	0.7877	0.8201	0.8637	0.8560
14	0.9596	0.9098	0.8782	0.8912	0.8051	0.7874	0.8205	0.8502	0.8471
15	0.9356	0.8696	0.8640	0.8670	0.7789	0.7656	0.8037	0.8425	0.8387
20	0.8065	0.7549	0.7519	0.7638	0.7463	0.7275	0.8110	0.8108	0.8008
25	0.7717	0.7303	0.7215	0.7186	0.6996	0.7124	0.7818	0.7842	0.7723
30	0.7651	0.7337	0.7270	0.7392	0.6873	0.7212	0.7687	0.7749	0.7801

11.2. *Key Exchange*

We also ran numerous examples of the key exchange protocols described in [7], again using both real and imaginary curves and $\mathbb{F}_p$ (p prime) and $\mathbb{F}_{2^n}$ as base fields. The genus of our curves ranged from 2 to 6 and the underlying

Table 11.8. Exponentiation ratios (NUCOMP / Cantor) over $\mathbb{F}_p$, unusual.

g	$\log_2 p$								
	2	4	8	16	32	64	128	256	512
2	1.0438	1.1079	1.0435	1.0444	0.9607	0.9345	0.9257	0.8955	0.8749
3	1.0500	1.0649	1.0506	1.0377	0.9102	0.8970	0.8815	0.8443	0.8417
4	1.1220	1.0905	1.0576	1.0565	0.9182	0.9301	0.9102	0.8782	0.8698
5	1.0824	1.0539	1.0030	1.0216	0.8861	0.8631	0.8382	0.8535	0.8465
6	1.1224	1.0404	1.0259	1.0419	0.9115	0.8951	0.8580	0.8769	0.8744
7	1.0081	0.9179	0.9309	0.9179	0.8604	0.8461	0.8158	0.8424	0.8312
8	0.9882	0.9695	0.9654	0.9900	0.8668	0.8647	0.8342	0.8627	0.8526
9	0.9340	0.8902	0.8883	0.8784	0.8286	0.8274	0.7986	0.8336	0.8274
10	0.9245	0.9151	0.9224	0.9308	0.8523	0.8396	0.8182	0.8484	0.8424
11	0.9641	0.9075	0.8642	0.8858	0.8033	0.8117	0.8016	0.8177	0.8164
12	0.9570	0.8997	0.8892	0.9055	0.8215	0.8265	0.8166	0.8258	0.8338
13	0.9615	0.8977	0.8662	0.8828	0.7821	0.7650	0.7964	0.8115	0.8026
14	0.9448	0.8719	0.8652	0.8858	0.7839	0.7699	0.8104	0.8328	0.8251
15	0.8945	0.8203	0.8111	0.8306	0.7663	0.7389	0.7822	0.7945	0.7858
20	0.8458	0.8234	0.8108	0.8250	0.7220	0.7456	0.8074	0.7927	0.7996
25	0.7964	0.7660	0.7606	0.7656	0.7252	0.7429	0.7942	0.7901	0.7822
30	0.7781	0.7591	0.7575	0.7622	0.7055	0.7307	0.7769	0.7766	0.7682

Table 11.9. Exponentiation ratios (NUCOMP / Cantor) over $\mathbb{F}_{2^n}$, imaginary.

g	$\log_2 p$								
	2	4	8	16	32	64	128	256	512
2	1.0068	0.9696	0.9498	0.9257	0.9185	0.8919	0.8824	0.8433	0.8205
3	0.9857	0.9757	0.9401	0.9244	0.9244	0.9251	0.8789	0.8951	0.8855
4	0.9725	0.9638	0.9448	0.9301	0.9056	0.9204	0.9285	0.9102	0.9110
5	0.9916	0.9705	0.9404	0.9360	0.9115	0.9153	0.9192	0.9035	0.9008
6	0.9632	0.9479	0.9248	0.9155	0.8915	0.9025	0.9132	0.9045	0.9035
7	0.9688	0.9248	0.9083	0.9050	0.8855	0.8761	0.9181	0.9161	0.9158
8	0.9305	0.9110	0.8928	0.8903	0.8866	0.9096	0.9263	0.9237	0.9247
9	0.9245	0.8985	0.8799	0.8902	0.8766	0.8926	0.9061	0.9079	0.9098
10	0.8890	0.8843	0.8809	0.8907	0.8715	0.8823	0.8937	0.8971	0.8996
11	0.8932	0.8695	0.8777	0.8780	0.8640	0.8776	0.8865	0.8938	0.8955
12	0.8744	0.8581	0.8621	0.8593	0.8666	0.8798	0.8811	0.8852	0.8905
13	0.8551	0.8623	0.8401	0.8469	0.8537	0.8696	0.8678	0.8759	0.8778
14	0.8407	0.8298	0.8202	0.8332	0.8492	0.8669	0.8676	0.8751	0.8802
15	0.8220	0.8171	0.8041	0.8173	0.8430	0.8609	0.8649	0.8750	0.8805
20	0.7449	0.7362	0.7399	0.7625	0.7950	0.8106	0.8316	0.8481	0.8536
25	0.7015	0.7089	0.7146	0.7263	0.7585	0.7847	0.8059	0.8174	0.8270
30	0.6744	0.7118	0.6993	0.7078	0.7419	0.7632	0.7839	0.7996	0.8131

Table 11.10. Exponentiation ratios (NUCOMP / Cantor) over $\mathbb{F}_{2^n}$, real.

g	$\log_2 p$								
	2	4	8	16	32	64	128	256	512
2	0.9277	1.0896	1.0930	1.0854	1.1152	1.0998	1.0653	1.0787	1.0661
3	0.8948	0.9597	0.9860	0.9622	0.9695	0.9791	0.9640	0.9824	0.9814
4	1.0213	1.0360	1.0375	1.0274	1.0267	1.0252	1.0289	1.0421	1.0440
5	0.9587	0.9672	0.9630	0.9505	0.9372	0.9295	0.9499	0.9516	0.9491
6	0.9989	0.9776	0.9743	0.9838	0.9718	0.9715	0.9689	0.9777	0.9790
7	0.9370	0.9193	0.9025	0.9126	0.8990	0.9041	0.9311	0.9413	0.9424
8	0.9534	0.9439	0.9222	0.9379	0.9365	0.9476	0.9650	0.9785	0.9842
9	0.9008	0.8771	0.8685	0.8928	0.8707	0.8727	0.8891	0.8954	0.8963
10	0.9053	0.8863	0.8854	0.9142	0.9098	0.9115	0.9252	0.9384	0.9491
11	0.8601	0.8518	0.8504	0.8713	0.8624	0.8595	0.8755	0.8838	0.8870
12	0.8878	0.8679	0.8589	0.8705	0.8938	0.9006	0.9154	0.9220	0.9301
13	0.8377	0.8230	0.8171	0.8281	0.8478	0.8476	0.8675	0.8729	0.8770
14	0.8393	0.8258	0.8206	0.8384	0.8659	0.8783	0.8863	0.8957	0.9031
15	0.7970	0.7775	0.7800	0.7942	0.8204	0.8423	0.8548	0.8594	0.8659
20	0.7221	0.7313	0.7312	0.7552	0.7968	0.8291	0.8311	0.8548	0.8638
25	0.6565	0.6298	0.6678	0.6954	0.7907	0.7356	0.7520	0.7756	0.8010
30	0.6576	0.6649	0.6722	0.6936	0.7353	0.7663	0.7823	0.8043	0.8187

finite field was chosen so that the size of the Jacobian (approximately q^g where the finite field has q elements) was roughly 2^{160}, 2^{224}, 2^{256}, 2^{384}, and 2^{512}. Assuming only generic attacks with square root complexity, these curves offer 80, 112, 128, 192, and 256 bits of security for cryptographic protocols based on the corresponding discrete logarithm problem. NIST [9] currently recommends these five levels of security for key establishment in U.S. Government applications. Although the use of curves with genus 3 and larger for cryptographic purposes is questionable, we nevertheless included times for higher genus as our main goal is to provide a relative comparison between our formulation of NUCOMP with Cantor's algorithm.

For curves defined over $\mathbb{F}_p$, we chose a random prime p of appropriate length such that p^g had the required bit length, and for curves over $\mathbb{F}_{2^n}$ we chose the minimal value of n with gn greater than or equal to the required bit length. For each genus and finite field, we randomly selected 2000 curves and executed Diffie–Hellman key exchange twice for each curve, once using Cantor's algorithm and once using our version of NUCOMP (Algorithm 9.1). We used Algorithm 9.1 as opposed to Algorithm 9.2, because our previous experiments indicated that it is more efficient for low genus curves. The random exponents used had $160, 224, 256, 384$, and 512 bits, respectively, ensuring that the number of bits of security corresponds to the five levels recommended by NIST (again, considering only generic attacks). In order to provide a fair comparison, the same sequence of random exponents was used for each run of the key exchange protocol.

Table 11.11 contains the average CPU time in seconds for each version of the protocol using real and imaginary curves over $\mathbb{F}_p$ and $\mathbb{F}_{2^n}$. The columns labeled "Cantor" contain the runtimes when using Cantor's algorithm, and those labeled "NC" the runtimes when using NUCOMP. The times for any precomputations, as described in [7], are not included. We also give the ratios of the average time spent for key exchange using NUCOMP versus that using Cantor's algorithm in Table 11.12. Clearly, in almost all cases, NUCOMP offers a fairly significant performance improvement as opposed to Cantor's algorithm, even for genus as low as 2.

12. Conclusions

Our results indicate that NUCOMP does provide an improvement for divisor arithmetic in hyperelliptic curves except for the smallest examples in terms of genus and finite field size. They also show that both versions of NUCOMP, Algorithm 9.1 and Algorithm 9.2, are useful. Nevertheless, a careful complexity analysis and further numerical experiments are required to compare NUCOMP and Cantor's algorithm more precisely.

There are a number of possible improvements to NUCOMP that need to be investigated. For example, our remarks at the end of Sec. 8 indicate that basis normalization need not be done when NUCOMP is used as a

Table 11.11. Key exchange timings over $\mathbb{F}_p$ and $\mathbb{F}_{2^n}$ (in seconds).

Security level (in bits)	g	$\mathbb{F}_p$ Imaginary Cantor	$\mathbb{F}_p$ Imaginary NC	$\mathbb{F}_p$ Real Cantor	$\mathbb{F}_p$ Real NC	$\mathbb{F}_{2^n}$ Imaginary Cantor	$\mathbb{F}_{2^n}$ Imaginary NC	$\mathbb{F}_{2^n}$ Real Cantor	$\mathbb{F}_{2^n}$ Real NC
80	2	0.0322	0.0290	0.0324	0.0306	0.0320	0.0282	0.0282	0.0291
	3	0.0382	0.0350	0.0390	0.0363	0.0342	0.0320	0.0322	0.0317
	4	0.0492	0.0438	0.0487	0.0438	0.0443	0.0404	0.0403	0.0382
	5	0.0466	0.0435	0.0483	0.0444	0.0611	0.0601	0.0560	0.0563
	6	0.0124	0.0124	0.0123	0.0122	0.0737	0.0705	0.0667	0.0658
112	2	0.0562	0.0498	0.0554	0.0520	0.0585	0.0505	0.0511	0.0522
	3	0.0737	0.0649	0.0707	0.0660	0.0692	0.0627	0.0624	0.0636
	4	0.0723	0.0651	0.0730	0.0648	0.0691	0.0630	0.0622	0.0598
	5	0.0938	0.0875	0.0937	0.0867	0.0846	0.0822	0.0776	0.0781
	6	0.1182	0.1076	0.1171	0.1048	0.1032	0.0977	0.0946	0.0919
128	2	0.0667	0.0593	0.0663	0.0625	0.0692	0.0594	0.0598	0.0611
	3	0.0870	0.0771	0.0847	0.0790	0.0807	0.0732	0.0730	0.0734
	4	0.0904	0.0806	0.0906	0.0806	0.0791	0.0723	0.0697	0.0667
	5	0.1129	0.1044	0.1124	0.1037	0.0989	0.0957	0.0899	0.0909
	6	0.1354	0.1224	0.1318	0.1181	0.1192	0.1129	0.1090	0.1063
192	2	0.1439	0.1235	0.1375	0.1290	0.1620	0.1348	0.1369	0.1395
	3	0.1617	0.1436	0.1577	0.1480	0.1652	0.1484	0.1472	0.1486
	4	0.1832	0.1609	0.1793	0.1615	0.1743	0.1642	0.1537	0.1505
	5	0.2313	0.2114	0.2210	0.2069	0.2190	0.2147	0.1964	0.1985
	6	0.2247	0.2053	0.2242	0.2019	0.1912	0.1795	0.1726	0.1677
256	2	0.2517	0.2127	0.2303	0.2182	0.3037	0.2556	0.2540	0.2593
	3	0.2920	0.2538	0.2825	0.2633	0.3417	0.3129	0.3025	0.3106
	4	0.2875	0.2537	0.2771	0.2505	0.3015	0.2815	0.2664	0.2622
	5	0.3662	0.3375	0.3557	0.3341	0.3693	0.3599	0.3338	0.3344
	6	0.3968	0.3577	0.3792	0.3446	0.3555	0.3456	0.3185	0.3120

Table 11.12. Key exchange ratios over $\mathbb{F}_p$ and $\mathbb{F}_{2^n}$.

	g	Security level 80	112	128	192	256
$\mathbb{F}_p$ imaginary	2	0.8999	0.8869	0.8890	0.8585	0.8454
	3	0.9153	0.8804	0.8866	0.8882	0.8693
	4	0.8916	0.9004	0.8919	0.8781	0.8825
	5	0.9329	0.9332	0.9242	0.9140	0.9214
	6	0.9984	0.9102	0.9038	0.9135	0.9015
$\mathbb{F}_p$ real	2	0.9435	0.9383	0.9429	0.9383	0.9477
	3	0.9305	0.9342	0.9323	0.9384	0.9321
	4	0.9000	0.8867	0.8895	0.9008	0.9041
	5	0.9197	0.9255	0.9229	0.9363	0.9391
	6	0.9905	0.8947	0.8961	0.9007	0.9088
$\mathbb{F}_{2^n}$ imaginary	2	0.8800	0.8621	0.8579	0.8320	0.8417
	3	0.9364	0.9066	0.9074	0.8984	0.9157
	4	0.9132	0.9125	0.9144	0.9420	0.9336
	5	0.9829	0.9718	0.9677	0.9803	0.9744
	6	0.9558	0.9467	0.9475	0.9388	0.9722
$\mathbb{F}_{2^n}$ real	2	1.0334	1.0222	1.0225	1.0190	1.0206
	3	0.9855	1.0181	1.0067	1.0097	1.0270
	4	0.9493	0.9616	0.9567	0.9796	0.9840
	5	1.0046	1.0065	1.0112	1.0106	1.0016
	6	0.9870	0.9707	0.9753	0.9717	0.9796

component for binary exponentiation, because the degree of $\hat{P}$ will generally be at most $g + 1$ at the end of NUCOMP. Not performing normalization saves one division with remainder at the cost of the inputs to subsequent applications of NUCOMP having slightly larger degrees. In addition, the results in Sec. 10 indicate that in some cases, it is possible to perform one extra NUCOMP step to guarantee that the output of NUCOMP is reduced without having to perform a continued fraction step. Further investigation and analysis is required to determine which of these options is the most efficient in practice.

Our data also indicate that using NUCOMP is more efficient than Cantor's algorithm for cryptographic key exchange using low genus hyperelliptic curves, for both imaginary and real models, However, explicit formulas based on Cantor's algorithm have been developed for divisor arithmetic on curves of genus $2, 3$, and 4 (see [5] for a partial survey and references). NUCOMP, as presented in this paper, is generic in the sense that it works for any genus and as such does not compete in terms of performance with these explicit formulas. Given that NUCOMP out-performs Cantor's algorithm, it is conceivable that some of the ideas used in NUCOMP can be applied to improve the explicit formulas. This, as well as the open problems mentioned above, is the subject of on-going research.

References

[1] E. Artin, Quadratische Körper im Gebiete der höheren Kongruenzen. *Math. Zeitschr.* **19** (1924), 153–206.

[2] D. G. Cantor, Computing in the Jacobian of a hyperelliptic curve, *Math. Comp.* **48** (1987), 95–101.

[3] A. Enge, How to distinguish hyperelliptic curves in even characteristic. *Public-Key Cryptography and Computational Number Theory.* De Gruyter (Berlin), 2001, 49–58.

[4] H. Hasse, *Algebraic Number Theory*, Springer, Berlin 2002.

[5] M. J. Jacobson, Jr., A. J. Menezes, and A. Stein, Hyperelliptic curves and cryptography, in *High Primes and Misdemeanors: Lectures in Honour of the 60th Birthday of Hugh Cowie Williams*, Fields Inst. Comm. **41**, American Mathematical Society, 2004, 255–282.

[6] M. J. Jacobson., Jr. and A. J. van der Poorten, Computational aspects of NUCOMP, *Proc. ANTS-V, Lect. Notes Comp. Sci.* **2369**, Springer (New York), 2002, 120–133.

[7] M. J. Jacobson, Jr., R. Scheidler, and A. Stein, Cryptographic protocols on real and imaginary hyperelliptic curves, submitted to *Advances Math. Comm.*, 2006.

[8] M. J. Jacobson, Jr., R. Scheidler and H. C. Williams, An improved real quadratic field based key exchange procedure. *J. Cryptology* **19** (2006), 211–239.

[9] National Institute of Standards and Technology (NIST), *Recommendation on key establishment schemes*, NIST Special Publication 800-56, January 2003.

[10] S. Paulus and H.-G. Rück, Real and imaginary quadratic representations of hyperelliptic function fields, *Math. Comp.* **68** (1999), 1233–1241.

[11] A. J. van der Poorten, A note on NUCOMP. *Math. Comp.* **72** (2003), 1935–1946.

[12] M. Rosen, *Number Theory in Function Fields*, Springer, New York 2002.

[13] D. Shanks, On Gauss and composition I, II. In *Proc. NATO ASI on Number Theory and Applications*, Kluwer Academic Press 1989, 163–204.

[14] V. Shoup, NTL: A library for doing number theory. Software, 2001. Available at `http://www.shoup.net.ntl`.

[15] A. Stein. Sharp upper bounds for arithmetics in hyperelliptic function fields. *J. Ramanujan Math. Soc.* **16** (2001), 1–86.

[16] A. Stein and H. C. Williams, Some methods for evaluating the regulator of a real quadratic function field. *Experiment. Math.* **8** (1999), 119–133.

[17] H. Stichtenoth, *Algebraic Function Fields and Codes.* Springer, Berlin 1993.

The number of inequivalent binary self-orthogonal codes of dimension 6

Xiang-dong Hou

Department of Mathematics, University of South Florida
Tampa, FL 33620, USA
E-mail: xhou@math.usf.edu

We announce an explicit formula for the number of inequivalent binary self-orthogonal codes of dimension 6 and arbitrary length. Formulas for dimension ≤ 5 have been obtained in a previous paper.

Keywords: Binary self-orthogonal code, Equivalence, General linear group, The symmetric group.

1. Introduction

Let $\mathbb{F}_2$ be the binary field and let $\langle \cdot, \cdot \rangle$ denote the usual inner product of $\mathbb{F}_2^n$. An $[n, k]$ binary code is a k-dimensional subspace C of $\mathbb{F}_2^n$; C is self-orthogonal if $C \subset C^\perp$, where $C^\perp = \{x \in \mathbb{F}_2^n : \langle x, y \rangle = 0 \text{ for all } y \in C\}$. Self-dual codes are $[2k, k]$ self-orthogonal codes.

Let $\mathfrak{S}_n$ denote the symmetric group on $\{1, \ldots, n\}$. $\mathfrak{S}_n$ acts on $\mathbb{F}_2^n$ by permuting its coordinates. Call two codes $C_1, C_2 \subset \mathbb{F}_2^n$ equivalent if there exists $\sigma \in \mathfrak{S}_n$ such that $C_2 = C_1^\sigma$. Self-orthogonality of codes is preserved under this equivalence. Classification of self-orthogonal codes, especially, of self-dual codes, has been a focus of attention since the early days of coding theory; see Pless [7], Pless and Sloane [8], Conway and Pless [1], Conway, Pless and Sloane [2], and Huffman [5]. Complete classifications of self-dual codes are known up to length 32 [2].

Let $\Psi_{k,n}$ denote the number of inequivalent $[n, k]$ self-orthogonal codes. We are interested in the computation of $\Psi_{k,n}$. More precisely, to what extent can $\Psi_{k,n}$ be explicitly determined? In a recent paper [4], we introduced an algorithm which essentially says that for a given moderate k, it is possible to find an explicit formula for $\Psi_{k,n}$ that holds for all n. In fact, formulas for $\Psi_{k,n}$ with $k \leq 5$ have been found in [4]. The main purpose of the present paper is to announce the formula for $\Psi_{6,n}$. The result consists of a master formula and forty sub formulas which are ingredients of the master formula. In Section 2, we describe the method of computation. The master formula is the Cauchy-Frobenius lemma, i.e., Burnside's lemma, applied to an action by the group $\mathrm{GL}(6,\mathbb{F}_2) \times \mathfrak{S}_n$ on a certain set of $6 \times n$ matrices over $\mathbb{F}_2$. The sub formulas count numbers of matrices fixed by representatives from the conjugacy classes of $\mathrm{GL}(6,\mathbb{F}_2) \times \mathfrak{S}_n$. We tabulate the sub formulas in Table 2 of the appendix. In Table 4 of the appendix, one can find the values of $\Psi_{k,n}$ with $k \le 6$ and $n \le 40$.

Knowing the value of $\Psi_{k,n}$ is a big advantage if we try to classify all $[n,k]$ self-orthogonal codes. Without knowing $\Psi_{k,n}$ beforehand, the algorithm to classify $[n,k]$ self-orthogonal codes relies on the mass formula, see [6, §9.7.1]. This algorithm consists of two types of steps: (i) search for an inequivalent code, (ii) computation of the order of the automorphism group of the newly found inequivalent code. The purpose of type (ii) steps is to check if the list of inequivalent codes already found is complete. However, if $\Psi_{k,n}$ is known beforehand, type (ii) steps are not needed. All we have to do is to find $\Psi_{k,n}$ pairwise inequivalent $[n,k]$ self-orthogonal codes (by whatever method). For example, classification of $[16,6]$ self-orthogonal codes seems quite an undertaking. (To the author's knowledge, this classification is not known.) However, the problem becomes more feasible when reformulated as "find 153 ($= \Psi_{6,16}$) pairwise inequivalent $[16,6]$ self-orthogonal codes".

2. Method of Computation

The method for computing $\Psi_{k,n}$ has been laid out in [4]. Since all details are available in 4, we only outline the approach here.

Let $\Psi_{\le k,n}$ be the number of inequivalent self-orthogonal codes in $\mathbb{F}_2^n$ with dimension $\le k$. Since

$$\Psi_{k,n} = \Psi_{\le k,n} - \Psi_{\le k-1,n},$$

it suffices to compute $\Psi_{\le k-1,n}$ and $\Psi_{\le k,n}$. (Since $\Psi_{\le 5,n}$ has been determined in [4], for the purpose of this paper, we only need to determine $\Psi_{\le 6,n}$.) Let $M_{k\times n}$ be the set of all $k \times n$ matrices over $\mathbb{F}_2$ and let $S_{k\times n} = \{X \in M_{k\times n} : XX^T = 0\}$. The group $\mathrm{GL}(k,\mathbb{F}_2) \times \mathfrak{S}_n$ acts on $S_{k\times n}$ by

$$X^{(A,P)} = A^{-1}XP, \qquad (A,P) \in \mathrm{GL}(k,\mathbb{F}_2) \times \mathfrak{S}_n,\ X \in S_{k\times n}.$$

(Here, a permutation in $\mathfrak{S}_n$ is treated an $n \times n$ permutation matrix.) The number $\Psi_{\le k,n}$ is precisely the number of $\mathrm{GL}(k,\mathbb{F}_2) \times \mathfrak{S}_n$-orbits in $S_{k\times n}$.

By the Cauchy-Frobenius lemma,

$$\Psi_{\leq k,n} = \frac{1}{|\mathrm{GL}(k,\mathbb{F}_2) \times \mathfrak{S}_n|} \sum_{\substack{A \in \mathrm{GL}(k,\mathbb{F}_2) \\ P \in \mathfrak{S}_n}} |\mathrm{Fix}(A,P)|, \tag{1}$$

where

$$\begin{aligned} \mathrm{Fix}(A,P) &= \{X \in S_{k\times n} : X^{(A,P)} = X\} \\ &= \{X \in M_{k\times n} : AX = XP,\ XX^T = 0\}. \end{aligned}$$

Equation (1) can be reduced to

$$\Psi_{\leq k,n} = \sum_{A \in \mathcal{C}(\mathrm{GL}(k,\mathbb{F}_2))} \frac{1}{|\mathrm{cent}_{\mathrm{GL}(k,\mathbb{F}_2)}(A)|} \sum_{\lambda=(\lambda_1,\lambda_2,\dots)\vdash n} \frac{|\mathrm{Fix}(A,P_\lambda)|}{\lambda_1!\lambda_2!\cdots 1^{\lambda_1}2^{\lambda_2}\cdots}. \tag{2}$$

In (2), the symbol $\lambda = (\lambda_1, \lambda_2, \dots) \vdash n$ means that λ is a partition of n, i.e., $\lambda_i \geq 0$ and $\sum_{i\geq 1} i\lambda_i = n$; $P_\lambda \in \mathfrak{S}_n$ is a permutation with cycle type λ; $\mathcal{C}(\mathrm{GL}(k,\mathbb{F}_2))$ is the set of all rational canonical forms in $\mathrm{GL}(k,\mathbb{F}_2)$; $\mathrm{cent}_{\mathrm{GL}(k,\mathbb{F}_2)}(A)$ is the centralizer of A in $\mathrm{GL}(k,\mathbb{F}_2)$. The cardinality $|\mathrm{cent}_{\mathrm{GL}(k,\mathbb{F}_2)}(A)|$ is given by the following two facts.

- If $A = A_1 \oplus A_2$ ($= \left[\begin{smallmatrix} A_1 & \\ & A_2 \end{smallmatrix}\right]$), where $A_i \in \mathrm{GL}(k_i,\mathbb{F}_2)$ and every elementary divisor of A_1 is prime to every elementary divisor of A_2, then

$$|\mathrm{cent}_{\mathrm{GL}(k,\mathbb{F}_2)}(A)| = |\mathrm{cent}_{\mathrm{GL}(k_1,\mathbb{F}_2)}(A_1)|\,|\mathrm{cent}_{\mathrm{GL}(k_2,\mathbb{F}_2)}(A_2)|.$$

- ([3, Theorem 3.6]) Assume that $A \in M_{k\times k}$ has elementary divisors $\underbrace{f^1,\dots,f^1}_{\mu_1}, \underbrace{f^2,\dots,f^2}_{\mu_2}, \dots$, where $f \in \mathbb{F}_2[x]$ is irreducible of degree d. Then

$$\begin{aligned} &|\mathrm{cent}_{\mathrm{GL}(k,\mathbb{F}_2)}(A)| \\ &= \prod_{i\geq 1} 2^{d\mu_i(1\mu_1+2\mu_2+\cdots+i\mu_i+i\mu_{i+1}+i\mu_{i+2}+\cdots)} \prod_{j=1}^{\mu_i}(1-2^{-dj}). \end{aligned}$$

The cardinality $|\mathrm{Fix}(A,P_\lambda)|$ is given by the following theorem.

Theorem 2.1 ([4, Theorem 3.3]). *Let $\lambda = (\lambda_1,\lambda_2,\dots) \vdash n$ and let $A \in \mathrm{GL}(k,\mathbb{F}_2)$ with multiplicative order $o(A) = t$. For each $d \mid t$, let $s_d = k - \mathrm{rank}(A^d - I)$, let $B_d \in M_{k\times s_d}$ such that its columns form a basis of $\{x \in \mathbb{F}_2^k : (A^d - I)x = 0\}$, and let*

$$\alpha_d = \sum_{\substack{i\geq 1,\, \nu(i)\leq \nu(t) \\ \gcd(i,t)=d}} \lambda_i,$$

where ν is the 2-adic order. Then

$$|\mathrm{Fix}(A, P_\lambda)| = 2^{\sum_{\nu(i)>\nu(t)} s_{\gcd(i,t)}\lambda_i} \cdot \mathfrak{n}(A),$$

where $\mathfrak{n}(A)$ is the number of sequences of matrices $(Y_d)_{d|t}$ with $Y_d \in M_{s_d \times \alpha_d}$ and

$$\sum_{d|t} \sum_{j=0}^{d-1} A^j B_d Y_d Y_d^T B_d^T (A^j)^T = 0.$$

The computation of $\mathfrak{n}(A)$ is complicated but is not out of reach for a moderate k. An algorithm for computing $\mathfrak{n}(A)$ is detailed in [4, Algorthim 3.4]. The algorithm is based on the predictable behavior of binary quadratic forms. In [4], one can also find many examples of step-by-step execution of this algorithm.

The following additional facts greatly simplify the computation of $|\mathrm{Fix}(A, P_\lambda)|$ for many A.

- ([4, Lemma 3.7 and Eq. (3.25)]) $|\mathrm{Fix}(I_k, P_\lambda)|$ is known.
- ([4, Corollary 3.10]) Let $f_1, \dots, f_t \in \mathbb{F}_2[x] \setminus \{x\}$ be irreducible such that $\{f_1, f_1^*\}, \dots, \{f_t, f_t^*\}$ are pairwise disjoint, where f_i^* is the reciprocal polynomial of f_i. Let $A = A_1 \oplus \cdots \oplus A_t \in \mathrm{GL}(k, \mathbb{F}_2)$, where $A_i \in \mathrm{GL}(k_i, \mathbb{F}_2)$ and the elementary divisors of A_i are powers of f_i or f_i^*. Then

$$|\mathrm{Fix}(A, P_\lambda)| = \prod_{i=1}^{t} |\mathrm{Fix}(A_i, P_\lambda)|. \tag{3}$$

- ([4, Lemma 3.11]) Let $f \in \mathbb{F}_2[x] \setminus \{x\}$ be an irreducible polynomial which is not self-reciprocal. Let t be the smallest positive integer such that $f \mid x^t - 1$. Let $A \in \mathrm{GL}(k, \mathbb{F}_2)$ have elementary divisors $\underbrace{f^1, \dots, f^1}_{\mu_1}, \dots, \underbrace{f^s, \dots, f^s}_{\mu_s}$ and let $\lambda = (\lambda_1, \lambda_2, \dots) \vdash n$. Then

$$|\mathrm{Fix}(A, P_\lambda)| = 2^{\deg f \sum_{j \geq 1} \lambda_{jt} \sum_{l \geq 1} \mu_l \min\{l, 2^{\nu(j)}\}}.$$

We now turn to the case $k = 6$. $\mathrm{GL}(6, \mathbb{F}_2)$ has 60 rational canonical forms $A_1, \dots, A_{60}$. Their elementary divisors and the cardinalities of their centralizers are given in Table 1 of the appendix. Formulas for $|\mathrm{Fix}(A_i, P_\lambda)|$, $1 \leq i \leq 60$, are computed by the method outlined above. The results are contained in Table 2 of the appendix. It is common that for several A_i, $|\mathrm{Fix}(A_i, P_\lambda)|$ share the same formula. As a result, Table 2 contains 40 formulas instead of 60. All computations in this project were done using Mathematica [9].

Remarks.

(i) Since the amount of computation (mostly symbolic) in this project is very large and since the result is very intricate, there is a natural concern about the possible errors in the computation. To have a strong assurance for the correctness of the formulas in Table 2, we have used them to compute the values of $\Psi_{\le 6,n}$ for $n \le 40$ (Table 3 of the appendix). If any of those formulas for $|\mathrm{Fix}(A_i, P_\lambda)|$ had gone wrong, most likely, the results of $\Psi_{\le 6,n}$ coming out of the master formula would not have been integers. The results of $\Psi_{\le 6,n}$ have turned out to be integers.

(ii) It is possible to simplify the computation and presentation of $\Psi_{\le k,n}$. The idea is to further exploit (3) in order to express $\Psi_{\le k,n}$ in terms of functions that are reusable. We will discuss this idea in details elsewhere. In fact, there do not seem to be insurmountable obstacles in the computation of $\Psi_{\le k,n}$ with reasonably larger k's.

Appendix. Tables

The appendix consists of four tables. Table 1 contains the information of $\mathcal{C}(\mathrm{GL}(6,\mathbb{F}_2)) = \{A_1,\dots,A_{60}\}$, the set of rational canonical forms in $\mathrm{GL}(6,\mathbb{F}_2)$. Table 2 contains the formulas for $|\mathrm{Fix}(A_i, P_\lambda)|$, $1 \le i \le 60$. In Table 2, $\lambda = (\lambda_1, \lambda_2, \dots)$ and

$$\lambda_{a,b} = \sum_{i \equiv a \pmod b} \lambda_i \qquad \text{for } 0 \le a < b.$$

The function $\delta : \{0,1,\dots\} \to \{0,1\}$ is defined by

$$\delta(x) = \begin{cases} 0 & \text{if } x = 0, \\ 1 & \text{if } x > 0. \end{cases}$$

Tables 3 and 4 give the values of $\Psi_{\le k,n}$ and $\Psi_{k,n}$ with $k \le 6$ and $n \le 40$. Portions of Tables 3 and 4 with $k \le 5$ are from [4].

Table 1. Information about $\mathcal{C}(\mathrm{GL}(6,\mathbb{F}_2) = \{A_1,\ldots,A_{60}\}$

	elementary divisors	$\lvert\mathrm{cent}_{\mathrm{GL}(6,\mathbb{F}_2)}(\ )\rvert$
A_1	$x+1,\ x+1,\ x+1,\ x+1,\ x+1,\ x+1$	$2^{15}\cdot 3^4\cdot 5\cdot 7^2\cdot 31$
A_2	$x+1,\ x+1,\ x+1,\ x+1,\ (x+1)^2$	$2^{15}\cdot 3^2\cdot 5\cdot 7$
A_3	$x+1,\ x+1,\ x+1,\ (x+1)^3$	$2^{11}\cdot 3\cdot 7$
A_4	$x+1,\ x+1,\ x+1,\ x+1,\ x^2+x+1$	$2^6\cdot 3^3\cdot 5\cdot 7$
A_5	$x+1,\ x+1,\ x+1,\ x^3+x+1$	$2^3\cdot 3\cdot 7^2$
A_6	$x+1,\ x+1,\ x+1,\ x^3+x^2+1$	$2^3\cdot 3\cdot 7^2$
A_7	$x+1,\ x+1,\ (x+1)^2,\ (x+1)^2$	$2^{14}\cdot 3^2$
A_8	$x+1,\ x+1,\ (x+1)^2,\ x^2+x+1$	$2^6\cdot 3^2$
A_9	$x+1,\ x+1,\ x^2+x+1,\ x^2+x+1$	$2^3\cdot 3^3\cdot 5$
A_{10}	$x+1,\ x+1,\ (x+1)^4$	$2^8\cdot 3$
A_{11}	$x+1,\ x+1,\ (x^2+x+1)^2$	$2^3\cdot 3^2$
A_{12}	$x+1,\ x+1,\ x^4+x^3+x^2+x+1$	$2\cdot 3^2\cdot 5$
A_{13}	$x+1,\ x+1,\ x^4+x+1$	$2\cdot 3^2\cdot 5$
A_{14}	$x+1,\ x+1,\ x^4+x^3+1$	$2\cdot 3^2\cdot 5$
A_{15}	$x+1,\ (x+1)^2,\ (x+1)^3$	2^{11}
A_{16}	$x+1,\ (x+1)^2,\ x^3+x+1$	$2^3\cdot 7$
A_{17}	$x+1,\ (x+1)^2,\ x^3+x^2+1$	$2^3\cdot 7$
A_{18}	$x+1,\ x^2+x+1,\ (x+1)^3$	$2^4\cdot 3$
A_{19}	$x+1,\ x^2+x+1,\ x^3+x+1$	$3\cdot 7$
A_{20}	$x+1,\ x^2+x+1,\ x^3+x^2+1$	$3\cdot 7$
A_{21}	$x+1,\ (x+1)^5$	2^6
A_{22}	$x+1,\ x^5+x^2+1$	31
A_{23}	$x+1,\ x^5+x^3+1$	31
A_{24}	$x+1,\ x^5+x^3+x^2+x+1$	31
A_{25}	$x+1,\ x^5+x^4+x^2+x+1$	31
A_{26}	$x+1,\ x^5+x^4+x^3+x+1$	31
A_{27}	$x+1,\ x^5+x^4+x^3+x^2+1$	31
A_{28}	$(x+1)^2,\ (x+1)^2,\ (x+1)^2$	$2^{12}\cdot 3\cdot 7$
A_{29}	$(x+1)^2,\ (x+1)^2,\ x^2+x+1$	$2^5\cdot 3^2$
A_{30}	$(x+1)^2,\ x^2+x+1,\ x^2+x+1$	$2^3\cdot 3^2\cdot 5$
A_{31}	$(x+1)^2,\ (x+1)^4$	2^8
A_{32}	$(x+1)^2,\ (x^2+x+1)^2$	$2^3\cdot 3$
A_{33}	$(x+1)^2,\ x^4+x^3+x^2+x+1$	$2\cdot 3\cdot 5$
A_{34}	$(x+1)^2,\ x^4+x+1$	$2\cdot 3\cdot 5$
A_{35}	$(x+1)^2,\ x^4+x^3+1$	$2\cdot 3\cdot 5$
A_{36}	$x^2+x+1,\ x^2+x+1,\ x^2+x+1$	$2^6\cdot 3^4\cdot 5\cdot 7$
A_{37}	$x^2+x+1,\ (x+1)^4$	$2^3\cdot 3$
A_{38}	$x^2+x+1,\ (x^2+x+1)^2$	$2^6\cdot 3^2$
A_{39}	$x^2+x+1,\ x^4+x^3+x^2+x+1$	$3^2\cdot 5$
A_{40}	$x^2+x+1,\ x^4+x+1$	$3^2\cdot 5$
A_{41}	$x^2+x+1,\ x^4+x^3+1$	$3^2\cdot 5$
A_{42}	$(x+1)^3,\ (x+1)^3$	$2^9\cdot 3$
A_{43}	$(x+1)^3,\ x^3+x+1$	$2^2\cdot 7$
A_{44}	$(x+1)^3,\ x^3+x^2+1$	$2^2\cdot 7$

Table 1. continued

	elementary divisors	$\lvert \mathrm{cent}_{GL(6,\mathbb{F}_2)}(\ )\rvert$
A_{45}	$x^3+x+1,\ x^3+x+1$	$2^3 \cdot 3^2 \cdot 7^2$
A_{46}	$x^3+x+1,\ x^3+x^2+1$	7^2
A_{47}	$x^3+x^2+1,\ x^3+x^2+1$	$2^3 \cdot 3^2 \cdot 7^2$
A_{48}	$(x+1)^6$	2^5
A_{49}	$(x^2+x+1)^3$	$2^4 \cdot 3$
A_{50}	$(x^3+x+1)^2$	$2^3 \cdot 7$
A_{51}	$(x^3+x^2+1)^2$	$2^3 \cdot 7$
A_{52}	x^6+x+1	$3^2 \cdot 7$
A_{53}	x^6+x^3+1	$3^2 \cdot 7$
A_{54}	$x^6+x^4+x^2+x+1$	$3^2 \cdot 7$
A_{55}	$x^6+x^4+x^3+x+1$	$3^2 \cdot 7$
A_{56}	x^6+x^5+1	$3^2 \cdot 7$
A_{57}	$x^6+x^5+x^2+x+1$	$3^2 \cdot 7$
A_{58}	$x^6+x^5+x^3+x^2+1$	$3^2 \cdot 7$
A_{59}	$x^6+x^5+x^4+x+1$	$3^2 \cdot 7$
A_{60}	$x^6+x^5+x^4+x^2+1$	$3^2 \cdot 7$

Table 2. Formulas for $\lvert\mathrm{Fix}(A_i, P_\lambda)\rvert$, $1 \le i \le 60$

i	$\lvert\mathrm{Fix}(A_i, P_\lambda)\rvert$
1	$2^{6\lambda_{0,2}}[2^{6\lambda_{1,2}-21}(914068-914067\,\delta(\lambda_{1,2}))+2^{5\lambda_{1,2}-21}\,651\,(3+(-1)^{\lambda_{1,2}})$ $+2^{4\lambda_{1,2}-18}\,4557\,(5+3(-1)^{\lambda_{1,2}})+2^{3\lambda_{1,2}-13}\,217\,(9+7(-1)^{\lambda_{1,2}})]$
2	$2^{6\lambda_{0,4}}[2^{5\lambda_{1,2}+6\lambda_{2,4}-16}[2920-1459\,\delta(\lambda_{1,2})-1459\,\delta(\lambda_{1,2}+\lambda_{2,4})-\delta(\lambda_{2,4})]$ $+2^{4\lambda_{1,2}+6\lambda_{2,4}-16}\,35(3+(-1)^{\lambda_{1,2}})(2-\delta(\lambda_{2,4}))$ $+2^{3\lambda_{1,2}+6\lambda_{2,4}-13}\,7(5+3(-1)^{\lambda_{1,2}})(2-\delta(\lambda_{2,4}))$ $+2^{5\lambda_{1,2}+5\lambda_{2,4}-10}\,525(1-\delta(\lambda_{1,2}))+2^{4\lambda_{1,2}+5\lambda_{2,4}-12}\,15(3+(-1)^{\lambda_{1,2}})$ $+2^{3\lambda_{1,2}+5\lambda_{2,4}-11}\,105(5+3(-1)^{\lambda_{1,2}})]$
3	$2^{6\lambda_{0,8}}[2^{4\lambda_{1,2}+5\lambda_{2,4}+6\lambda_{4,8}-11}[200-35\,\delta(\lambda_{1,2})-64\,\delta(\lambda_{1,2}+\lambda_{2,4})$ $-64\,\delta(\lambda_{1,2}+\lambda_{4,8})-\delta(\lambda_{2,4}+\lambda_{4,8})-35\,\delta(\lambda_{1,2}+\lambda_{2,4}+\lambda_{4,8})]$ $+2^{3\lambda_{1,2}+5\lambda_{2,4}+6\lambda_{4,8}-11}\,7(3+(-1)^{\lambda_{1,2}})(2-\delta(\lambda_{2,4}+\lambda_{4,8}))$ $+2^{4\lambda_{1,2}+4\lambda_{2,4}+6\lambda_{4,8}-7}\,21(2-\delta(\lambda_{1,2})-\delta(\lambda_{1,2}+\lambda_{4,8}))$ $+2^{3\lambda_{1,2}+4\lambda_{2,4}+6\lambda_{4,8}-9}\,7(3+(-1)^{\lambda_{1,2}})(2-\delta(\lambda_{4,8}))$ $+2^{2\lambda_{1,2}+4\lambda_{2,4}+6\lambda_{4,8}-8}\,7(5+3(-1)^{\lambda_{1,2}})(2-\delta(\lambda_{4,8}))]$
4	$2^{4\lambda_{0,2}+2\lambda_{0,6}-1}[2^{2\lambda_{3,6}}+(-1)^{\lambda_{3,6}}2^{\lambda_{3,6}}]\,[2^{4\lambda_{1,2}-10}(436-435\,\delta(\lambda_{1,2}))$ $+2^{3\lambda_{1,2}-10}\,35(3+(-1)^{\lambda_{1,2}})+2^{2\lambda_{1,2}-7}\,7(5+3(-1)^{\lambda_{1,2}})]$
5 6	$2^{3\lambda_{0,2}+3\lambda_{0,7}}[2^{3\lambda_{1,2}-6}(36-35\,\delta(\lambda_{1,2}))+2^{2\lambda_{1,2}-6}\,7(3+(-1)^{\lambda_{1,2}})]$
7	$2^{6\lambda_{0,4}}[2^{4\lambda_{1,2}+6\lambda_{2,4}-13}[976-27\,\delta(\lambda_{1,2})-3\,\delta(\lambda_{2,4})-945\,\delta(\lambda_{1,2}+\lambda_{2,4})]$ $+2^{3\lambda_{1,2}+6\lambda_{2,4}-13}(3+(-1)^{\lambda_{1,2}})(76-75\,\delta(\lambda_{2,4}))$ $+2^{4\lambda_{1,2}+5\lambda_{2,4}-13}[4(237+7(-1)^{\lambda_{2,4}})-27(35+(-1)^{\lambda_{2,4}})\delta(\lambda_{1,2})]$ $+2^{3\lambda_{1,2}+5\lambda_{2,4}-13}(3+(-1)^{\lambda_{1,2}})(75+(-1)^{\lambda_{2,4}})+2^{4\lambda_{1,2}+4\lambda_{2,4}-4}\,3(1-\delta(\lambda_{1,2}))$ $+2^{3\lambda_{1,2}+4\lambda_{2,4}-6}(3+(-1)^{\lambda_{1,2}})+2^{2\lambda_{1,2}+4\lambda_{2,4}-7}\,7(5+3(-1)^{\lambda_{1,2}})]$
8	$2^{4\lambda_{0,4}+2\lambda_{0,6}-1}[2^{2\lambda_{3,6}}+(-1)^{\lambda_{3,6}}2^{\lambda_{3,6}}]\,[2^{3\lambda_{1,2}+4\lambda_{2,4}-7}[24-\delta(\lambda_{2,4})-11\,\delta(\lambda_{1,2})$ $-11\,\delta(\lambda_{1,2}+\lambda_{2,4})]+2^{2\lambda_{1,2}+4\lambda_{2,4}-7}(3+(-1)^{\lambda_{1,2}})(2-\delta(\lambda_{2,4}))$ $+2^{2\lambda_{1,2}+3\lambda_{2,4}-5}\,3[7+(-1)^{\lambda_{1,2}}-4\,\delta(\lambda_{1,2})]]$
9	$2^{2\lambda_{0,2}+4\lambda_{0,6}-3}[2^{2\lambda_{1,2}-3}(4-3\,\delta(\lambda_{1,2}))+2^{\lambda_{1,2}-3}(3+(-1)^{\lambda_{1,2}})]$ $\cdot[2^{4\lambda_{3,6}-1}+2^{3\lambda_{3,6}-1}\,5(-1)^{\lambda_{3,6}}+2^{2\lambda_{3,6}}\,5]$
10	$2^{6\lambda_{0,8}}[2^{3\lambda_{1,2}+4\lambda_{2,4}+6\lambda_{4,8}-8}[24-\delta(\lambda_{4,8})-3\,\delta(\lambda_{1,2})-3\,\delta(\lambda_{1,2}+\lambda_{4,8})$ $-8\,\delta(\lambda_{1,2}+\lambda_{2,4})-8\,\delta(\lambda_{1,2}+\lambda_{2,4}+\lambda_{4,8})]$ $+2^{2\lambda_{1,2}+4\lambda_{2,4}+6\lambda_{4,8}-8}(3+(-1)^{\lambda_{1,2}})(2-\delta(\lambda_{4,8}))$ $+2^{3\lambda_{1,2}+3\lambda_{2,4}+6\lambda_{4,8}-4}\,3(1-\delta(\lambda_{1,2}+\lambda_{4,8}))$ $+2^{2\lambda_{1,2}+3\lambda_{2,4}+6\lambda_{4,8}-6}\,3(3+(-1)^{\lambda_{1,2}})(1-\delta(\lambda_{4,8}))$ $+2^{3\lambda_{1,2}+4\lambda_{2,4}+5\lambda_{4,8}-8}(1+(-1)^{\lambda_{4,8}})[12-\delta(\lambda_{2,4})-3\,\delta(\lambda_{1,2}+\lambda_{2,4})-8\,\delta(\lambda_{1,2})]$ $+2^{2\lambda_{1,2}+4\lambda_{2,4}+5\lambda_{4,8}-8}(3+(-1)^{\lambda_{1,2}})(1+(-1)^{\lambda_{4,8}})(1-\delta(\lambda_{2,4}))$ $+2^{3\lambda_{1,2}+3\lambda_{2,4}+5\lambda_{4,8}-4}\,3(1-\delta(\lambda_{1,2}))+2^{2\lambda_{1,2}+3\lambda_{2,4}+5\lambda_{4,8}-6}\,3(3+(-1)^{\lambda_{1,2}})]$
11	$2^{2\lambda_{0,2}+4\lambda_{0,12}-1}[2^{2\lambda_{1,2}-3}(4-3\,\delta(\lambda_{1,2}))+2^{\lambda_{1,2}-3}(3+(-1)^{\lambda_{1,2}})]$ $\cdot[2^{2\lambda_{3,6}+4\lambda_{6,12}-1}+2^{2\lambda_{3,6}+3\lambda_{6,12}-1}(-1)^{\lambda_{6,12}}+2^{\lambda_{3,6}+2\lambda_{6,12}}(-1)^{\lambda_{3,6}}]$
12	$2^{2\lambda_{0,2}+4\lambda_{0,10}-2}[2^{2\lambda_{1,2}-3}(4-3\,\delta(\lambda_{1,2}))+2^{\lambda_{1,2}-3}(3+(-1)^{\lambda_{1,2}})]$ $\cdot[2^{4\lambda_{5,10}}+2^{2\lambda_{5,10}}\,3(-1)^{\lambda_{5,10}}]$
13 14	$2^{2\lambda_{0,2}+4\lambda_{0,15}}[2^{2\lambda_{1,2}-3}(4-3\,\delta(\lambda_{1,2}))+2^{\lambda_{1,2}-3}(3+(-1)^{\lambda_{1,2}})]$
15	$2^{6\lambda_{0,8}}[2^{3\lambda_{1,2}+5\lambda_{2,4}+6\lambda_{4,8}-9}[80-\delta(\lambda_{1,2})-\delta(\lambda_{2,4})-43\,\delta(\lambda_{1,2}+\lambda_{2,4})$ $-4\,\delta(\lambda_{1,2}+\lambda_{4,8})-2\,\delta(\lambda_{2,4}+\lambda_{4,8})-26\,\delta(\lambda_{1,2}+\lambda_{2,4}+\lambda_{4,8})]$ $+2^{2\lambda_{1,2}+5\lambda_{2,4}+6\lambda_{4,8}-7}(3+(-1)^{\lambda_{1,2}})[3-\delta(\lambda_{2,4})-2\,\delta(\lambda_{2,4}+\lambda_{4,8})]$ $+2^{3\lambda_{1,2}+4\lambda_{2,4}+6\lambda_{4,8}-9}[2+2(13+2(-1)^{\lambda_{2,4}})(1-\delta(\lambda_{1,2}))$ $+(1+(-1)^{\lambda_{2,4}})(1-\delta(\lambda_{4,8}))+(43+3(-1)^{\lambda_{2,4}})(1-\delta(\lambda_{1,2}+\lambda_{4,8}))]$ $+2^{2\lambda_{1,2}+4\lambda_{2,4}+6\lambda_{4,8}-7}(3+(-1)^{\lambda_{1,2}})(3-\delta(\lambda_{4,8}))$ $+2^{3\lambda_{1,2}+3\lambda_{2,4}+5\lambda_{4,8}-2}(1-\delta(\lambda_{1,2}))+2^{2\lambda_{1,2}+3\lambda_{2,4}+5\lambda_{4,8}-4}(3+(-1)^{\lambda_{1,2}})]$

Table 2. continued

i	$\|\mathrm{Fix}(A_i, P_\lambda)\|$
16 17	$2^{3\lambda_{0,4}+3\lambda_{0,7}}[2^{2\lambda_{1,2}+3\lambda_{2,4}-4}[8-3\,\delta(\lambda_{1,2})-\delta(\lambda_{2,4})-3\,\delta(\lambda_{1,2}+\lambda_{2,4})]$ $+2^{\lambda_{1,2}+2\lambda_{2,4}-3}(3+(-1)^{\lambda_{1,2}})]$
18	$2^{2\lambda_{0,6}+4\lambda_{0,8}-1}[2^{2\lambda_{3,6}}+(-1)^{\lambda_{3,6}}2^{\lambda_{3,6}}]\,[2^{2\lambda_{1,2}+3\lambda_{2,4}+4\lambda_{4,8}-4}[8-\delta(\lambda_{2,4}+\lambda_{4,8})$ $-\delta(\lambda_{1,2})-2\,\delta(\lambda_{1,2}+\lambda_{4,8})-2\,\delta(\lambda_{1,2}+\lambda_{2,4})-\delta(\lambda_{1,2}+\lambda_{2,4}+\lambda_{4,8})]$ $+2^{\lambda_{1,2}+2\lambda_{2,4}+4\lambda_{4,8}-4}(3+(-1)^{\lambda_{1,2}})(2-\delta(\lambda_{4,8}))]$
19 20	$2^{\lambda_{0,1}+2\lambda_{0,6}+3\lambda_{0,7}-2}(2-\delta(\lambda_{1,2}))[2^{2\lambda_{3,6}}+(-1)^{\lambda_{3,6}}2^{\lambda_{3,6}}]$
21	$2^{6\lambda_{0,16}}[2^{2\lambda_{1,2}+3\lambda_{2,4}+5\lambda_{4,8}+6\lambda_{8,16}-5}[8-\delta(\lambda_{1,2})-2\,\delta(\lambda_{1,2}+\lambda_{2,4})-\delta(\lambda_{4,8})$ $-\delta(\lambda_{1,2}+\lambda_{4,8})-2\,\delta(\lambda_{1,2}+\lambda_{2,4}+\lambda_{4,8})]$ $+2^{\lambda_{1,2}+2\lambda_{2,4}+5\lambda_{4,8}+6\lambda_{8,16}-4}(3+(-1)^{\lambda_{1,2}})(1-\delta(\lambda_{4,8}+\lambda_{8,16}))$ $+2^{2\lambda_{1,2}+3\lambda_{2,4}+4\lambda_{4,8}+6\lambda_{8,16}-5}(1+(-1)^{\lambda_{4,8}})[4-2\,\delta(\lambda_{1,2}+\lambda_{8,16})$ $-\delta(\lambda_{2,4}+\lambda_{8,16})-\delta(\lambda_{1,2}+\lambda_{2,4}+\lambda_{8,16})]$ $+2^{\lambda_{1,2}+2\lambda_{2,4}+4\lambda_{4,8}+6\lambda_{8,16}-4}(3+(-1)^{\lambda_{1,2}})]$
22 | 27	$2^{\lambda_{0,1}+5\lambda_{0,31}-1}(2-\delta(\lambda_{1,2}))$
28	$2^{6\lambda_{0,4}}[2^{3\lambda_{1,2}+6\lambda_{2,4}-12}[1184-35\,\delta(\lambda_{2,4})-7\,\delta(\lambda_{1,2})-1141\,\delta(\lambda_{1,2}+\lambda_{2,4})]$ $+2^{2\lambda_{1,2}+6\lambda_{2,4}-7}\,7(3+(-1)^{\lambda_{1,2}})(1-\delta(\lambda_{2,4}))$ $+2^{3\lambda_{1,2}+5\lambda_{2,4}-12}\,7(3+(-1)^{\lambda_{2,4}})(8-7\,\delta(\lambda_{1,2}))$ $+2^{3\lambda_{1,2}+4\lambda_{2,4}-5}\,7(1-\delta(\lambda_{1,2}))+2^{2\lambda_{1,2}+4\lambda_{2,4}-7}\,7(3+(-1)^{\lambda_{1,2}})]$
29	$2^{4\lambda_{0,4}+2\lambda_{0,6}-1}[2^{2\lambda_{3,6}}+(-1)^{\lambda_{3,6}}2^{\lambda_{3,6}}]$ $\cdot[2^{2\lambda_{1,2}+4\lambda_{2,4}-6}[16-3\,\delta(\lambda_{2,4})-3\,\delta(\lambda_{1,2})-9\,\delta(\lambda_{1,2}+\lambda_{2,4})]$ $+2^{2\lambda_{1,2}+3\lambda_{2,4}-6}(3+(-1)^{\lambda_{2,4}})(4-3\,\delta(\lambda_{1,2}))+2^{\lambda_{1,2}+2\lambda_{2,4}-3}(3+(-1)^{\lambda_{1,2}})]$
30	$2^{\lambda_{1,2}+2\lambda_{0,2}+4\lambda_{0,6}-5}[4-\delta(\lambda_{1,2})-\delta(\lambda_{2,4})-\delta(\lambda_{1,2}+\lambda_{2,4})]$ $\cdot[2^{4\lambda_{3,6}-1}+2^{3\lambda_{3,6}-1}\,5(-1)^{\lambda_{3,6}}+2^{2\lambda_{3,6}}\,5]$
31	$2^{6\lambda_{0,8}}[2^{2\lambda_{1,2}+4\lambda_{2,4}+6\lambda_{4,8}-7}[16-\delta(\lambda_{1,2})-\delta(\lambda_{2,4})-5\,\delta(\lambda_{1,2}+\lambda_{2,4})-\delta(\lambda_{4,8})$ $-\delta(\lambda_{1,2}+\lambda_{4,8})-\delta(\lambda_{2,4}+\lambda_{4,8})-5\,\delta(\lambda_{1,2}+\lambda_{2,4}+\lambda_{4,8})]$ $+2^{2\lambda_{1,2}+3\lambda_{2,4}+6\lambda_{4,8}-6}[(1+(-1)^{\lambda_{2,4}})(1-\delta(\lambda_{4,8}))$ $+(5+(-1)^{\lambda_{2,4}})(1-\delta(\lambda_{1,2}+\lambda_{4,8}))]$ $+2^{2\lambda_{1,2}+4\lambda_{2,4}+5\lambda_{4,8}-6}(1+(-1)^{\lambda_{4,8}})[4-\delta(\lambda_{1,2})-\delta(\lambda_{2,4})-2\,\delta(\lambda_{1,2}+\lambda_{2,4})]$ $+2^{2\lambda_{1,2}+3\lambda_{2,4}+5\lambda_{4,8}-5}[1+(2+(-1)^{\lambda_{2,4}})(1-\delta(\lambda_{1,2}))]$ $+2^{\lambda_{1,2}+2\lambda_{2,4}+4\lambda_{4,8}-3}(3+(-1)^{\lambda_{1,2}})]$
32	$2^{\lambda_{1,2}+2\lambda_{0,2}+4\lambda_{0,12}-3}[4-\delta(\lambda_{1,2})-\delta(\lambda_{2,4})-\delta(\lambda_{1,2}+\lambda_{2,4})]$ $\cdot[2^{2\lambda_{3,6}+4\lambda_{6,12}-1}+2^{2\lambda_{3,6}+3\lambda_{6,12}-1}(-1)^{\lambda_{6,12}}+2^{\lambda_{3,6}+2\lambda_{6,12}}(-1)^{\lambda_{3,6}}]$
33	$2^{\lambda_{1,2}+2\lambda_{0,2}+4\lambda_{0,10}-4}[4-\delta(\lambda_{1,2})-\delta(\lambda_{2,4})-\delta(\lambda_{1,2}+\lambda_{2,4})]$ $\cdot[2^{4\lambda_{5,10}}+2^{2\lambda_{5,10}}\,3(-1)^{\lambda_{5,10}}]$
34 35	$2^{\lambda_{1,2}+2\lambda_{0,2}+4\lambda_{0,15}-2}[4-\delta(\lambda_{1,2})-\delta(\lambda_{2,4})-\delta(\lambda_{1,2}+\lambda_{2,4})]$
36	$2^{6\lambda_{0,6}}[2^{6\lambda_{3,6}-9}+2^{5\lambda_{3,6}-9}\,21(-1)^{\lambda_{3,6}}+2^{4\lambda_{3,6}-8}\,105+2^{3\lambda_{3,6}-6}\,35(-1)^{\lambda_{3,6}}]$
37	$2^{2\lambda_{0,6}+4\lambda_{0,8}-1}[2^{2\lambda_{3,6}}+(-1)^{\lambda_{3,6}}2^{\lambda_{3,6}}]\,[2^{\lambda_{1,2}+2\lambda_{2,4}+4\lambda_{4,8}-3}[4-\delta(\lambda_{4,8})$ $-\delta(\lambda_{1,2}+\lambda_{2,4})-\delta(\lambda_{1,2}+\lambda_{2,4}+\lambda_{4,8})]$ $+2^{\lambda_{1,2}+2\lambda_{2,4}+3\lambda_{4,8}-3}(1+(-1)^{\lambda_{4,8}})(2-\delta(\lambda_{1,2})-\delta(\lambda_{2,4}))]$
38	$2^{6\lambda_{0,12}}[2^{4\lambda_{3,6}+6\lambda_{6,12}-5}+2^{3\lambda_{3,6}+6\lambda_{6,12}-5}(-1)^{\lambda_{3,6}}$ $+2^{4\lambda_{3,6}+5\lambda_{6,12}-5}(-1)^{\lambda_{6,12}}+2^{3\lambda_{3,6}+5\lambda_{6,12}-5}(-1)^{\lambda_{3,6}+\lambda_{6,12}}$ $+2^{3\lambda_{3,6}+4\lambda_{6,12}-2}(-1)^{\lambda_{3,6}}+2^{2\lambda_{3,6}+4\lambda_{6,12}-3}\,5]$
39	$2^{2\lambda_{0,6}+4\lambda_{0,10}-3}[2^{2\lambda_{3,6}}+(-1)^{\lambda_{3,6}}2^{\lambda_{3,6}}][2^{4\lambda_{5,10}}+2^{2\lambda_{5,10}}\,3(-1)^{\lambda_{5,10}}]$
40 41	$2^{2\lambda_{0,6}+4\lambda_{0,15}-1}[2^{2\lambda_{3,6}}+(-1)^{\lambda_{3,6}}2^{\lambda_{3,6}}]$

Table 2. continued

i	$\lvert\mathrm{Fix}(A_i, P_\lambda)\rvert$
42	$2^{6\lambda_{0,8}}\,[2^{2\lambda_{1,2}+4\lambda_{2,4}+6\lambda_{4,8}-6}[16-3\,\delta(\lambda_{1,2}+\lambda_{2,4})-3\,\delta(\lambda_{1,2}+\lambda_{4,8})$ $-3\,\delta(\lambda_{2,4}+\lambda_{4,8})-6\,\delta(\lambda_{1,2}+\lambda_{2,4}+\lambda_{4,8})]$ $+2^{2\lambda_{1,2}+3\lambda_{2,4}+5\lambda_{4,8}-6}[3+(-1)^{\lambda_{2,4}+\lambda_{4,8}}$ $+3(1-\delta(\lambda_{1,2}))(2+(-1)^{\lambda_{2,4}}+(-1)^{\lambda_{4,8}})]$ $+2^{\lambda_{1,2}+2\lambda_{2,4}+4\lambda_{4,8}-3}(3+(-1)^{\lambda_{1,2}})]$
43 44	$2^{\lambda_{1,2}+2\lambda_{2,4}+3\lambda_{0,4}+3\lambda_{0,7}-2}[4-\delta(\lambda_{1,2}+\lambda_{2,4})-\delta(\lambda_{1,2}+\lambda_{4,8})-\delta(\lambda_{2,4}+\lambda_{4,8})]$
45 47	$2^{6\lambda_{0,7}}$
46	$2^{6\lambda_{0,14}}[2^{6\lambda_{7,14}-3}+2^{3\lambda_{7,14}-3}\,7]$
48	$2^{6\lambda_{0,16}}\,[2^{\lambda_{1,2}+2\lambda_{2,4}+4\lambda_{4,8}+6\lambda_{8,16}-3}[4-\delta(\lambda_{1,2}+\lambda_{2,4}+\lambda_{4,8})$ $-\delta(\lambda_{1,2}+\lambda_{2,4}+\lambda_{8,16})-\delta(\lambda_{4,8}+\lambda_{8,16})]+2^{\lambda_{1,2}+2\lambda_{2,4}+3\lambda_{4,8}+5\lambda_{8,16}-3}$ $\cdot[(1+(-1)^{\lambda_{4,8}+\lambda_{8,16}})(1-\delta(\lambda_{1,2}))+((-1)^{\lambda_{4,8}}+(-1)^{\lambda_{8,16}})(1-\delta(\lambda_{2,4}))]]$
49	$2^{6\lambda_{0,24}}[2^{2\lambda_{3,6}+4\lambda_{6,12}+6\lambda_{12,24}-2}+2^{2\lambda_{3,6}+3\lambda_{6,12}+5\lambda_{12,24}-2}(-1)^{\lambda_{6,12}+\lambda_{12,24}}$ $+2^{\lambda_{3,6}+2\lambda_{6,12}+4\lambda_{12,24}-1}(-1)^{\lambda_{3,6}}]$
50 51	$2^{3\lambda_{0,7}+3\lambda_{0,14}}$
52 55 | 59	$2^{6\lambda_{0,63}}$
53	$2^{6\lambda_{0,18}}[2^{6\lambda_{9,18}-3}+2^{3\lambda_{9,18}-3}(-1)^{\lambda_{9,18}}\,7]$
54 60	$2^{6\lambda_{0,21}}$

Table 3. Values of $\Psi_{\leq k,n}$, $k \leq 6$, $n \leq 40$

$n\backslash k$	0	1	2	3	4	5	6
1	1	1	1	1	1	1	1
2	1	2	2	2	2	2	2
3	1	2	2	2	2	2	2
4	1	3	4	4	4	4	4
5	1	3	4	4	4	4	4
6	1	4	7	8	8	8	8
7	1	4	7	9	9	9	9
8	1	5	11	16	18	18	18
9	1	5	11	17	20	20	20
10	1	6	16	28	37	39	39
11	1	6	16	30	42	46	46
12	1	7	23	49	77	92	95
13	1	7	23	53	89	112	118
14	1	8	31	82	157	218	245
15	1	8	31	89	187	281	329
16	1	9	41	133	323	551	704
17	1	9	41	144	389	740	1,016
18	1	10	53	210	654	1,447	2,244
19	1	10	53	229	804	2,059	3,602
20	1	11	67	325	1,324	4,029	8,330
21	1	11	67	354	1,651	6,032	15,012
22	1	12	83	490	2,654	11,774	36,548
23	1	12	83	534	3,356	18,581	75,207
24	1	13	102	727	5,291	36,239	194,365
25	1	13	102	793	6,759	59,798	454,191
26	1	14	123	1,058	10,433	116,020	1,238,014
27	1	14	123	1,154	13,444	198,489	3,196,838
28	1	15	147	1,515	20,363	382,272	9,024,639
29	1	15	147	1,651	26,384	670,031	24,685,875
30	1	16	174	2,136	39,229	1,276,454	70,478,121
31	1	16	174	2,329	51,025	2,267,431	196,702,836
32	1	17	204	2,972	74,574	4,260,828	557,194,708
33	1	17	204	3,237	97,143	7,596,889	1,547,951,716
34	1	18	237	4,078	139,660	14,050,410	4,299,971,583
35	1	18	237	4,439	181,923	24,965,555	11,732,683,283
36	1	19	274	5,532	257,592	45,384,782	31,774,581,057
37	1	19	274	6,017	335,029	79,965,507	84,618,649,911
38	1	20	314	7,418	467,600	142,792,476	222,909,144,028
39	1	20	314	8,061	606,613	248,697,834	577,998,702,214
40	1	21	358	9,843	835,392	497,412,483	1,480,493,480,646

Table 4. Values of $\Psi_{k,n}$, $k \leq 6$, $n \leq 40$

$n\backslash k$	0	1	2	3	4	5	6
1	1	0	0	0	0	0	0
2	1	1	0	0	0	0	0
3	1	1	0	0	0	0	0
4	1	2	1	0	0	0	0
5	1	2	1	0	0	0	0
6	1	3	3	1	0	0	0
7	1	3	3	2	0	0	0
8	1	4	6	5	2	0	0
9	1	4	6	6	3	0	0
10	1	5	10	12	9	2	0
11	1	5	10	14	12	4	0
12	1	6	16	26	28	15	3
13	1	6	16	30	36	23	6
14	1	7	23	51	75	61	27
15	1	7	23	58	98	94	48
16	1	8	32	92	190	228	153
17	1	8	32	103	245	351	276
18	1	9	43	157	444	793	797
19	1	9	43	176	575	1,255	1,543
20	1	10	56	258	999	2,705	4,301
21	1	10	56	287	1,297	4,381	8,980
22	1	11	71	407	2,164	9,120	24,774
23	1	11	71	451	2,822	15,225	56,626
24	1	12	89	625	4,564	30,948	158,126
25	1	12	89	691	5,966	53,039	394,393
26	1	13	109	935	9,375	105,587	1,121,994
27	1	13	109	1,031	12,290	185,045	2,998,349
28	1	14	132	1,368	18,848	361,909	8,642,367
29	1	14	132	1,504	24,733	643,647	24,015,844
30	1	15	158	1,962	37,093	1,237,225	69,201,667
31	1	15	158	2,155	48,696	2,216,406	194,435,405
32	1	16	187	2,768	71,602	4,186,254	552,933,880
33	1	16	187	3,033	93,906	7,499,746	1,540,354,827
34	1	17	219	3,841	135,582	13,910,750	4,285,921,173
35	1	17	219	4,202	177,484	24,783,632	11,707,717,728
36	1	18	255	5,258	252,060	45,127,190	31,729,196,275
37	1	18	255	5,743	329,012	79,630,478	84,538,684,404
38	1	19	294	7,104	460,182	142,324,876	222,766,351,552
39	1	19	294	7,747	598,552	248,091,221	577,750,004,380
40	1	20	337	9,485	825,549	496,577,091	1,479,996,068,163

References

[1] J. H. Conway and V. Pless, *On the enumeration of self-dual codes*, J. Combin. Theory A **28**, 26 – 53 (1980).

[2] J. H. Conway, V. Pless, and N. J. A. Sloane, *The binary self-dual codes of length up to 32: a revised enumeration*, J. Combin. Theory A **60**, 183 – 195 (1992).

[3] X. Hou, $\mathrm{GL}(m,2)$ *acting on* $R(r,m)/R(r-1,m)$, Discrete Math. **149**, 99 – 122 (1996).

[4] X. Hou, *On the number of inequivalent binary self-orthogonal codes*, IEEE Trans. Inform. Theory, to appear.

[5] W. C. Huffman, *On the classification and enumeration of self-dual codes*, Finite Fields Appl. **11**, 451 – 490 (2005).

[6] W. C. Huffman and V. Pless, *Fundamentals of Error-Correcting Codes* (Cambridge University Press, Cambridge, 2003).

[7] V. Pless, *A classification of self-orthogonal codes over* GF(2), Discrete Math. **3**, 209 – 246 (1972).

[8] V. Pless and N. J. A. Sloane, *On the classification and enumeration of self-dual codes*, J. Combin. Theory A **18**, 313 – 335 (1975).

[9] *Mathematica*, Wolfram Research, Inc. Champaign, IL, `http://www.wolfram.com/`